# MINDING THE BEES

## *A Vision for Apiculture at Douai Abbey*

Gabriel Wilson
*Order of St. Benedict*

2023

Northern Bee Books

**MINDING THE BEES** - *A Vision For Apiculture at Douai Abbey*
Copyright © 2023 Gabriel Wilson, Order of St. Benedict

ISBN 978-1-914934-59-9


Published by Peacock Press, 2023
Scout Bottom Farm
Mytholmroyd
Hebden Bridge HX7 5JS (UK)


Design and artwork by DM Design and Print

# MINDING THE BEES

## *A Vision for Apiculture at Douai Abbey*

Gabriel Wilson
*Order of St. Benedict*

2023

# Contents

# MINDING THE BEES

## Prologue
### A Stone Bee

'PURE BLACKS – Healthy Natural Swarms
10s 6d, 15s ready shortly. Please book early.
ALSFORD, Expert, Blandford.' [1]

*British Bee Journal 1896*

I knew I had to visit the grave, the moment I saw that strange bee.

From the photograph on the computer screen it looks about six inches long, a relief-carving within an oblong recess; a queen bee fossilised in stone, or a tarnished version of those chunky insect brooches you sometimes see worn on a favourite coat. Once grey and perfect, a century has marked her with small rings of yellow and pale-green lichen growing across the entire headstone. The design reminds me of the illustrated queen in an oblong advert for Italian imports of 1921, typical of bee adverts of that time, 'Guinea Gold Queens from D. ALLBON & Co., Hitchin, the Goldens That Breed True To Colour'.(2) Or the BKA medals awarded in Victorian times at honey shows, embossed with a silver queen.

*BBJ 1921*

With difficulty I decipher *Annie Woodley* at the top of the stone - I think she died in her early fifties, in 1904. Beneath her name is the carved bee. Below: *William Woodley died 1923,* aged 78. Expansive and spreading lichen stains the rest of the lettering, like exploding cauliflowers of wet watercolour. The bee is disfigured and partially obscured by the lichens. If I hadn't recognised it as a queen bee, I wouldn't quite know what it was. With careful observation I might discern the long legs and the distinctive, slender body tapering to a point, the wings folded across her back. It isn't what most people would recognise as a bee. This is a honeybee queen, part of the hidden, inner mechanism of a beehive - the mainspring that powers the whole colony. Most people never see a queen honeybee so they wouldn't necessarily recognise the image, and many wouldn't have done, I suspect, even when she was new and perfectly-carved.

*The stone bee*

Beekeepers like me would have identified her straight away though, before time and nature obscured her beneath the lichens. I only know her now because I stumbled upon William Woodley's *Notes By The Way,* in archived editions of *The British Bee Journal,* and she is perhaps more poignant in her imperfect state than when she was a pristine jewel; for she seems to me like those dead queens we beekeepers sometimes find among strewn detritus on the bottom of a hive in early spring. Those shapes of honey-coloured lichen are fragments of broken brace comb and nibbled store cappings among which we might find our wrecked queen's remains, the bright enamel with which we marked her thorax last season the only remnant of her glorious, short reign.

It was that first long lockdown of the Covid pandemic in 2020 that I discovered him, or rather when he emerged from the ochre pages of an archive, the way this stone bee emerges now from the excrescences of time. I'd been reading through online editions of the *British Bee Journal* to unearth more information about the mysterious *Isle of Wight disease* that was thought by many until fairly recently to have killed the entire population of our *native Black bees* (*Apis Mellifera mellifera*) in three major episodes between 1904 and the years of the First World War. A description of what happened is epitomised by J. Vincent, writing in the *British Bee Journal* 11 March 1915:

'The time for the autumn driving from skeps in the surrounding villages arrived, and I determined - in spite of all that seemed to oppose me - to work up as many stocks as possible from this source, but all my work, hopes, and bees were doomed to failure, and before Christmas more than twenty out of thirty-one stocks were dead, and I felt that I could not go near the hives after I found out how things were. I concluded I had no bees left until now I find five or six colonies (driven bees last autumn) have been busy to-day. I know they were not well supplied with either combs or stores, and I found out, when too late, they had been clearing out stores from hives affected with Foul Brood and "Isle of Wight" disease. As sugar was so dear I decided to let the lot take their chance, and it is almost a miracle that they have pulled through up to the present. My aim has been to keep bees enough to pay my rent, and last spring I was proud to think I had got the best lot of bees I had possessed since starting bee- keeping twenty years ago — this spring 1 feel as though my bee-keeping has come to an end. It is a £40 loss to me, as all my hives, supers, shallow frames, &c., are simply useless lumber. What can I do? I cannot afford to throw more money after that I have lost.' (3)

T.A.R. of Glasgow described what beekeepers had been experiencing since the phenomenon's first appearance on the Isle of Wight in 1904, writing in the *British Bee Journal* 8 April 1915, 'The first symptoms are the bees loafing about on the alighting board and the colony showing very little inclination to work. Here and there "crawlers" may be noticed; these increase in numbers as the disease progresses.' (4)

The mysterious Isle of Wight disease is usually referred to as a discrete crisis in beekeeping, but I have begun to question this. Beekeeping had undergone a revolution since the invention of the modern frame hive by Lorenzo Langstroth (1818-1895). His invention in 1851 had opened the door to beekeepers, not just to new beekeeping methods but to the possibilities of keeping and breeding different races of the western honeybee in much larger colonies than the native bee, which meant more honey. Foreign races such as the Ligurian, Carniolan and the Cyprian bee, which were becoming available to British beekeepers, promised to unlock the full potential of the frame hive by increasing honey production and commercialising apiculture. By the time Isle of Wight disease first appeared beekeepers had for half a century already experimented with cross breeding different races to propagate their most desired characteristics such as docility and fecundity; thereby maximising production from bees in the same way that farming had for centuries selectively bred domesticated animals and plants for food and other products.

When Isle of Wight disease emerged in 1904 half a century of British apiculture had already gone down the path of an increasing use of foreign races of honeybee, and hybridisation between them and the native British bee. Even before the appearance of this disease the bacterial disease of foul brood had been around for decades, though William Woodley and others claimed it had never been known when bees had been kept in traditional skeps. Was this mere coincidence, or were modern, untried methods with foreign subspecies and hybridisation responsible - (if not for causing emerging new diseases) for leaving bees more susceptible to them than before due to what we now understand as epigenetic modification? Epigenetics is the comparatively recent science of heritable changes in gene function that can't be explained by changes in an organism's DNA, but are caused instead by something 'on top of ' the DNA, such as environmental influences and diet. Was Isle of Wight disease just another disease, or a complex crisis synthesising half a century of experimentation with new methods, of which cross breeding of different races was one?

At the time there was one group who had very clear and evangelical ideas about apiculture and bee diseases, and they were the new Beekeeping Association. The first Beekeeping Association (BKA) had been started as a London club in 1874 by Thomas William Cowan (1840-1926) and Charles Nash Abbot. By 1890 it had grown, with its own governing body and Presidents in twenty-six affiliated County Associations. Whereas bees had for centuries been associated with myths and legends, a little political theory and a lot of religious symbolism, by the late nineteenth century the goals the new BKA had begun to outline something approximating a particular ideology of beekeeping. This set of ideas and assumptions appeared in the editors' introduction to the very first edition of the *British Bee Journal* in May 1873 in which the first page disparaged the superstitious cottage skeppist while focusing their efforts squarely on those who were more enlightened. Interestingly, before any text, there appeared a number of large adverts for Ligurian bees, illustrating that from the beginning the BKA's emphasis on modern methods included the foreign race of bee and not the British Black bee. W. J. Pettitt's Ligurian queens were advertised, along with adverts throughout the journals that year for Ligurian queens at 12s, 6d or Ligurian queens crossed with English drones at 5s each.

The word *Ideology*, from the French Ideologie, is rooted in the Greek *ideo* (idea) a logos (the study of), and was first coined in 1801 by the French philosopher Destrutt de Tracy who was influenced by the earlier English philosopher Francis Bacon. Bacon stated that the destiny of science was to improve knowledge and the life of humanity. Ideology then proposes a coherent set of ideas and beliefs based upon a set of assumptions. These assumptions were clearly embedded in the early aims of the first BKA, which were to improve and advance apiculture alongside the philanthropic cause of improving the conditions of cottagers and the labouring classes. They believed from the start in the application of new methods based on science, which meant that from their origins they regarded traditional skep beekeeing as primitive, and cottager skeppists as ignorant, superstitious and in need of education. Clearly, a foundational assumption of the organisation was that the traditional straw skep was an obstacle to progress in the craft and to the advancement of the beekeeper. Beekeeping was now set on a particular course towards improving the performance of honeybees and beekeepers for economic gain.

As with all ideologies, their basic assumptions developed into new patterns of thought and practice through the subjective choices beekeepers began to make; the experimentation with different races of honeybee, cross-breeding and the application of new sciences such as Mendel's laws of heredity. These assumptions were accepted passively by most, with fervent advocacy by others, but with resistance by some. William Woodley represented this third group of people who do not accept new ideologies uncritically because they question or deny some or all of its basic assumptions. Arguably, he remains important today, not just to help us understand an important historical moment in British beekeeping, but to remind us of the need to question and examine the basic assumptions that underpin our own beekeeping, and indeed to apply critical thinking to ideologies presenting themselves today, in beekeeping, politics and elsewhere.

What is clear to me is that beekeeping for decades after Isle of Wight disease was set on a trajectory from which it could not escape because of a particular ideology it had developed. From 1904 -1919 beekeepers had searched in vain for the cure to Isle of Wight disease, as they had also struggled for years before eventually overcoming foul brood. By 1919, with British stocks decimated by disease and the need to avert a crisis in food production after World War I, the only short term answer was to import even more foreign stocks which seemed to be resistant to the disease. Beekeepers who had started keeping foreign races of imported bees as a subjective choice were now locked into a culture of keeping imported bees out of sheer necessity in their search for bees that were resistant to Isle of Wight disease and to address the urgent need to restock in order to meet the pollination demands of British agriculture. Arguably, the problem all along had been neither traditional skep beekeeping nor the modern frame hive, but just as much (if not more) the arrival and indiscriminate hybridisation of imported stocks. After 1919 the Government restocking programme inadvertently sent apiculture down a cul de sac to find the perfect, disease-resistant and commercially profitable bee so that nothing like Isle of Wight disease could happen again.

After Isle of Wight disease a long term solution to the problem of susceptibility to known bee diseases was needed. One solution came from a Benedictine monk called Brother Adam Kehrle (1898-1996) at Buckfast Abbey in Devon, England, who believed (in the BKA's tradition of Enlightenment ideals) that progress in beekeeping could be made by using Mendel's laws of heredity in the relatively new field of bee breeding. By cross-breeding different races and strains of honeybee, Brother Adam's holy grail was to produce the Buckfast strain – a superbee that he hoped might answer the commercial needs of the modern beekeeper and prevent a repetition of Isle of Wight disease. The foundation of the new strain was bred from survivor stocks of this popular cross (*Ligustica* and *AMm*) at Buckfast and elsewhere, and (as evidenced in the first pages of Vol 1 of the journal in 1873) the fashion of *Ligurianizing* skeps of English Black bees.

The path British beekeeping had taken, however, would turn out to be a dead-end little over half a century later when the Buckfast bee was finally ready for commercial use. By the early 1990s the mite *Varroa destructor* had arrived in Britain, first seen in Devon. Breeders had taken European bees to Asia in the mid-twentieth century, where the mite had jumped from *Apis cerana*, the Asian honeybee, which was resistant to the parasite. It was a clear example of the spread of modern bee diseases and pests by the movement of alien subspecies and strains to new areas of the world.

With varroa, British beekeeping had reached the bottom of the sack, with nowhere to go until a treatment, rather than a cure, was found in 1997. Even then there was no way back from the cul de sac, except to continue treating, to keep the mites under control. Unfortunately, while treatments might help keep stocks alive, varroa mites only became stronger, acquiring resistance to medication. Medicated bees, on the other hand, were progressively weakened, with compromised resistance and immunity not only to varroa but to the range of viruses transmitted by the mites and to other viruses and pathogens to which bees were exposed.

This discovery of the history of Isle of Wight disease ignited my interest in increasing reports of surviving stocks of *native* Black bees *(Apis Mellifera mellifera)* or *Amm* that were thought to have been wiped out by the mysterious plague. It was there, in the pages of those old editions of the *British Bee Journal* that William Woodley emerged, a well-known beekeeper in his day, and a regular contributor to beekeeping journals. In particular, I began to appreciate a man who had embraced modern beekeeping methods during the First Industrial Revolution with its opportunities to commercialise apiculture, but also someone who still had one foot firmly in the pre-industrial world of the traditional skep and the agricultural cottage labourer. It was like finding treasure. William Woodley's *Notes By The Way* correspondence transported me back, through the pages of the old journals in which he wrote, into an age in the history of beekeeping of a long-running debate about controversies in apiculture that would completely change the way I approach the craft.

The same basic controversies are still very much alive and unresolved today. They raise questions about whether the honeybee can ever be domesticated; how we define progress in beekeeping; what constitutes the perfect honeybee; how we keep honeybees in ways that are beneficial to them as well as to us; our care for the environment and farming methods; bee-breeding and the use of foreign races of bee, bee legislation and the freedoms and responsibilities that underpin western civilisation itself. It also raises specific questions about fundamental freedoms British beekeepers have enjoyed for centuries, enshrined by Roman law and *Magna Carta*. More disturbingly, the same trajectory of a particular ideology of bee improvement and the perfect bee has arrived at a new controversy today – that of the very real possibility of a transgenic bee created by the controversial technology of Precision Breeding.

On 23 January 2023 the House of Lords voted on *The Genetic Technology (Precision Breeding Bill)*, aimed to remove existing EU measures that prevent the development, marketing and release of precision-bred animals and plants. The Bill was passed in its report stage. Precision breeding involves a range of technologies, including gene editing, that allow DNA to be edited more precisely than with traditional breeding methods. What Brother Adam took fifty years to achieve with his Buckfast bee could, in the not too distant future, be achieved in a laboratory, with greater precision, in only a couple of years.

William Woodley has helped me to question the underlying assumptions that have brought us to this stage in beekeeping and which could take us into a development in bee culture that could be disastrous for bees, for beekeepers and for the environment. Unquestioned ideologies, whether in beekeeping, or in other areas, such as extreme environmentalism, if unexamined and accepted uncritically, could take us passively towards some very dark places – even those we might think are green, but which could turn out to be a much darker shade of green.

As I read through those old journals, there were apparent contradictions about William Woodley which further intrigued me. The great debate into which he was drawn, about the rise of bee diseases and the need for Bee Disease legislation to control them, became a fascinating area of research. Not only did I become immersed in the story of a Victorian beekeeper, but also in the enigma of Isle of Wight disease and the complex reasons why William Woodley was blamed by many beekeepers at the time for its destructive and uncontrollable consequences for British bees and beekeeping because of his opposition to Bee Disease legislation.

I began to see a strange and poignant parallel between the demise of William Woodley, the respected, famous beekeeper, and the demise of our native British bee. Moreover, I was increasingly struck by the alarming parallels between that crisis in bee keeping a century ago and the crises facing modern apiculture, comprising many of the same issues and underlying assumptions that existed when William Woodley kept his bees in Beedon, only a dozen miles down the road from Douai Abbey. Most alarming of all was the discovery that Isle of Wight disease was possibly (at least partially) caused by a reemerging disease today, known as Chronic Bee Paralysis Virus (CBPV), which has been growing again exponentially since 2007. History, it would appear, is repeating itself, and beekeepers are still uncertain, as they were a century or more ago, about the best way forwards for the craft. A further parallel is that, as then, we tend today to polarise the debate between two different methods of beekeeping. For them the focus of debate was between the traditional skep and the modern frame hive, while for us it's treatment-free beekeeping or medicating as a way of dealing with varroa mites. Then, as now, the debate rarely focused sufficiently on the type of bee kept and the consequences of selective breeding and cross-breeding of different races of the honeybee. Then, as now, there was little debate about the underlying assumptions that form the culture (or even the prevailing ideology) of beekeeping.

My research has made me examine and question this culture and its received ideas and beliefs, together with the wisdom of many beekeeping books and videos whose underlying assumptions are often in continuity with the ones William Woodley opposed. Many books will tell you, for example, that Isle of Wight disease wiped out the native Black bee, which I have come to believe is increasingly doubtful. Many books will quote Brother Adam of Buckfast, who believed that there were no native Black bees left in Britain after1915 and that any Black bees were imported stock from France or Holland after the Great War. It is assumed that all these imported stocks were resistant to Isle of Wight disease, but historical evidence in the old bee journals suggests otherwise. There is also a prevailing belief in some beekeeping circles and many books that unless we treat our colonies against the contemporary threat posed by varroa mites we will invariably lose them in a year or two, or three at the most. Isle of Wight disease is often also retrospectively discussed as a specific ailment (such as the mite *Acarapis woodii*), though there are arguments against this diagnosis. On the topic of other pests, some books also tell you that wax moths will not attack the stored combs in your supers, but only in your brood frames – but not in my experience. Other books tell you that if pollen is coming into the hive then you must have brood and a laying queen – not in my experience. There are many more examples of beekeeping dogma that is often unquestioned. Above all, an underlying assumption is still the idea that the honeybee can be domesticated and perfected like any other species, to maximise its productive potential and commercial value – ideas which I now dispute and discourage. This has been for me a radical point of departure from the mindset of Brother Adam and a century or more of modern beekeeping.

Something similar has happened in beekeeping, exemplified by a quotation I found attached to an article in a recent edition of the British Beekeepers' Association (BBKA) magazine. The writer quoted Socrates' advice that the secret of change is to be focused on building the new rather than on fighting the old. Most might ascribe the advice to Socrates, the Greek philosopher, whereas it's actually from a fictional character, also called Socrates, in a novel from 1980. There's a real danger that in beekeeping we receive too much of our information in a similar way, relying on second hand, received wisdom that can be misleading or even inaccurate, whereas we ought to question what we read and test it against our own experience.

I began to ask my own questions about modern beekeeping a few years ago when I decided to go *cold-turkey* with the apiary here at Douai Abbey (a Benedictine monastery in rural West Berkshire), by stopping chemical medication for *varroa mites*. I continue to ask many questions and to experiment as I journey further into treatment-free apiculture, raising my own queens and advocating locally adapted bees - especially after researching the Isle of Wight phenomenon.

Many questions also remain unanswered about the discovery of that carved queen bee on William Woodley's grave stone, such as how she came to be there. Was she put there on the instructions of William, as a touching tribute to Annie who had helped him run the bee farm? Or did his second wife have the bee carved when William died in 1923 and his memory joined Annie's on the headstone? If the latter was the case, then the strange, sculpted queen bee on an obscure rural grave in Beedon parish is all the more poignant. If it came to be there after William's death, it seems not only a fitting tribute to a beekeeper and his wife, but an enduring symbol of decline and death from which we might still learn; for, like her Master, the stone bee lies in her cold, hard cell like the grave itself, a memorial not only to a beekeeper but to a type of bee and to the vestiges of an ancient way of beekeeping. She and William Woodley have faded together into the past - not just a man and his bees, but the loss of a type of beekeeper and his native Black bees from a bygone age that can never return.

Despite being a man of the First Industrial Revolution, Mr. Woodley was, I think, aware with Thomas Hardy, the poet and novelist, that they were living at a time when the remnants of the old Victorian world were passing away. One of its most enduring symbols was the straw skep, introduced by the Saxons in the fifth century when they established the ancient Kingdom of Wessex where Woodley and Hardy both lived. Even in 1915, over half a century after the invention of the modern frame hive, the *British Bee Journal still* carried photographs and articles on skep beekeeping and what must have been the last of the old agricultural cottagers; those like a certain J. Vincent who relied on skeps as part of a cottager economy; though by then the editors were the first detractors of the traditional old ways. Their mocking (or at best, patronising) tone was perhaps a foreshadowing of the virtue-signalling commonplace today: 'With all his stubbornness and assertion that "nobody can teach him nowt "and that "book larning beant no good. I went to wuk when I wure foure I did, no waste o' time at skule," we love him because there is a great deal of practical knowledge of Nature and her ways crammed into his simple brain.' (5)

*William Woodley BBJ 1892*

The fading image of the stone bee defaced by time and neglect also seems to me a poignant metaphor for what has happened to honeybees in the last one hundred and fifty years since beekeeping adopted its modern methods to make life more convenient and more profitable for the beekeeper. Consequently, the western honeybee has had its distinct genotypes and ecotypes blurred by constant hybridization and importation. Its resistance to emerging and reemerging diseases and pests have been consistently compromised by chemicals, miticides and the disturbance of its adaptive genetics by rising numbers of imported queens; while many of the beekeeping methods and approaches that continue to cause its decline have fossilised like Mr. Woodley's sculpted queen. Lastly, that stone bee seems to me a symbol of what has already happened to our *native* Black bees. Like the stone bee, their genetic identity is still faintly discernible in some of our wild bee population, though obscured by a century or more of out-crossing with other races and strains. But there is a lot of recent evidence that the genetics are still there, like treasure buried in the minefield of modern apiculture, lying beneath the surface of our locally adapted wild and survivor bees - waiting to come to the surface.

The patina of lichens on the Woodley gravestone might also be seen as representing the dilemma faced by modern British beekeepers about the next step to take in the craft. Just as there are two conflicting views about lichen on gravestones, there would appear to be two main conflicting views about the way beekeeping should progress. There are those who would clean the lichens from the headstone, to reveal the fading memorial inscribed beneath.

This, it might be argued, runs the risk of damaging the stone where water has gone into the substrate and where the lichens have attached to it. Is it better instead to leave the lichen, with the damage it has done to a memorial, or to remove it and risk damaging the stone even further, along with the environmental damage of losing important lichens?

The problem is analogous with the dilemma faced by British beekeeping: do we stop importing queens, cease medicating our stocks with chemicals and start again with a monoculture of our original *native* bee stocks, but risk the initial collapse of our local bee population not seen since 1919; or do we change nothing about our approach and continue compromising our bees at a physiological and genetic level which will eventually lead to a crisis down the line anyway – perhaps with the emergence of a new pest or crisis equivalent to Isle of Wight disease or varroa, for which there will similarly be no cure or treatment?

Or is there a middle way between two extremes in modern beekeeping that can keep everyone happy? I propose that there is – that we focus our efforts on keeping and raising locally adapted bees, including stocks from the wild that have built up resistance to varroa mites. These bees would become genetically as close as we can get to our native Black bees, with naturally-selected traits suited to our climate and environment. Not only would we eventually not need to medicate these stocks, but we would also eliminate the risk of introducing new pests and emerging diseases from abroad. In the meantime, those who can't go cold-turkey could still medicate their bees as needed rather than slavishly following the treatment routines taught by the books. Perhaps in a decade or so we might not need to medicate our bees at all.

There was probably a middle way in Mr. Woodley's day which neither he nor the beekeeping associations could see: for them it was presented as a choice between skeps or frame-hives, with little debate about the impact of imported and cross-bred bees that had escalated since the mid-nineteenth century.

Like them, we can never go backwards, but we can go forwards in a new direction. This book explores the history and some of the latest science, to understand where we are in beekeeping and how we got here. It documents a difficult season (2021) in beekeeping at Douai Abbey during our change of direction from Buckfast bees towards treatment-free beekeeping with locally adapted bees. This new chapter of our monastic beekeeping at Douai Abbey makes the case for a proposed middle way out of the crisis of our modern beekeeping, in an effort to avoid repeating the mistakes of a century ago.

William Woodley, and even the famous Brother Adam of Buckfast Abbey, with many others, made some of those mistakes, it's fair to suggest, as we all do. Their stories and my own as a monastic beekeeper can, I think, help us (as all history can) to learn the lessons of the past century in apiculture so that we might enjoy bees and beekeeping another hundred years from now. William Woodley especially serves as a reminder not just of a type of beekeeper, but a type of British man - rooted, as we all are in the west, in certain freedoms. Above all, his greatest warning to me is of the importance of critical thinking and freedom of thought and speech which are the foundations of every beekeeper's freedoms to practice their ancient and noble craft.

The ideas, beliefs and assumptions of much of modern beekeeping since the First Industrial Revolution have emphasised attaining the perfection of the honeybee and the advancement of apiculture, mainly through education and science. Up to a point these have been worthy aims, but the over emphasis on improving the performance of the bee (for its commercial potential and its resistance to diseases) is a trajectory that might well be setting us on a course towards the next big debate in beekeeping. In William Woodley's time the debate concerned the perceived meddling of government and the imposition of legislative powers that he regarded as a tyranny over the freedoms and responsibilities of the individual beekeeper. In modern times the same issue of meddling and tyranny might be looming on the horizon in an entirely new form; it could manifest in the power, ambition and finance of Technocrats and Big Business, upheld by laws and new technology, to alter irrevocably the genome of the honeybee. For William Woodley and the beekeepers of his age the fundamental question regarding bee diseases was whether to do something or nothing. We face the same question in the face of diseases, pathogens, parasites and other contemporary problems in our own beekeeping.

Whatever course beekeeping takes it will be important for the individual beekeeper to examine the ideas, beliefs, assumptions, subjective choices and responsibilities that underpin their own beekeeping, because it is my own view that the exploration of precision breeding and the prospect of a transgenic superbee would be irresponsible, wrong and an enormous mistake. Indeed it would ally itself potentially with the prospect of a post human ideology of the Fourth Industrial Revolution that even now threatens to unleash its own new tyranny by limiting the freedoms not only of the beekeeper but of the human race. We might then see further and darker manifestations of a truth we already know - that minding the bees is inseparably linked to minding humanity.

# MINDING THE BEES

## Chapter 1
### The End of a Simpler Life

'Imported Cyprians & Syrians. See December Journal.
Remittances and Letters posted
up to March 4th to LARNICA;
March 4th to 19th, BEYROUT;
after that MUNICH, GERMANY.' [1]


**British Bee Journal 1885**


It's good news and bad news. The bad news is that another colony has crashed, one of several so far this winter. There are many reasons why colonies can crash in the winter; some are down to bad luck or weak stocks, and some are due to the beekeeper's mismanagement or neglect. There's a certain wisdom perhaps in the old country tradition of *Telling the bees* when someone has died, because the person who does this not only observes a tradition (albeit a superstition) – but they also observe a duty beneficial to their bees' welfare.

George D. Leslie R.A, wrote in the *British Bee Journal* in October 1886 on *Superstitions about Bees* relating a 'lady informant' (2) who had given her own experience of Telling the Bees. George Leslie explained the old tradition: '... an old man told us that these bees had forsaken their regular hives for this place, because the bee- master had neglected to inform them on the occasion of a death in the family belonging to the manor. He said that it was a well-known fact that bees, if they were not told by knocking on their hive and by the voice of a death in the family they would either die or leave their hive.' He continued with the lady's experience, sent to him in a letter: ' ''On the death of my youngest son, came to ask me if he  might go tell the bees; not understanding the old custom of  that county I  marvelled, but said " Yes." It was done, and the bees  remained; but on the death of my brother, some time afterwards, though I  counselled the same, the matter was pooh-poohed, and the bees all died." ' (3)

I admit that in my case the reason for the losses is more prosaic and is indeed a type of neglect – though a strategic one. And the good news is that this situation is what I want - sort of. All right, maybe not exactly what I want, but what I need if my change of approach towards treatment-free beekeeping is going to work. I stopped treating the Douai Abbey apiary with chemicals and miticides a few years ago by simply going *'cold turkey'* and for the first year or two things were fine. Then the winter losses began. And you have to expect those losses. When the apiary is covered in snow and winter clenches the world in its most unforgiving grip, it can be hard to hold your nerve.

*Douai Abbey hives*

Treatment-free simply means not putting anything into the hives – not just chemicals, but even organic treatments such as essential oils, which beekeepers have used for many years to manage varroa mites, the scourge of modern beekeeping. It's not natural beekeeping, however, because keeping bees in frame hives isn't natural anyway. Neither is it organic beekeeping because you can never guarantee that your bees aren't foraging genuinely organic forage where pesticides are not used. Treatment-free begins from the premise that bees can solve many of their own problems without beekeepr intervention, given a chance, because that's what they've been doing for millions of years. Nor is treatment-free beekeeping another version of the obsession purely with varroa mites. Experiments in the US have found that feral colonies that have been treatment-free actually have greater immunity against many diseases than managed bees. They also have three times the immunity against a number of specific diseases.

It's important also to realise that treatment-free beekeeping is not some magical destination at which point problems no longer exist. We have to expect that there will always be problems with the weather, with poor seasons and with pests and diseases. Anyone who farms knows that this is just how it is in any farming, and beekeeping is another kind of farming. As far back as 1853 Lorenzo Langstroth described colonies that were suddenly deserted except for the queen. In 1891 and 1896 in the US large numbers of bees also began vanishing in an epidemic of May Disease. Throughout the decades of editions of the *British Bee Journal*, from the 1880s to the end of World War I, there is evidence of good seasons and bad seasons alike, the vagaries of the British weather, as well as the problems of foul brood and Isle of Wight disease. Perhaps beekeeping has arrived at its own version of the 'new normal' to which we are all adapting after the Covid pandemic?

It's not that easy in practice though. I've been watching the larger stocks and the nucleus hives going down like dominoes since December when the cold weather really started to set in. Until then they looked fine. Of course, every winter beekeepers expect to lose a small percentage of hives. It's why we carry nucleus hives (half hives) through the winter, as an insurance policy. The general rule of thumb is to carry a third of the number of your hives in 'nucs' or *nuc boxes* (short for nucleus boxes).This has served me well for the last couple of years when, out of six hives, average winter losses have been only one or two colonies. Last season I started nine nucs just to be safe, and four of them have also crashed. Now I don't feel safe at all. The apiary is on a knife-edge. I remind myself that when I made the decision to stop treating I also decided I would stop worrying. This pattern of collapse and recovery is nothing new. It's part of the territory of a treatment-free approach because it's fundamentally nature's way. In the wild colonies die off all the time under pressure of selection for survival. Even managed colonies that are treated can die off too.

It's important also to see this choice as a choice; sometimes not to do something is a choice too. But this approach is much more than simply choosing not to do something. It also involves the choice to have to do other things differently. For example, you have to begin looking for local wild bees because it's likely that some of these have been living in the wild, unmanaged by beekeepers, for some years and have already adapted  for survival. You want the genetics of these survivor stocks, especially if you start off with a recognised race or strain with an imported queen because they will almost certainly die quite quickly when you go treatment-free.

Local bees have been found to have many important microbes living in their gut, as well as having about 8,000 micro-organisms that have been identified living in the ecosystem of the hive. Treatments build up in the comb, killing off the friendly microbes on the comb and in the bees' guts, but local bees will re-introduce these microbes and build clean, natural comb, improving the all-round health of the colony in ways not yet fully understood. In short, local bees don't just live on comb; the comb is a whole ecosystem for organisms the bees have not only evolved to live alongside but require if they are to survive.

Bright sunshine illuminates the subtlety and variety of colour in winter trees – copper washes in the tops of the line of trees around the meadow; pink and green branches; silver of the birches. Others are holding their dried leaves, such as the oaks, copper-coloured with sparse dabs of paint like the marks of an Edward Wesson Watercolour. The afternoon is filled with loud birdsong – great tits and robins taking the starring roles.

Beekeeping opens your eyes to the natural world. If you stop and look for just a few minutes something always happens. Among the conifers next to the apiary I find a patch of wood pigeon feathers on the ground – mostly soft, white and grey body feathers strewn like an accident with a pillow. On top, several flight or tail feathers edged with black. Probably the remains of a sparrowhawk attack, as they often hunt through the trees at the edges of woodland – our only woodland raptor. Flying at fifty miles per hour, reaching attack speed in two seconds and hugging close to the ground, the sparrowhawk creates a high pressure cushion between its wings and the earth. Males take smaller prey; this meal was had by a female.

I empty another *dead-out* of old frames and notice a tell-tale ball of hay at the bottom of the box. Two field mice, also called wood mice or long-tailed field mice *(Apodemus sylvaticus)* stare up at me, motionless, eyes like blackcurrants. I decide they're doing no damage at the moment and leave them in the empty nucleus box, replacing its lid. I'm a softy.

This aside, the main reason I expected and wanted the colonies to collapse is because of my treatment-free approach to the problem of varroa mites that have plagued beekeepers ( metaphorically) since the 1990s in Britain, and have plagued bees (literally) since they became a global pandemic in the late 1980s. In short, the idea is that putting chemicals into hives to kill the mites is making the mites more resistant and the bees weaker, so it doesn't make any sense. Research suggests that unmanaged colonies in the wild, without beekeeper interventions, can develop resistance to *varroa* within five years through Darwinian natural selection. The downside to this approach is that initially you have to expect heavy losses (you're telling me!) My losses so far this winter exceed my worst expectations.

The first officially UK-approved varroa treatment became available in 1997, which allowed beekeepers to begin combating varroa mites in managed hives. The increased danger of these wild, infected colonies drifting into or robbing managed colonies led beekeepers to begin treating twice a year. In the wild, however, colonies were increasingly affected as much as managed ones, but were dying out because of lack of treatment. Despite the decreasing problem of wild bees spreading the mites, UK beekeepers and those abroad had become committed to treating their bees twice a year with chemicals and miticides – a situation that persists to this day. According to recent BBKA news, 40% of UK beekeepers treat twice a year, though 51% who treat annually or not at all would now maintain that one treatment a year is all that is necessary.

Just as during the mysterious Isle of Wight disease a century ago many beekeepers never saw the disease, there are beekeepers today who have never seen varroa, or have come across it only rarely. The longest known treatment-free beekeeper in Britain has not put chemicals in his hives for twenty-five years, while there is now evidence that there are many others who have reached six years or more without treating, or even as long as ten years. According to the BBKA annual Winter Survival Survey, about 25% of British beekeepers don't treat. 10% of those registered in the four UK associations have managed to maintain treatment-free beekeeping for at least six years and their colonies have developed natural mite-resistance

Alongside this change, wild colonies have also returned. As wild bees received no treatment they either died out or adapted to cope with varroa mites. Consequently, these wild colonies have been able to spread again and are now a key resource in developing managed colonies that are treatment-free. Resistant colonies have developed a range of traits that have been identified by researchers, disrupting the breeding cycle of the mites. This development has been replicated in resistant colonies in other countries too.

There are even whole beekeeping associations that are treatment-free, such as the Gwynedd experience of Lleyn and Eifionydd BKA in North Wales who have about five hundred colonies in north-west Wales that are treatment-free. Four other UK associations now have 70% of members no longer medicating their bees.

The problem with going treatment-free in any area is whether there are enough other beekeepers who are also treatment-free and the kind of bees kept by other beekeepers in the same area. The introduction of external bees, such as queens from abroad or from different areas of the country, floods the local area with their drones and disturbs the evolution of the adaptive genetics of local bees that are treatment-free. The importance of drones can also be overlooked by beekeepers who don't raise their own queens and who remove the drone brood favoured by the mites, but drones are 50% of the genetics in any area.

I began beekeeping at Douai Abbey with the Buckfast strain developed by Brother Adam, a bee bred to meet the commercial needs of beekeepers at a particular time. The bees were bred to be hardy in our climate, docile, high-yielding, non-swarming and disease-resistant. They were lovely to work with, especially for a beginner who was nervous about being stung, while they never swarmed and they produced a decent crop of honey. But I was faced with a problem once the original queens died off and were superseded; you lose the genetics of a particular pure strain or subspecies very fast – 50% in the first generation as they hybridise with local drones. I could either allow this process, in which case the bees would become local mongrels within only a few seasons, or I could replace them with more Buckfast queens, or queens of a different race, such as Carniolans.

Replacing them with pure queens was a problem for me because it meant buying imported queens. Even Buckfasts are no longer bred at Buckfast Abbey. I made a choice then to raise my own open-mated queens, but this meant losing the Buckfast genetics over a few years, with no guarantee that I would be left with mongrel bees with any desirable traits. The first consequence of this decision was that I had to lower my expectations of the amount of honey I could expect from each stock. Rather than keep a few Buckfast stocks I'd need to double this number of mongrel stocks to get the same amount of honey.

There was another unexpected consequence, however, that showed up in the first couple of years, which was that I noticed the bees in our apiary becoming noticeably darker as the Buckfast genes became more diluted with each new generation. What happened was that natural selection was de-selecting the Buckfast genes in favour of another kind of bee that was more adapted to survive under the pressure of selection for being treatment-free. This interested me and was the beginning of my passion for locally adapted bees – bees that are adapted for survival in a particular environment.

I glance up from the derelict hive thoughtfully. A couple of kites circle above, mewling, eyeing death. An herbaceous clot of mistletoe hangs high in one of the skeletal trees in the Abbey grounds. Mistletoe is parasitic, and this clump has been there since I have been at the monastery (but was probably here before I arrived), growing larger each year. Mistletoe might be parasitic, but there's *parasitic* and *pathogenic*, and *Varroa destructor*, the plague mite of honey bees, comes into the second category, vectoring all kinds of viruses to the bees as it sucks their fat, weakening their immunity to disease - and generally compromising them.

In clear blue sky the moon is a partial potato-stamp in white paint. I wonder if my strategy is the real deal, or a facsimile of Darwinian beekeeping. Am I just a child making potato prints, or is this really going to work, I wonder? The fact is that I have neglected for three years now to do what most modern beekeepers have come to regard as essential for keeping their bees alive. I'm suddenly tempted to buy some *varroa* treatments to save the remaining colonies at all costs, but decide I've come this far and must stick to my principles. I must hold my nerve.

It would be too easy to go back to a simpler life and to medicate the bees, but that simpler life is now an illusion.  Even if I have one hive left at the end of winter, this is Nature telling me something important and I have to listen. One hive left standing through natural selection is clearly survival of the fittest in action, and I can make increase of the stocks from that colony, knowing it's the strongest colony.

Using the J-end of my hive tool, I loosen the lugs of the top bars of a couple of frames towards one end-wall of the collapsed hive. Red propolis cracks like hard toffee as a frame comes loose. I turn the frame, holding both lugs, to gather information. There's plenty of sealed honey on this frame, and white cells that are stuffed with the winter fondant they've had as an emergency food. I pull a couple more frames, noticing a pile of dead bees on the hive floor beneath the bottom bars of the frames. At last, I pull a frame that has a small cluster of dead bees, their heads inside empty cells. On the reverse side of the frame another smaller cluster tells the same story - heads in empty cells. Some cells have only a back-end protruding - mini mortuaries in which the bees crawled, perhaps for food or heat, or both. In this frame there is no food. The evidence points to isolation starvation, which fits the profile of the recent weather pattern alternating between mild and Arctic-cold. In mild weather the bees move away from their stores, the cluster breaking up. A sudden cold-snap hits and the bees go torpid, unable to regroup or to move back to their stores. The bees on the hive floor probably fell there as they died. In short, it doesn't look like *varroa* is the main suspect here. It's been a similar story in some of the other colonies. Perhaps in a milder winter, losses from varroa would account for a smaller percentage of crashed colonies. The coldest winter for eight years has thrown in its own share of hardships for colonies that might already be stressed by varroa and the viruses that further weaken them. It's hard to arrive at any definite conclusions, except that varroa and this winter have both contributed to the heavy losses in ways I can't quite piece together.

I sigh. There's still March to get through. We could have another very cold spell, with alternating patterns, which could mean more isolation starvation. Then there could still be other hives that crash due to varroa. Another scenario is that in March winter bees begin to die off. If the queen isn't laying from late January, producing spring bees, the colony contracts before the spring expansion kicks in, and this makes temperature- regulation difficult. The bees can't then keep brood warm, to raise spring bees, or keep the cluster warm if the temperature plummets below freezing.

It's not only the cold we have to worry about either. A pair of silhouetted kites dog-fight in the gusting half-light of grey dawn over the monastic graveyard; flexing blades of danger as they bank and turn. By late afternoon the wind is roaring, battering the Abbey Church like ocean waves, making the wood-ribbed roof of the nave creak like a wooden ship. I begin to worry now about the hives blowing over.

After Compline (Night prayer) I take a torch down to the apiary and inspect all the hives. One has its roof partially blown off and the perspex inner cover is dislodged. No rain has got in, and it's not too cold, though the colony will have lost some valuable heat. I look below the perspex cover and see bees moving. Hoping they're all right and that I've caught the situation before disaster strikes, I replace the lid and put a brick on top. Crisis averted – I hope, as I walk back to bed in squalls of rain and battering wind.

I'm seized by a sudden romantic yearning for the seemingly simple, old world of nineteenth century black and white photographs in the old bee books and journals, depicting cottage gardens in full summer with skep hives and Bee Masters in tweed jackets, smoking pipes; a world depicted by Tickner Edwardes' The Bee-Master of Warrilow, first published in 1907: 'Among the beautiful things of the countryside which are slowly but surely passing away must be reckoned the old Bee-Gardens – fragrant, sunny nooks of blossom, where the bees are housed only in the ancient straw- skeps, and have their own way in everything, the work of the beekeeper being little more than a placid looking-on at events of which it would have been heresy to doubt the finite perfection.' (4)

This world was vanishing, and was indeed doomed to extinction by the BKA before bee diseases, even as the beekeeper, William Woodley of Beedon, was writing his fortnightly *Notes By The Way*. Among his records of a disappearing way of life on the Berkshire Downs he related its traditions and superstitions, such as the old Wessex custom of Telling the bees, or *Waking The Bees* in 1908:

'Little brownies, wake.

Your master's dead –

Another you must take.' (5)

He continued, lamenting, 'But old customs have nearly died out, and the apiaries that graced the cottage gardens are dying out also in the village of Beedon.' (6)

It's tempting to think of modern diseases and pandemics in agriculture and apiculture and environmental issues as something unknown in the good old days. The truth is that bee diseases and their related issues have been around longer than we might think. *The British Bee Journal* of 1893 published documents showing the Beekeeping Association exam results in diseases such as foul brood. The minutes of the Bristol, Somersetshire and Gloucestershire Association for 2 March 1893 record that, '...it was proposed that napthaline should be supplied to each expert free of cost for the purpose of suppressing foul brood' (7) and that, 'Unless we get legislation shortly it will stop bee keeping in this parish as every hive is more or less affected.' (8)

Clearly, even before the early twentieth century and what came to be called generically Isle of Wight disease, nineteenth century beekeepers in Britain were already struggling with foul brood, the disease of that time. Indeed, so serious was the situation that in 1891 the BBKA syllabus was amended to include a section on foul brood, with minimum marks required in that section to obtain a pass. It was considered so important and such an issue in beekeeping that previously successful candidates were required to be examined in the new part of the syllabus in order to retain their expert certificates.

One expert created at the turn of the century was William Herrod, who rose to become secretary of the BBKA 1909 - 1930. He was initially allocated to Lancashire for eight-week periods during the season, with a salary of £3.00 per week. His job was to enrol new members to the association and to collect subscriptions. His 1896 diary details the 10 April when he began at 7:00 am, cycling forty miles to visit ten beekeepers. Having finished at 9:30 pm, he had examined twenty-nine colonies, one of which was a skep hive. Foul brood was found in four colonies. His diary entry includes a quaint detail of tea, bed and breakfast for 2s and 6d.

Those were the days! But they clearly weren't. As foul brood began to spread, the controversy about Bee Diseases legislation raised its head increasingly in the pages of the *British Bee Journal*. By January 1905, when Isle of Wight disease was also an emerging problem, William Woodley recalled in his *Notes By The Way*, 'I wrote ten years ago in BBJ these words:- "The more I go into the subject of foul brood the less I fear it", and I am of the same opinion today.'

At the same time he called for more supporters of his stance against legislation to write in to the *British Bee Journal*. I. Weston, Vice-Chair of BBKA, Hook, Winchfield, wrote to the journal, opposing Woodley's biased views on legislation, while Allen Sharp of Huntingdon supported Mr. Woodley. T.W. Swabey, Linc. , however, accused him of failing to come up with any alternative to the suggested legislation.

Even then, and further back, beekeepers would have lost a certain number of colonies in the winter, even with the efforts they made in those days to protect their stocks from winter weather. When you look at the traditional straw-woven skep covered in cow dung, sitting in the recess of a wall, known as a bee-bole, such conditions must have been much closer to the cavity of a tree trunk, the natural home of European honeybees, than today's frame hives such as the British National or the American Langstroth with their comparatively thin walls and larger volume. It's also too easy to romanticise the old-fashioned bee skep that was used, virtually unchanged, from Saxon times until the mid-nineteenth century and the invention of modern hives with moveable frames.

The skep, fashioned from coils of straw rope, sewn together with split bramble, dogwood or willow, is a beautiful object belonging to the age of rustic crafts that has all but gone. When I was a boy there was a television programme called *Out of Town* with a slow-talking Jack Hargreaves who visited cottage crafters such as skep-makers. Those programmes had the feel of a last glimpse at the same passing world Thomas Hardy captured in his novels.

The word *skep* comes from the Norse *skeppa*, meaning a half-bushel, which indicates that the first skeps were probably baskets used to measure out grain. They varied in design and size across Europe, but the basic design was standard. Bundles of wheat or rye straws, from winter-sown crops, were twisted through a short length of hollow cow horn to give an even width, then sewn together in coiled ropes, usually with split bramble. The skep-makers used two tools - a knife and an awl, the latter made traditionally from a piece of carved and sharpened bone. It would take many hours to make a skep, ensuring it was an even shape, with a lip that sat level on a base. A good skep often lasted up to a hundred years, reinforced with its waterproof covering of cow dung.

The skep design predates the arrival of the Saxons in Britain. There is evidence from lower Saxony of the top part of a woven wicker skep, dated 200 AD, found in 1977, which is the earliest known example. For all its associated romance, however, the history of beekeeping suggests that it was the issue of bee diseases - notably *foul brood*, that brought about the demise of the skep, together with the decline of the agricultural cottager economy during the First Industrial Revolution.

A few miles from the Abbey is the Berkshire parish of Beedon, Newbury. In the village churchyard is a gravestone marking the resting place of William Woodley. Carved into the gravestone is a queen bee. Why? Because William Woodley was at one time among the most well-known beekeepers in Britain, having at one stage the largest bee farm in Britain, with apiaries of one hundred hives at Stanmore and the same at World's End - hamlets within Beedon parish. Though now an obscure character in the history of beekeeping, his writing for publications such as the *British Bee Journal* document the revolution in beekeeping methods that he came to regard eventually as the cause of modern bee diseases. In his *Notes By The Way* in the *British Bee Journal* in 1905 he wrote: 'I have handled bees nearly all my life. I hived my first swarm in June 1856, and I have kept bees in straw skeps till some 27 years ago, and till then did not know foul brood even by name.'(10)

A sketch of William Woodley's biography appeared in the *British Bee Journal* on 17 March 1892 in an article on 'Our Prominant Beekeepers'. It informed the reader that he was born in Oxford on 9 March 1846. After losing his mother at a young age, he went to stay with a great aunt in Stanmore, a hamlet of Beedon in Newbury, on the edge of the Berkshire Downs. She gave him the task, when he had turned seven, of *mindin' the bees* for several weeks during the swarm season, which meant watching them and alerting her if they swarmed. She insisted he was kept occupied while doing this, by weeding the garden and reading the Gospels from his Bible. This experience gave him an early exposure to beekeeping, although his first choice of career in 1859 was as an apprentice grocer in nearby Chieveley. Later, after a passing interest in photography, he turned an amateur interest in clock-making into a second career, until his cousin, A. D. Woodley of Reading, started him off in beekeeping. By 1878, having begun keeping bees in skeps, he adopted the new technology of the frame hive and began exhibiting his honey at shows, where he became a prize-winning exhibitor. In 1889 he was presented to Queen Victoria at Windsor.

Woodley's career in beekeeping developed into at one time the largest bee farm in Britain, with 200 hives. It was a time when beekeeping was undergoing a revolution that paralleled the Industrial Revolution. Both revolutions were closely connected with the passing of a way of life exemplified by the rural cottage economy. Like many hamlets, Beedon's inhabitants were mostly tenant agricultural workers. The article on Mr. Woodley in 1892 informs us that he was an adviser, will-maker and counsellor to his neighbours and that he was often consulted about neighbourhood matters. We also learn that he managed a benefit society. These biographical details help provide useful information with which to understand his fierce defence of the cottager and their skeps.

Cottages and a piece of land often came with a job in agriculture, and cottagers would work that piece of land to supplement their poor income from agricultural work. Beekeeping was a common feature of the cottager economy: '...nearly every cottager in the neighbourhood kept bees', (11) Woodley wrote in 1913. 'The cottagers used to winter five or seven as a rule but some kept more for winter stock.' (12)

Beekeeping therefore was a vital part of the cottage economy. In *The Cottage Economy* and *The Poor Man's Friend* 1821 William Cobbett explained: 'A good swarm of bees, that is to say, the produce of one, is always worth about two bushels of good wheat. The cost is nothing to the labourer. He must be a stupid countryman indeed who cannot make a beehive, and a lazy one if he will not.' (13) He further suggested that, '...suppose a man get three stalls of bees in a year. Six bushels of wheat give him bread for an eighth part of the year.'(14)

Woodley was part of the beekeeping revolution that happened around 1860-70. At the same time as industrialisation and the disappearance of a whole way of life on the rural Berkshire Downs, workers from agriculture began moving to towns and cities for work as mechanisation took over on farms. In the beekeeping world a similar transition was happening, as beekeepers began to move from skep beekeeping to the new frame hives. But it was more than just the equipment that was changing. From the 1850s beekeepers also began experimenting with other races of honeybee, imported from abroad, such as Italians of the Ligurian type and Carniolans. Having experimented with them, and hybrids, many beekeepers returned to the Black bee, which William Woodley never abandoned, believing it beyond compare in our climate and equal, if not superior, to any Italian bee. Indeed, on January 26 1911 a letter appeared in the *British Bee Journal* referring to a recent edition of the *American Bee Journal* in which reference was made to a meeting of Maine beekeepers who had overwhelmingly favoured the Black bee over any Italian.

It was inevitable that British beekeepers would also begin cross-breeding the different races. There are many letters in the early years of the twentieth century in the *British Bee Journal* from beekeepers, asking advice about the outcomes of certain crosses between races of bee and from those who had experience of the most favourable crosses they had made in their apiaries. There were also adverts for different races of bees at least as far back as 1886; 'pure', 'Blacks' and 'English Bees', 'Pure Natives' 'Pure Black swarms' and 'Ligurians', (15) with some breeders specialising in English native bees and foreign races such as the Ligurian or the Carniolan, evidenced in this advert from 1896:

'RELIABLE QUEENS (Ligurian and English).

Prolific laying Queens, 5s. 6d. Sent post free in my introducing cage.

Safe arrival guaranteed. Orders filled in rotation.

Henry W. Brice, The Apiary, Thornton Heath.' (16)

Among the foreign races available were F. Benton's Cyprian bee, 'The race of the island of Cyprus is the noblest and most valuable of all bees which up to this time have become publicly known.' (17)

William Woodley, however, along with many others, preferred the native Black or English bee, 'I still stand by "ye olde Englishe bee".' (18) He was dismissive of foreign races of bees, writing in June 1908 that other factors than race were important, such as the skill of the beekeeper and the location in which the bees were kept. In August that year he also warned against hybridisation, clearly aware of what we now understand as *F2 aggression* in cross-breeds of bee: 'I say do not hybridise your and your neighbour's bees by the introduction of foreign blood without consulting him, or in a year or two the latter may feel unneighbourly if you turn your erstwhile docile stock of bees into veritable demons. ' (19) In fact, in 1908 he seems to have regarded foreign bees as no more than pretty (which they were, compared to native bees), and that they were nothing more than a craze: '...I attach far less value to the golden colour of the abdomen in the bees bred by "faddists" for beauty than I do to the energy displayed...'(20)

In all his writings for the *British Bee Journal* I have only come across one piece in which Mr. Woodley seriously engaged with the idea of foreign bees, when he wrote in 1896 regarding the bee-breeder, Mr. Brice's bees:

'Mr. Brice, I notice, has both "native" and "hybrid" queens for sale; may I ask how the mating of these queens is managed, and if all are reared in the same apiary? I should suppose that the mothers are English or Ligurian, as the case may be, and the drones? Would the daughters of an English queen mating with a Ligurian drone be considered English or native, or how are the queens classified? I should expect that if I reared English queens, and there were Ligurians in the neighbourhood, that I should get a mixture. It will be very interesting if Mr. Brice will give us his modus operandi in securing pure natives, if his hybrids are reared and mated in his home apiary, and if apart, how far apart?' (21)

The virtues of the native Black bee were well-known to beekeepers like William Woodley, despite how many dismissive opinions appeared in the letters and columns of the beekeeping journals. Even as late as 1922, when Isle of Wight disease was burning itself out, there were letters extolling its many good points as a prolific and hardy bee. On 22 January a correspondent signing himself ROBIN HOOD of Pickering was one such advocate of the native bee, claiming that in 1916 a single hive had given him two-hundred and eighty pounds of honey, excluding stores left to them for overwintering. Mr. HOOD also observed that during the Isle of Wight disease the first to show disease were not the natives but Italian hybrids, (presumably hybridised with Black bees) and therefore the same hybrid that survived the disease at Buckfast Abbey after 1915-16 and from which Brother Adam began to build his Superbee. This casts doubt too on Brother Adam's claim that on the basis of his experience alone this particular hybrid was categorically resistant to Isle of Wight disease in every case. It's more likely perhaps that similar or identical crosses could still have had very different genetics, accounting for why some were more resistant than others.

At this time many other beekeepers were not only experimenting with particular races, but were also consciously beginning to cross-breed them, as the monks were at Buckfast Abbey. It had been known for centuries that crossing plant and animals species could create hybrids which in the first generation displayed superior characteristics to their parents. Hybridisation in honeybees would often result, as with chickens, in increased productivity. Unfortunately, hybridisation requires not only the original cross but also line-breeding (mating within a closely-related line) and inbreeding (mating individuals that are closely-related), to fix certain desired traits and to remove others. This can lead to problems, as seen in other domestic animals, in which undesirable traits (such as deafness in Dalmatians and skin problems in Red Setters) is often the price paid for fixing a desired trait.

Honeybees are especially susceptible to inbreeding, which is the breeding of closely-related individuals. Their biology has developed to guard against inbreeding through polyandry in mating, which maintains a diverse gene pool and multiplies the number of possible recombinations of genes. Inbreeding narrows this gene pool. A known risk with inbreeding is that the cost of fixing desirable traits is that undesirable ones can become attached to a trait the breeder actually wants to fix, while other desirable characteristics can be lost. One of the biggest risks of such inbreeding, or any cross-breeding, is also that an organism can suddenly become more vulnerable to new diseases. Bee breeding therefore is always a trade-off between the desired traits and those that might be lost, and breeders have to balance the gene pool of desirable traits with the need to keep introducing new blood, to maintain genetic diversity. This is complicated by the fact that bee breeding is not about individual bees but about whole colonies, compounded by a biology and way of breeding unlike other species we have domesticated.

It is not a loud voice in the *British Bee Journal* in these years of controversy about beekeeping methods, but there were a few who wondered if the rise in bee diseases such as foul brood and Isle of Wight disease might have had something to do with the increasing number of foreign bees entering the country. Brother Adam's mentor at Buckfast Abbey – Brother Columban, even wrote to the journal with his own opinion on the issue. He responded to speculation that foul brood might have originated abroad with imported foreign bees, suggesting that the sale of driven bees (bees cleared from skeps to harvest honey) was a more likely cause. This was in sharp contrast to William Woodley's view, expressed in June 1905, and repeated at other times, describing himself as 'Expert' (22) in skep beekeeping for twenty-seven years before changing to frame hives, and having kept bees all his life,'...with practical experience, who says FB practically does not exist in straw skeps.' (23)

We don't know, but half a century of selective breeding since the 1850s, abroad and in the UK, and hobbyist beekeepers' own experiments with crossing bee races, might have resulted in hybrids which were losing the resistance to foul brood that native British bees had developed over centuries; which is why beekeepers such as William Woodley had never seen the disease in bees traditionally kept in skeps. These bees would be swarming, some of which had been cross-breeding with native bees (managed and wild) for half a century by the time foul brood and Isle of Wight disease emerged. No one at the time would have known enough about bee breeding or more recent areas of genetics to know to what extent decades of mostly random and uncontrolled cross-breeding might have made bees more vulnerable to new or reemerging diseases.

It was the industrialisation of beekeeping, which William Woodley used to his advantage, allowing increased and more efficient management and production of bees and their products such as section honey. Unfortunately, it happened just at the time when rural cottagers found it hard to retain tenancies; with agricultural work becoming less secure, cottage labourers had to move about much more. Mr. Woodley observed that over twenty or thirty years he had observed agricultural labourers changing their employer annually so that they had to stop beekeeping. In May 1913 his *Notes By The Way* record: '...nearly every cottager in the neighbourhood kept bees...the cottagers use to winter five or seven as a rule, but some kept more for winter stocks.' (24) In an earlier edition that year he had lamented similarly that: '...50 years ago nearly every cottager and every farmer kept bees in this district. Now I know of two farmers only who keep bees... The cottager beekeepers can be counted on the fingers of one hand in two or three villages.'(25) By 1913 he could only recall one agricultural labourer in Beedon who still kept bees.

This situation doubtless compounded the cottagers' increasing poverty. In the past, on top of reasonably secure wages, Mr. Woodley's *Notes By The Way* inform us, '...four or five skeps in a good season would produce enough to pay the rent.' (26) The poverty of the cottager at this time of great change not only meant they were unable to have the stability of tenure that enabled them to own stocks, but they were also unable to afford the transition from skeps to expensive frame hives with the related equipment. Whereas in the past many cottagers might have had the skill and the basic materials to craft their own skeps, frame hives required more science than art in their construction, not to mention more expensive materials, while the decline of the cottage labourer no doubt coincided with the decline in traditional craft skills such as skep-making.

Since their beginnings in 1873 and 1874 the *British Bee Journal* and the British Beekeepers Association had also become evangelical about frame hives and foreign races of bees, promoting them at places such as county shows. In an ironic entry in the *British Bee Journal* of 1910 Woodley stated: '...if I remember aright, the main object of the BBKA was to help the cottage beekeeper to be a more humane and profitable system of beekeeping.' (27) Clearly, from Mr. Woodley's observation of the decline of cottage beekeeping and the social and economic reasons that contributed to the decline, the BBKA was not succeeding in this aim. Beekeeping was even in danger of becoming elitist. In fairness, however, events had overtaken that agenda as foul brood and Isle of Wight disease became more prevalent in the first two decades of the twentieth century until the Government was petitioned by beekeepers to introduce a *Bee Disease Prevention Bill*, the aim of which (among other goals) was nothing short of the extinction of the traditional straw skep, a goal that seems to have been intended from the start by the BKA.

In 1911 Tickner Edwardes wrote to the *British Bee Journal*, criticising William Woodley's objection to this stated aim and denying that it would become an explicit part of any legislation: 'Mr. Woodley's remarks in "B.B.J." for Dec. 7: (p'. 486) are, I think, liable to create a false impression as to the nature of the petition to the Board of Agriculture which the "Smallholder" has for some time past been putting forward amongst its readers. The petition does not, as Mr. Woodley states, ask for the abolition of the straw-skep, nor does it advocate any particular measure what-ever.' (28)

Mr. Woodley was, however, correct on this occasion, and it was Mr. Edwardes whose facts were wrong. The editors responded with a footnote, probably hoping to shut down further controversy on the subject rather than with any intention of defending Mr. Woodley, such was their own evangelical view with regard to the shortcomings of the primitive skep:

'Our correspondent has overlooked the fact that the letter accompanying the petition asks for "support in bringing about the passing of a Bee (Prevention of Diseases) Act as drafted in the 'Small- holder,' and in the first clause of this it is distinctly stated that its object is the abolition of the skep. The exact wording is "Keeping of bees in anything but movable comb hives prohibited. Abolition of old skeps, box-hives, and the like to be complete in one year from date when the Act comes into force." We therefore cannot see how the petition can be dissociated from the proposed Bill, and the signing of the former would be a pledge of support of the latter.' (29)

The decline of the Douai Abbey apiary's stocks has halted for now. I check the stock that had almost lost its roof in bad weather. I look under the roof and beneath the perspex cover I see a black mass of jostling bees. Relieved, I replace the roof and the brick. This stock is one of a small number of survivor stocks from which I will continue to make increase each year. It's a valuable stock that I can't afford to lose.

A jay hops across the garden among a dozen foraging pigeons, tossing leaves aside and staring, motionless, at an area of lawn it has cleared. At the end of the meadow the farmer creeps along in machinery flashing sunlight, cutting his side of the old blackthorn hedge.

Winter's iron grip continues. It turns fridge-cold. A kite glides over sunshine and frost against pale-blue sky, turning and banking like something made from balsa wood and fabric, or floats behind the black, lino-printed trees, huge and terrible and Jurassic. Mallard ducks are back on the monastery pond. Between them and my room pink cherry-blossom, delicate as flowers made by a confectioner. Just after 2.00 pm a tawny owl hoots in the garden.

At this time of year, as in all nature, it's usually all happening beneath the surface in a bee hive, the queen beginning to lay eggs and producing brood from January or February. A whole world of mysteries is happening in every colony of bees, and one mystery I hope is happening is that the bees in the Douai apiary are beginning to develop a mechanism to cope with varroa mites. The mechanism has been developing for some years but has only recently been discovered by scientists, and involves the uncapping of affected brood where the mites reproduce. Affected larvae are thrown out and the mites usually abandon the cell.

A century ago the greatest mystery in beekeeping was Isle of Wight disease, a phenomenon in about three separate waves that spanned 1904 – 1919 and decimated hive populations across Britain. To this day the phenomenon is not fully understood and remains something of a mystery, with scientists disagreeing about the catalyst and the many contributory factors that fed into what seems to have been primarily a virus. Brother Adam of Buckfast Abbey was adamant that Isle of Wight disease was caused by acarine, also known as the mite *Acarapis woodi*, '...in the space of a mere 12 years the whole race was exterminated by this disease.'(30)

Writing about the native Black bee later in his career, he continued to believe that, 'She lives today only in memory. Some 50 years ago she fell a victim to the acarine epidemic and was completely wiped out.' (31) There are good reasons to challenge Brother Adam's assertion of these facts. Not every one (scientists included) agree, even today, that acarine was the underlying cause, and there is plenty of evidence that the native bee was not entirely wiped out by the Isle of Wight phenomenon. Brother Adam could not have known with any certainty that there were no survivors of the native race in parts of the country, and was dismissive of the hopes many entertained that there might have been survivor colonies in remote areas of the British Isles. Just as there are many things we still don't understand about honeybees, there were many more unsolved mysteries about their lives in Brother Adam's time.

*Brother Adam Kerhle -Buckfast Abbey archive*

For me personally, it's humbling to approach the bees with respect and awe for their mysteries. As I smoke them, when opening the hives, it often occurs to me that the smoker is a reminder that there are still many unsolved mysteries about honeybees. I often recall what the late Joseph Ratzinger (Pope Benedict XVI) referred to as *smoke in the eyes of the mind* when explaining why Roman Catholics use incense. For monks, smoke reminds us of the divine mysteries into which we enter as we celebrate, and the mystery of a transcendent God we neither see nor fully comprehend. Similarly, even for those who are not religiously minded, a beekeeper's smoker can be a reminder of the many mysteries of the honeybee, which in turn allude to the great mystery of Creation itself and the evidence for intelligent design. It would seem that at the fingertips of every beekeeper is the evidence monks chant in the psalms: '...all His works are to be trusted.' (32) Even those who are not religious can hardly deny that there is a compelling mystery about bees.

There are many mysteries still to be explained in the lives of this fascinating little insect. For example, we don't quite know how scout bees, who find a nest site for a swarm, know when to trigger the swarm. We don't know why all drones (males) in a geographical region for miles around make their way to places known as Drone Congregation Areas (DCAs) to mate, and how these remain stable areas for mating generation after generation. Indeed one DCA is known to exist at Selborne, Hampshire, discovered over two hundred years ago by the naturalist Gilbert White. What is remarkable is that drones only live a matter of weeks, and no one knows how they find these DCAs when there are no surviving drones or other worker bees from previous years to show them the way. Even more incredible and mysterious is that new queens from a particular colony will go to different DCAs from those to which drones in their own colony will go – in order not to end up inbreeding with their own sons.

There is one mystery in particular though which appears throughout the secret and strange lives of honeybees and indeed elsewhere in nature. It is called the Golden Ratio (sometimes referred to as the Greek letter PHI), containing a number-mysticism that seems to be profoundly purposeful and which suggests a kind of mathematical order in the Universe. The Golden Ratio, or Divine Proportion, has a numerical form, called the Fibonacci sequence, which occurs in remarkable ways in the lives of bees.

To fully understand this we must know something of the details of bee biology and reproduction, and it's a fun way to introduce the complexities of bee biology. What is then revealed is a mystical number at work, called PHI (1.618034...) and an associated mystical sequence of numbers deep in the lives of honeybees, and indeed throughout nature, manifesting in galaxies and hurricanes, the double helix, and even the molecules of our own DNA. In fact, our whole bodies, along with the wonder of bees, are part of a cosmic symphony of the Golden Ratio, a universal law that applies to all nature and which we find inherently satisfying to the eyes. There has to be a reason why this number occurs so universally, but as yet no one knows.

In 1202 Leonardo Pisano Bogollo (Fibonacci) introduced to the West a number sequence that had already been known by Indian mathematicians for a thousand years. The Fibonacci sequence, as it is now called, is a string of numbers, beginning with 0, that can be extended infinitely by adding up any two consecutive numbers to find the next number in the sequence.

Eg

0+1=1

1+1=2

1+2=3

2+3=5

3+5=8

etc...

The Fibonacci sequence emerges thus: 0, 1, 2, 3, 5, 8...

The relationship between the 2 numbers along the sequence can be described as 1.6 (The Golden Ratio), and the further along the sequence you go, the closer the relationship of the two numbers to the Golden Ratio (1.618034). To put it another way, the Fibonacci sequence expands at a 1:1.6 ratio, so the Fibonacci sequence is a numerical form of the Golden Ratio.

The number PHI is also unique because if you multiply or divide it by itself it goes up by 1 or down by 1. No other number does this.

We can depict this pictorially **Fig 1**:

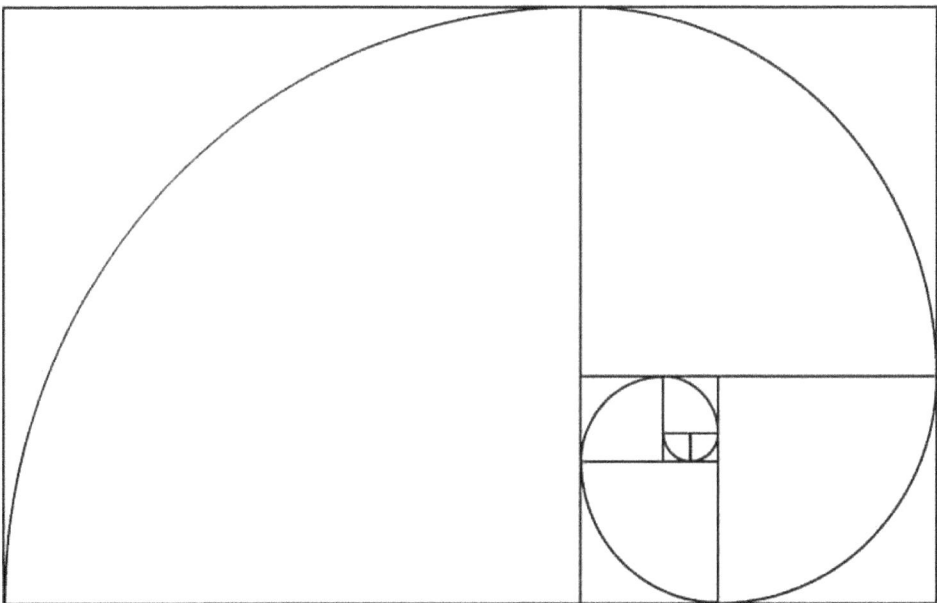

*Fig 1 Golden Ratio Rectangles*

if we draw a rectangle with sides a and b and divide them (a/b) it will give us the number PHI. When a rectangle of sides in this ratio is geometrically added to a square dimensioned to the long side of the rectangle, the resulting larger rectangle has the same ratio. If we keep repeating this process of adding a square dimensioned to the long side of the rectangle, smaller rectangles will keep emerging of the same ratio of 1:1.6. Every time you take a square off a golden rectangle it leaves a smaller golden rectangle.

We can then draw a spiral within the rectangles that looks like the spirals of a nautilus shell. This spiral pattern is found throughout nature; in many plants, the spiral of seed-heads such as the sunflower, shells, the curl of an elephant's tusk, storms, hurricanes and galaxies. What then has this to do with honeybees? To explore this question we need to understand something about bee biology.

Bee reproduction is very different from our own because the biology of their reproduction is unique. A queen produces two kinds of egg; fertile eggs which all produce worker females and infertile eggs which produce male drones, referred to as *parthenogenesis* in the queen. This means a drone only inherits one set of chromosomes from its mother, as it comes from an infertile egg.

Now pay attention at the back – this could get complicated: the drone egg is not fertilised (*haploid*), as opposed to the (*diploid*) fertile egg of a female worker or a queen. The latter have two sets of chromosomes, one from each parent (32 in total). The haploid drone has one copy of chromosomes (16) – from his mother only. Gametes are also haploid cells, which is why a drone is basically a flying gamete.

With one set of chromosomes from its mother, a drone has no father, but he does have a mother and a grandmother. For the same reason, he cannot have sons, but only grandsons. Therefore, a single drone has one parent and two grandparents. The grandfather has one parent because he is a drone, but the grandmother has two parents. So, a drone has three great grandparents. In the next generation, because he is a male and will have only one parent, the females will have four parents, making a total of five great, great grandparents… and so on…

1 parent

2 grandparents

3 great grandparents

5 great great grandparents

8 great great great grandparents

Each worker (all females) gives you a different number sequence from the drone sequence:

2 parents

3 grandparents

5 great grandparents

8 great great grandparents

13 great great great grandparents

Mathematicians will have already spotted something extraordinary in this sequence – it's a Fibonacci sequence, in which each number is the sum of the previous two. The ratios produce the number 1.61803 which is known as the Golden Ratio, or what Johanne Kepler called the Divine Proportion. It's found in the petals of a rose, the body ratios in humans and the spirals of snail shells, to name a few more examples. Just consider human hands; one thumb on a hand; and 5 digits per hand; eight fingers – all numbers in the Fibonacci sequence.

Returning to our honeybee example of the number of parents in each generation: if you divide the numbers of parents of a female bee by the number of parents of a male bee, you also end up with the golden ratio. No one really knows why. But without the drones in a honeybee colony, this golden ratio would not exist.

If we call drones **M** and Queens **F**, a five-generation family tree may be seen in **Fig 2**:

**A Drone's Parents**

*Drone family tree (Fig 2)*

In the first generation a drone has just one parent – his mother, because he only receives chromosomes from a queen who lays an infertile egg. Paradoxically, in this first generation then there are therefore no F1 drones.

By the second generation the same drone now has two ancestors: his grandfather and grandmother. In the third generation he has three ancestors – his great grandmother on the paternal side, his great grandfather on the maternal side and another great grandmother on the maternal side.

And so on... If you write down the numbers of ancestors in the five generations, the *Fibonacci sequence* begins to show up:

*1, 2, 3, 5...........*

We can do the same exercise with a queen or worker in **Fig 3,** which will always have two parents in the first generation, as they come from fertilised eggs.

## A Queen or Worker's Parents

*Queen or Worker family tree (Fig 3)*

In the next generation they have three grandparents,

then five great grandparents in the following generation,

with eight great great grandparents in the fourth generation,

and thirteen great great great grandparents in the fifth generation.

Again, for females, the *Fibonacci sequence* shows up in the number of ancestors in each generation:

*2, 3, 5, 8, 13........*

If you look at the number of male and female parents in each generation, again the *Fibonacci sequence* shows itself. In a drone's ancestry, there is one female in the first generation, two in the next, three in the following generation and five in the next.

*1, 2, 3, 5...etc*

Counting the male parents over five generations gives a similar result:

*1, 1, 2, 3, etc*

We can also see the sequence at work in a little mathematical puzzle with two rows of hexagon cells representing a section of bee comb ***Fig 4.***

A-B = 1 possible route
A-C = 2 possible routes
A-D = 3 possible routes (1+2)
A- E = 5 possible routes (2+3)

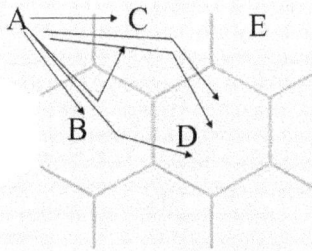

*Hexagonal cells puzzle (Fig 4)*

If we start from cell **A** and want to move to another cell to the right only, we can do it by going straight across or diagonally up or down. If we want to reach cell **B** there will only be 1 way to do this. If we want to get from cell **A** to cell **C** there are 2 possible pathways. From cell **A** to cell **D** there will be 3 ways. To reach cell **E** we now know that there are already 2 pathways + 3 pathways, giving 5 possible ways to get there. If we count cell **A** as 1 because to get there you simply stay where you are, then we arrive at a sequence of pathways:

## 1, 1, 2, 3, 5, ...etc

It has also been found that the elliptical form of the early stage of honeybee comb is not random but contains geometrical rules related to the *Golden Ratio* or *Divine Proportion*. Moreover, the number of drones to workers follows the *Golden Ratio*, with 10 drones for every 16 females. The larger the hives becomes, the closer it gets to the *Golden Ratio* 1.618034.

If this doesn't have people scratching their heads in wonder, one of the many mysteries which has only recently been discovered about honeybees is that they are developing unexpected ways to combat *varroa* mites. It was known that there were colonies surviving *varroa* in various parts of the world: the Arnot forest of New York State, USA; Mexico and Costa Rica; Primorsky, Russia; Gotland, Sweden; Avingnon, France; Tunisia and other places; but it wasn't understood how they were adapting. The breakthrough came in 2018 when a Norwegian PhD student (Melissa Oddie) discovered cell recapping behaviour was elevated in resistant colonies. Other key *varroa* traits found in honeybee colonies were then linked together by Izzy Grindrod, another PhD student, at Salford University.

I remind myself then that if the world of beekeeping seems to have been so much simpler a hundred years ago, it's important to remember that it wasn't. We simply had a simpler understanding back then. It's little wonder no one understood the possible repercussions of imported bees, of cross-breeding, of viral and bacterial diseases and the pests that importing might have introduced.

Right up to the early years of the twentieth century there were even letters in the *British Bee Journal* from beekeepers who still believed queens were mated within the hive by drones from their own colony. Indeed the fact of multiple matings was not really proven until the 1940s. Even Mendel, the father of the modern science of genetics, had failed to explore bee genetics because he had been unable to understand or control their mating. Most hobbyists during the late nineteenth and early twentieth centuries had even less understanding of the complex mechanisms of bee breeding, let alone the genetic consequences of cross-breeding and inbreeding and the long term detrimental effects that might come hand in hand with the short term gains from hybridisation of the honeybee.

Beekeeping was heading by the end of the nineteenth century in a direction that would lead to a crisis by the first decade of the twentieth century. Beekeepers would soon be scratching their heads for years, not to mention pulling out their hair in despair, at what began happening to their bees – a crisis that hit the apiaries of many beekeepers, including William Woodley of Beedon and Brother Adam of Buckfast Abbey. It helped finish off William Woodley's career, but was the beginning of a life's extraordinary work for Brother Adam who began breeding a new kind of bee that he hoped might provide the answer to everyone's problems.

# MINDING THE BEES

## Chapter 2
### Worthy Aims

'Consignment of Carniolan, Ligurian
& Cyprian Queens expected shortly.
CHARLES T. OVERTON,
LOWFIELD APIARY, CRAWLEY, SUSSEX.' [1]

*British Bee Journal 1890*

Meteorological spring begins 1 March.

20 March – the Vernal Equinox, when the hours of daylight and night are now equal. It's also the first official day of astronomical spring (though you wouldn't think so by the weather at the moment!)

From the Latin *aequus – equal* and *nox* – night, Equinox means *equal* night. According to folklore, you can stand a raw egg on its end on the equinox, popularised by a *LIFE* magazine article in 1945. Apparently several Almanac editors have tried; seventeen out of twenty-four of the eggs in their attempts did stand on end. Three days later the experiment was repeated; the result was the same.

Since 6 March I've been watching for a different kind of egg - frog spawn accumulating in the garden pond. It always appears in an area of the shallows, among the green daggers of emerging irises, speckled masses of pond tapioca several feet long and at least two feet wide, always on the western side of the pond. I've also watched from my cell window a few times this month a heron fishing for frogs, usually early in the morning. It stabs and I catch a momentary glimpse of silvery-white before the heron's throat bulges and the contents sink down its long, grey neck. There are no fish in the pond though - not with herons around, but there always seem to be plenty of frogs and newts.

I've also seen a male and female Mallard duck on the pond at the beginning of the month. Our Call ducks have started laying, a sign that the wild ducks will soon be nesting too, perhaps nearby. Around the apiary site bracken has begun sending up its first furry coils of new growth. Bracken *(Pteridium aquilinum)* from the Old English *braken* – probably from Old Norse *brakni*, meaning undergrowth, is a native fern, poisonous to livestock by producing hydrogen cyanide in its leaves. Its eventual growth to two metres tall, however, screens the apiary from the meadow-side in summer, sending bees upward, as they fly off over the meadow, minimising the risk to passers-by on the meadow. Right now it serves as a hopeful reminder to me of nature's cycle of collapse and recovery, its new growth emerging through last year's flattened, rusty leaves.

Spring is pushing back the winter slowly, but the bees have been flying regularly on sunny afternoons. March is also the best month for hearing the sheer joy of the dawn chorus. Three geese fly past my window, and over the pond, at first light, and I hear the whir of a woodpecker through nature's Matins.

One afternoon I walk across the neighbouring field into Eight Acre Gully, towards Greyfield Wood. Bright yellow celandine is flowering along the whispering little stream meandering through the trees. A nuthatch calls incessantly, high on a tree trunk in front of me. The flash of white wing-feathers – a jay through sunlit trees. Squadrons of redwings pass over the tops of the trees, calling *pseep...pseep*.

I'm always relieved during these sunny advances of the new season to see the bees bringing in bright-yellow pollen. There are four surviving stocks in the apiary, and all are bringing in pollen, a good sign that a queen is probably present and laying, as pollen is used to feed the young brood, though this rule of thumb doesn't always hold true. If it does, it means the spring expansion of each colony has begun, which not only means the stocks are probably queen-right, but that new spring bees are replacing the winter caste of bees now dying off. Winter bees are physiologically different from summer bees and live for several months, whereas summer bees live for four to six weeks.

It's still too early to open the hives, except to check under the roof that they still have a slab of fondant. March is the time when many stocks fail, simply through starvation, so I'm always vigilant this month. This year I supplement with pollen fondant, giving them extra food for brood production as well as for the adult bees. In the past I've opened the hives without protection, but consequently can now admit the shame and indignity of having being stung on the hand as early as 13 February when the bees aren't even active. My remedy for stings is a tube of yellow paraffin wax dabbed on the affected area as soon as possibly after being stung. Half an hour later there's hardly any swelling; an hour later you wouldn't know you've been stung at all. I first discovered it by accident when I had a horsefly bite a few years ago and picked up the wrong tube, thinking it was sting-relief. Now I swear by the stuff, which means I'm less likely to swear about being stung! Lesson learnt though – always be suited and booted, and safe rather than sorry.

There's not much else to do outside though, so I turn my attention to some jobs in the workshop. So far this winter I've made new frames for brood and honey, repainted the spare brood boxes and honey supers, and all the nucleus boxes in which winter stocks have not survived. But I've also decided that my new approaches to beekeeping are going to require swarm collection - swarms escaping from local apiaries and wild swarms, but preferably the latter. I already know that my stocks are depleted at the start of this season, while I could still lose others before inspections can begin in April, so I need a plan B. I've caught swarms before and have hived them successfully, but I've never actually baited them, which is the plan. In particular, I want wild swarms. Not only will they boost my stocks, but they are more likely to have developed some resistance to *varroa* mites, which will give natural selection a head start on my own efforts to select for resistance. As Brother Adam of Buckfast Abbey pointed out in *Breeding The Honeybee*, 'It is well known that in the battle against disease it is the wild forms of plant which play a decisive role.' (2) In the same way, it is honeybees living in the wild that are now playing a decisive role in developing resistance to *varroa* mites.

You can see varroa mites with the naked eye. They're about the size of a pin head and are brick-red. They usually enter a hive when a worker or drone carrying a mite drifts from its own colony or when workers rob another colony. Once inside the hive the female mite detaches itself from the host bee and finds its way into a brood cell where the queen bee has laid an egg in a pool of larval food in the base of the cell. Once the cell is capped the mite feeds on the developing larva and lays a small number of eggs. The first egg laid is male, which takes six days to develop. The following female eggs take eight days to develop, the larval mites passing through a couple of stages before becoming adults. These mites mate within the cell, the adult females leaving attached to an adult worker or drone bee as it emerges from the brood cell. The mites continue being parasitic upon the adult bee, leaving it after two weeks when they enter a new cell and repeat the cycle.

Wild stocks that have not been treated with chemicals are now known to develop resistance to the varroa mite, which is why I've suddenly become very interested in wild swarms. It's a complete change in my mentality as a beekeeper, because modern beekeeping puts a very heavy emphasis on swarm prevention, but this is a very new idea that was developed with the commercialisation of beekeeping. You find it in important books such as E. L. Snelgrove's *Swarming: Its Control and Prevention* and it is a foundational principle in the development of Brother Adam's Buckfast strain. Modern beekeepers are encouraged consistently by the beekeeping association to raise queens from non-swarming stocks. Apart from the inconvenience and possible danger of swarms, this is because modern beekeeping has emphasised increasing honey production, which relies on keeping a large colony intact during the main nectar flow. As the swarm season happens just before, or during, the main nectar flow, swarming results in lower honey harvests. This was fine for the cottager with a few skeps who could still take perhaps thirteen or fourteen pounds of honey in a good year from a single skep. Increasing productivity was a good intention when the British Bee

Keepers Association (BBKA) encouraged cottagers to abandon their skeps for frame hives, but it began to be very detrimental for bees.

Skeppists for hundreds of years, however, were indirectly selecting for the swarm impulse. Stocks that didn't swarm would not produce swarms for wintering, which then produced honey the following year or stocks that could be sold for profit, so non-swarming bees were not selected, but discarded.

Swarming also benefits bees as a strategy for avoiding disease, and might be a key factor in achieving resistance to varroa mites. It can certainly help manage the mites. Firstly, a swarming colony stops producing brood. As varroa reproduce on the brood and can only survive six days without a host, this reduces the mite load in the parent hive. When the prime (main) swarm then issues, a percentage of the bees carrying mites will drop them during swarming. There will also be a break in brood-rearing by this prime swarm which has to spend time building comb in the new nest site before the queen can start laying again. The parent hive will also experience a brood-break for at least four to six weeks until a new queen emerges, mates and begins laying. Swarming therefore benefits both the parent hive and the swarmed stock in its new site.

I have several swarm lures, of different designs. A couple are brewers' five gallon buckets with a convenient bee-sized hole in the bottom for a tap, which suffices as a bee entrance. I have a conventional hive cobbled together from parts of the old equipment left in the apiary by my predecessor. Lastly, I have made a couple of boxes from old packing cases or crates, to which I added a home-made removable lid, and an entrance hole. These hives were then coated inside with a melted concoction of old wax, propolis and a little honey, all gathered when cleaning the hive boxes and frames in storage for the winter; or old bits of comb that broke off the frames during movement back from the apiary room or during honey extraction. This coating inside the lures gives them the scent of derelict hives, which bees seem to find attractive. With the addition of a couple of pieces of wax foundation or even a little section of brace-comb cemented with a line of melted wax to the ceiling of the hives, additional incentive is given to a swarm which will build new comb from these foundations. The last ingredient when I set them out will be few drops of lemongrass oil which mimics the scent of the Nasonov gland in swarming bees' abdomens From my research, the best place to site these lures is along the edge of a tree line, facing South; ideally several feet off the ground, although there is plenty of evidence that bees will readily move into hives placed near the ground. From my own experience, swarms will happily settle on a nucleus box on a stand that's only a foot off the ground.

As I work, there are at least two conflicting beekeepers' voices in my mind. I recall the three beekeepers who delivered my first two swarms several years ago. Two of them sat me down to give me a few pointers before they left. Within minutes they were disagreeing with each other, the disagreements that I have continued proficiently ever since, on my own in an empty workshop. Beekeepers, I've found, frequently don't agree on anything to do with beekeeping. It was ever thus, as I've discovered in the ochre-aged pages of the *British Bee Journal*.

It isn't any wonder when there are so many more kinds of beekeeper today, with such varied reasons for keeping bees. The old skeppists had one simple reason for keeping stocks of bees, which was the cottage economy. Honey helped make ends meet for poor country folk. Today there are some beekeepers who simply want to make as much honey as possible, while others like to hug trees (and will try to hug a swarm as well in an effort to be natural and organic). And that's fine by me! There are beekeepers who want to make their life easy and efficient, and others who want to make it as comfortable as possible for the honeybee. Good! Still others are fascinated by the way bees work, and enjoy handling them, while there are others who see beekeeping as a way of saving the bees and looking after the environment and as a good excuse to work outdoors on a fine summer day. No problem with that either! With such varied approaches and motives, however, such a diverse group couldn't be expected to agree on everything about beekeeping.

In recent years, however, there seems to have been a more distinct division between two rather different approaches to the craft. On the one hand, there are those who adhere to an integrated pest management regime (IPM) which includes the use of chemicals and miticides. On the other hand, there are others who aim to be treatment-free, natural and environmentally aware. Each would accuse the others' approaches as deficient in some way, but in fairness, I can see the benefits of both approaches.

Natural beekeepers, however, don't necessarily all want to abandon frame hives, but there are many who want to keep bees in a more natural way that is more bee-friendly. Some might see that as a log hive that replicates the natural nest site in the cavity of a woodland tree. Many natural beekeepers might simply long to return to skep hives because they are less invasive and don't require inspections and pulling out or rearranging frames on a regular basis. In general, the opposing groups could be described as 'hands on' and 'hands off' beekeepers, and I suspect each of those groups could and would start an argument even over these definitions on the grounds that the different ways of keeping bees are probably much more nuanced than my generalised explanation. I accept that criticism. I venture to suggest that perhaps there are almost as many ways to keep bees as there are beekeepers, and that this was beginning to develop during William Woodley's career as a beekeeper when the Industrial age hobbyist superseded the agricultural cottager.

I admit to being torn, however, between two overall ideas of what constitutes responsible beekeeping. The advantages of modern frame hives include hive hygiene, because combs and bees can be inspected on a regular basis for diseases, and interventions can be made, where necessary, with a variety of treatments; everything from cutting out drone brood to reduce the varroa load, to introducing oxalic acid in January before the first brood appears.

It is a perfectly reasonable position to regard 'hands off' beekeeping, in contrast, as deficient or even irresponsible, because certain diseases and pests really are extremely serious, while some are notifiable, and that's absolutely as it should be. If we find avian flu in the country in wild birds or domestic ones, we have a responsibility to all poultry keepers and anyone who keeps pet birds, to mitigate the risk of infection, just as we had a collective responsibility to keep each other safe from the Covid-19 pandemic or other human diseases of the past. Even then, there are those who are against vaccinations or face masks, or restrictions and lockdowns. Just as there are those who will challenge someone in a shop during lockdown for refusing to wear a mask, there are beekeepers who might challenge other beekeepers for deciding not to treat their bees for varroa mites. Bees fly miles from their hives, meet other bees, and visit the same forage; they can transfer diseases and pests to other bees. They're also inclined (depending on the subspecies and time of year) in varying degrees, towards robbing other hives. A weak or crashing colony that has an intolerable mite load will often be robbed by stocks from the same apiary and from other people's apiaries. Robbing bees pick up not only easy honey and pollen, but whatever pests and diseases are in the collapsing colony. In the few minutes a robbing bee is stationary on the comb while it sips free honey, a *varroa* mite seizes its opportunity to find a new host and climbs aboard the visiting bee.

Conversely, 'hands off' beekeepers can see that modern beekeeping has locked itself into a position as disturbing as the need for farmers to use pesticides in order to make a reliable living. Pouring chemicals into the soil and the environment has all kinds of damaging consequences, which is why so many people have reacted by demanding organic fruit and vegetables, often grown themselves on allotments and in their back gardens. Similarly, pouring chemicals into bee stocks has wide-reaching consequences for bees. In the case of the varroa mites, the evidence seems to be that the parasites are becoming more resistant to treatments, just as bacteria are becoming resistant to modern antibiotics. Meanwhile, as bees are becoming more dependent on chemical treatments, they are more weakened by the viruses transmitted by *varroa* mites.

While we are nudging towards the topic, let me also address the fallacious slogan *saving the bees*. There's no such thing as saving the bee. There are approximately twenty thousand species of bee in the world, two hundred of which are found in Britain, all of which need saving from the effects of pesticides and crop spraying, pollution, loss of environment and monoculture in farming. Having a beehive in your garden or growing a few wild flowers through your lawn are worthy ambitions and have their place, but *the generic bee* doesn't exist. What exists is a complex and delicate ecology that has to be approached as a whole, in which honeybees play their part. But the idea that my stock of bees in a log hive down the garden is going to save the bee is nonsense, albeit well-intentioned nonsense. It's like suggesting that by keeping backyard chickens I'm helping the birds. Not that keeping chickens isn't environmentally friendly on its own terms though. I read recently of a Belgian town where two or three thousand residents were offered a couple of backyard chickens in 2010, to cut down on organic waste. In a year the town saved a fortune and cut down on landfill by 100 tons. In 2012 Pince, a town in northern France, introduced a similar scheme. Keeping chickens is clearly good for the environment, even if it doesn't save the birds; in the same way that keeping bees is good for the environment even if it doesn't save the bees.

I suppose I'm suspicious of both extreme positions in the beekeeping world, because the world we live in is complex. In a way I have great sympathy for every kind of beekeeper because they're all motivated by a love of bees, and that's good common ground. But I'm reluctant to dictate to the person who has two backyard hives, when the consequences of going treatment-free will be the probable loss of their bees and the repeated cost of replacing them. That's apart from the upset. That's just as unsustainable for some hobbyists as the need to continue treating stocks, but what else can a small hobbyist or a commercial beekeeper do when they don't really seem to have any choice? The beekeepers debating in my head are, in truth, great friends, just as were the three beekeepers who brought me my first stocks when they sat and argued. The truth is that since that day I've been conflicted, but on balance I can not see the sense in medicating the stocks any longer. Neither is that the kind of beekeeping I can enjoy.

If anyone had asked me in the first year of my beekeeping why I keep bees, the answer would not have been very complicated: I was asked to by my Abbot at the time (Geoffrey Scott OSB, now titular Abbot of Lindisfarne) but as is so often the case with obedience, a great blessing followed. I had never thought of keeping bees, or wanted to go anywhere near them. I wouldn't describe beekeeping as my vocation, though it has certainly become part of it, and like a vocation in a religious sense, it changes as we change and grow. Cardinal Basil Hume, a monk and Abbot of the English Benedictine Congregation, once wrote that he came to the monastery for one reason but found other reasons to stay, and I think this can also be applied to my beekeeping. Mine started as an act of obedience, with the idea that minding the bees would be about producing honey for the monastery. It hasn't remained completely that way though. To be honest, I'm really less interested in honey these days, but find myself fascinated by the way a honeybee colony works. My reason for keeping bees now is that it's an outdoor hobby which I enjoy, and a craft, but that the skills I continue to develop are now at the service of my monastic community but also the interests of bees and the environment they inhabit. Minding the bees is now more than just looking after the bees for my own benefit; it's also about *looking out* for the bees, for their benefit too.

When William Woodley's great aunt gave him the job of *mindin' the bees*, beekeeping for cottagers was not so much a hobby as an intrinsic and necessary part of the cottager economy, and this had been the way with beekeeping for centuries, when wax was needed for lighting and before sugar replaced honey as the standard sweetener in the human diet. But minding the bees has always been a craft, a learned set of skills and techniques that can be mastered through regular use and acquired experience. In that sense it's enjoyable in the same way golf is enjoyable for some people. Neither are an art; neither are forms of self-expression of emotions and experience, but getting good at minding the bees is just as satisfying, I imagine, as mastering a difficult swing in golf or learning to paint in Watercolour. So I suppose I keep bees quite simply for some of the the same reasons people might enjoy any hobby - it is satisfying to master a craft or a sport and have something to show for your efforts.

That is why experienced beekeepers have always been called *Masters*. It isn't a designation entitling us to dominate bees or the created order, but it denotes one who has mastered the craft, like a Master brewer or a Master builder. In Germany, even today, a Master is someone professionally qualified in their field who has reached the highest degree of skill.

For many cottagers in rural England and across Europe the craft of beekeeping was more than manipulating the bees or looking after them; there were also associated crafts, such as skep-making, using local materials that were readily available at low cost. Skeps would have been made through the winter months, probably occupying a cottager through dark winter evenings. Sitting by a fire, the skep-maker would weave their skeps from twists of rope or ropes of straw that were sewn together with split and soaked bramble into the coiled construction of a basket. The finished skep would then be covered with cow dung and dried before use.

A fascinating record exists in the Eva Crane Trust of an eighteenth century skep-maker and beekeeper called Thomas Owen. His records, in colloquial Welsh, written in two notebooks, record the weather, his family, local markets and his personal accounts, including information about his skep-making and the swarming of his stocks. These fascinating records are a valuable window into the life of a mid-eighteenth century rural cottager craftsman living in Denbighshire. It also gives us some idea of the lives of rural cottager beekeepers across Britain before the Industrial Revolution

Owen Thomas was a skilled maker of lip-work, or coiled straw basketry, which included skeps. His croft would have included making chairs, seed baskets, bushel measures and feed baskets, but his records show that he also made and sold skeps. Clearly, though a skilled craftsman, making a living was not easy and the family struggled to make ends meet. His wife, Elizabeth, and their children all supplemented the family income by gleaning barley, oats and wheat, from 1749 to 1774. They also burnt bracken and sold the ash between 1756 and 1765 and grew and sold potatoes. They were tenants in several places, which shows that even before the Industrial Revolution tenures could be difficult to secure.

Owen's main occupation towards the last thirty years of his life was skep-making, so it must have secured a reliable income. It suggests how common beekeeping was and that there was a demand for skeps, though his annual production varied, from 2 to 193, with an overall average of 97 skeps a year. Between 1745 to 1774 we can see from his records that he made 2,920 skeps and that some were made by his son.

Owen Thomas kept bees in the same traditional straw skeps that he made. Usually, he overwintered two or three stocks, with an average yield of thirteen pounds of honey from an overwintered stock. He overwintered medium-weight skeps from which he collected swarms the following year for overwintering. Heavy stocks were harvested for the honey and wax, while weak hives were united with other stocks. The process of harvesting that was common at this time was to kill the bees over a sulphur pit, but Owen drove his bees into another skep by rhythmic thumping on the upturned skep with another skep placed above it, driving the stock up into the empty skep. We know from his records that Owen drove his bees on the 14 July 1759, 3 July, 7 July 1772, and 29 June and 5 August 1774.

He left us detailed records of the swarming or *rising* of his stocks from 1757 to 1776. We know from these records not only the dates on which his swarms rose, but also second and third casts that swarmed. Although we don't know what he charged or therefore earned from selling skeps, the records tell us that he sold wax at local markets for fourteen to eighteen pence a pound.

Welsh law set out the value of swarms, a prime swarm being sixteen pence, a first after-swarm twelve pence and a third after-swarm eight pence. Of the old hives (meaning overwintered prime swarms) he recorded that thirty eight survived the winter and that a prime swarm issued from thirty of these. We know then that Owen lost some stocks in winter, as did every beekeeper; and that most of the British Black bees he was keeping swarmed. This was probably typical of the British Black bee, having for centuries been kept by skeppists in small skeps, which encouraged and naturally selected for early swarming and the swarm tendency. Early swarming would have been particularly favoured by labourers such as Owen Thomas because the swarmed stock then had more time in a good season to build up again and make some honey for harvesting.

Owen Thomas is an example of the cottage labourer who aspired to the vision set out by William Cobbett in 1830 in his introduction to *Cottage Economy:*'Better times, however, are approaching. The labourer now appears likely to obtain that hive of which he is worthy; and, therefore, this appears to me to be the time to press upon him the duty of using his best exertions for the rearing of his family in a manner that must give him the best security for happiness.' (3)

It was no hobby for Owen Thomas though. His crafts were vital to his survival, both the bee keeping and the skep-making, the latter being a substantial addition to his income, as evidenced by the number he made over three decades. It was hard-working, struggling people like him that the *British Beekeepers Association* wanted to improve when it was formed a hundred years after Owen Thomas:'...the encouragement, improvement and advancement of bee culture in the United Kingdom, particularly as a means of bettering the condition of cottagers and the agricultural labouring classes as well as the advocacy of humanity to that industrious labour the honey bee.'(4)

Past *BBKA* President, David Charles' modern objectives are unambiguous:'In short, the task was to effect a general change from the use of the straw skep to the management of the wooden movable frame hive.'(5) These were worthy aims, of course. Modern frame hives could increase production and efficiency, which would benefit labourers who wanted to supplement their income gained from labouring. It would also benefit the bees because honey could not until then be harvested without destroying the bees over sulphur pits.

But what we see here isn't only the modernisation or commercialisation of beekeeping. Another revolution was developing, that is perhaps happening again today: a revolution of the reasons for keeping honeybees among a more diverse group of people. Whereas bees had been kept for centuries as part of animal husbandry, for their produce, there were suddenly more hobbyists keeping them, who enjoyed the craft for its own sake. Some of those, like Sir John Lubbock, first president of the BBKA, had a scientific interest in bees. Still others, such as William Woodley of Beedon, were taking advantage of the revolution as industrialists had done in other business areas such as wool and iron. Beekeeping was splitting into two entities – a hobby and a livelihood. Both hobbyists and commercial beekeepers wanted efficiency, disease control and maximum yields of honey which could be better achieved with frame hives. Straw skeps could not be easily inspected or managed for swarm control, or easily harvested, and so they were falling out of favour. In short, why use a bicycle when you can use a motor car and reach more places more quickly? Of course, remaining with that analogy for a moment, we see today that people are getting back onto bicycles with their many benefits for health and the environment, so it isn't always the case that all modernization and mechanisation is always better. We also lose something by abandoning the old and simpler ways, whether they be the benefits of a push-bike or the many qualities of a traditional skep. And there are many advantages to skeps over frame hives that were perhaps too easily overlooked by the drive to modernise the craft.

An obvious advantages is that the skep shape suits the wild bee cluster. There are no corners that can get cold in a wild bee nest in a tree-cavity or skep, while both provide a more natural shape for comb-building, which is downwards rather than sideways. Skeps are also well-insulated, but breathable. Looking at my Nationals, I can't see how their comparatively thin wooden walls could be anywhere near as insulated as the thick straw walls of a skep that is then caked in a shell of mud or dung. As skeps are smaller than frame hives, this also makes for greater thermal-efficiency in the nest. Good, stable temperature regulation means winter bees use less energy and less food, while brood-rearing can begin in late winter, making for a rapid and early increase in spring so that swarming can occur early and swarms have time to build up stores for the winter.

Another feature of skep hives is that the bees make virgin comb each season, which reduces mites and the build up of pathogens, chemicals and insecticides when comb is kept for successive years. In a skep bees can also orientate their comb-building to manage their economy and ventilation. Skeppists observed that their bees naturally chose to build the cold way, the comb edges facing the entrance, which improves ventilation, especially in summer. When building their own comb, the stock in a skep hive can also determine their own cell size, rather than build to a size determined by the sheets or strips of foundation in a frame hive given by the beekeeper – which is often mostly worker-comb of a prescribed cell size..

One of the problems with frame hives and the modern approach to managing an apiary is that disease is easily spread. Movable frames can be taken out and transferred to other hives, to equalise stocks during the spring build-up, for example, or to donate brood and eggs to a queenless hive that has missed the chance to rear emergency queens. Because the combs are fixed in skeps, such manipulations are not possible, minimising the transfer of diseases between stocks. Propolis in fixed-comb skeps remains undamaged, whereas it is constantly cracked and removed by the beekeeper during frequent inspections of frame hives. Propolis contains natural antivirals and anti-bacterial properties, which means more propolis is actually good for bees.

The old skeps had the right amount of propolis, determined by the bees, whereas many modern beekeepers regard it as undesirable, as it glues the frames and boxes together, making it harder to take the hive apart for inspections. In the case of the Buckfast strain, which has been bred partly for its tendency not to produce large amounts of propolis, this can only expose them to greater risk of disease. One area that bees in the wild propolise is the nest entrance. In skeps the entrance is kept small, whereas in modern hives the entrance during the main season is deliberately enlarged. Not only does the latter lack the propolis door-mat that might disinfect workers' feet as they enter the hive, but it leaves colonies more exposed to robbing, with a much larger entrance to defend. And with robbing comes the danger of more varroa being imported into the hive. As skeps were not inspected as often as frame hives, neither were they smoked as much, which meant the bees didn't consume as much honey. It is estimated that every time a beekeeper smokes and opens a hive for inspection, they lose a day's honey because the natural instinct of a smoked bee is to gorge itself on honey, the smoke mimicking a forest fire, preparing the bees, if necessary, to evacuate the nest loaded up with honey, which means they are ready to set up home at a new site. Not least of all, the fewer inspections of skep hives also meant comb was not damaged or removed and fewer bees were squashed, leaving the stocks much less stressed. Lastly, in skeps stocks were actively encouraged to swarm, with natural requeening and natural brood breaks. Natural re-queening means the bees select the best queen, while the brood break reduces the *varroa* mite load.

Skeps would also have restricted the queen's laying, while the increased capacity of the frame hive encouraged the prolific laying more characteristic of the Italian bee *(Ligustica)* or the Carniolan *(Carnica)*. A consequence of this is that modern queens, stretched to maximum output in the current season, are not only less prolific the following season, but might even be weakened by this environmental stress in ways that we now know can be passed on, despite the bee's DNA being unchanged. Is this one of the complex reasons why many modern queens that seem to be laying well are superseded in their first season? The smaller capacity of the skep, however, was more suited to our indigenous Black bee, comparable to a tree cavity in the wild. Perhaps this was one reason why native bee queens were reputed to live up to several years in the wild. Although one trial does not constitute a scientific experiment, I have tried to keep a queen in the same five-frame nucleus hive for her entire life, to monitor the effect of restricting her laying. While other queens were usually superseded in their second season in a full National hive, the queen in the nucleus hive lasted into her fourth season before being superseded.

The ancient skep hive clearly had many qualities that commended it to the beekeeper, and yet it has all but gone. Although in Britain skeps could, in theory, be used, in some countries and many parts of the United States they are now illegal. The reason is that they cannot be routinely inspected for diseases and parasites, due to their fixed combs and the generally 'hands off' method of management by the beekeeper. I can understand why, and I have a lot of sympathy with the reasons skeps are regarded in such regions as retrograde and a danger to bees and humans alike. In many countries (none more so than the USA) Colony Collapse Disorder (CCD) has for many years been the greatest threat to honeybees, putting many bee farms out of business, and other farmers incurring huge financial losses from the reduction of pollinators each spring. The causes of *CCD* are not clearly understood, but seem likely to be a combination of factors, but it is understandable that the agricultural authorities and government legislation would want to do everything possible to reduce the threat to agriculture in these countries. Unmanaged hives and a lack of inspections would be a problem in this situation. Skeps, on the other hand, would risk worsening an already bad situation.

That aside, it seems to me that the virtue of the skep lies in the wisdom of the ancient art that it contains. There are many good lessons to be learned from what worked in skep hives that can be transferred to modern beekeeping in frame hives. A few modifications in the way we manage frame hives, borrowed from the skeppist, might go some way towards finding a better approach to keeping bees than the beekeeping books continued to propagate.

One of the changes this has made to my beekeeping is that I no longer want to put a queen under unnatural pressure to produce an enormous colony. Like the old skeppists who kept many skeps and hoped for a modest harvest from each stock, I prefer now to run more hives and to expect a smaller surplus from each one. I believe these queens can live longer. I also suspect that this approach might mitigate the environmental stress on many of our queens kept in unnaturally large hives in which they are under pressure to lay at full stretch.

In the last decades of the nineteenth century when Britain had one foot in skep beekeeping and the other in modern hives, or one foot in cottage crafts and the other in industrialisation, it was inevitable that the manifest destiny of 'progress' would win the day. The formation of experts by the BBKA, to counter increasing diseases such as foul brood, suggest that bee diseases were not a problem caused only by simple peasant skeppists, but worldwide were spread between managed frame hives too. One wonders to what extent the new frame hives were responsible for the spread of diseases such as foul brood and the mysterious Isle of Wight disease. Many people at that time were still learning how to use frame hives properly, and I'm sure there were many who had made the transition from skep hives who did not make regular inspections or changed old comb as regularly as they ought to have done. I've seen it even today; I know beekeepers who have never read a book on beekeeping and who know very little about it, even after keeping bees for several years, or more. In one beekeeper's largely unmanaged frame hive I've seen ancient black combs that must have harboured all kinds of diseases. Unsurprisingly, this beekeeper has great difficulty over-wintering a colony and almost annually loses all his bees.

Such a situation in Victorian times would have led to a polarisation of views. Skeppists could have maintained that bees had been kept for centuries in skeps, with relatively few problems. Frame hive beekeepers would have probably scapegoated the skeppist for the rise of bee diseases, maintaining that modern hives were more hygienic and more easily inspected for diseases; it was obvious that skeps were largely responsible, they argued, and the quicker they were superseded by frame hives, the better.

This view is indeed exemplified in the very first edition of the *British Bee Journal* on 1 May 1873 when the editors wrote: '...nor can we hope to induce the bee-keeping cottager (so called) to abandon at once the superstitions and obscure theories by which that class of beekeepers has been governed for so many generations. Our mission is to aid those enlightened members of the community who cultivate bees...' (6)

In fairness, perhaps today apiculture needs more knowledge about the natural conditions of a bee nest than it did in 1873. If you look at the changing trends in beekeeping at different times since the BBKA began, the current trend is characterised by uncertainty about the future of beekeeping. Perhaps it is time to look back at what we can learn from the past if we really want to advance apiculture and be able look forward to a more certain future for bee keeping.

A century after Thomas Owen's bees were *rising* and his skeps were making him a living, cottagers were abandoning beekeeping in large numbers. Some moved to towns and cities during the Industrial Revolution. Others could not secure stable agricultural work or tenure of land and had to abandon beekeeping. Those left were poor, and simply couldn't afford a modern wooden hive with all the equipment associated with it for extracting and filtering honey. William Woodley summed up their situation neatly in the *British Bee Journal*, 1 May 1913:'The frame hive is out of their reach. How can a cottager on 11s or 12s per week invest in hives and other appliances for modern beekeeping?'(7)

Despite this situation, after years of foul brood and Isle of Wight disease, pressure had grown to introduce legislation in the fight against bee diseases. On 17 March 1910 Lord Carrington asked the BBKA to obtain evidence that the majority of beekeepers were in favour of legislation, to strengthen his hand with the Board of Agriculture and Fisheries. By 1911 Tickner Edwardes had framed the *Apiaries Regulation Act* which appeared in draft form in *Smallholder* on 23 September 1911. The BBKA had already produced their own draft *Foul Brood Bill* as early as 1896, which was eventually published in the *British Bee Journal* on 14 July 1910, though Mr. Edwardes thought it deficient. Tickner Edwardes' Bill went much further, seeking to stop the importation of foreign queens from diseased areas, which some, such as G. Thomas of Stackpole, Pembroke, criticised as impractical: 'Clause VII. prohibits the importation of foreign bees from infected districts abroad. This will be of great benefit, but at present I do not see how it will be discovered whether a certain apiary abroad is infected or not.' (8)

On Friday 12 August 1912 Mr. Runcimann, President of the Board of Agriculture and Fisheries, introduced the Bill for its first reading in Parliament:

**'A BILL TO PROVIDE FOR THE PREVENTION OF THE INTRODUCTION AND SPREAD OF PESTS AND DISEASES AFFECTING BEES.'**

Be it enacted by the King's most Excellent Majesty, by and with the advice and consent of the Lords Spiritual and Temporal, and Commons in this present Parliament

assembled, and by the authority of the same as follows: — 1. — The Board of Agriculture and Fisheries (hereinafter referred to as the "Board") may make such orders as they "think expedient for preventing the introduction into England and Wales of any pest or disease affecting bees, and for that purpose any such order may prohibit and regulate the introduction or admission by post of bees, and of any articles or appliances used in connection with bee-keeping, and any other thing whereby any such pest or disease may be introduced, and any such order may direct or authorise the seizure, detention destruction, or disposal of any bees or things introduced or admitted in contravention of any such order.

2. — The Board may make such orders as they think expedient for preventing the spread in England and Wales of any pest or disease affecting bees, and any such order

may direct or authorise the destruction by the local authority of any colony of bees so affected, and any receptacle (other than a movable comb hive) in which there are or have been so affected bees, and the contents of any receptacle which is being used or has recently been used for bees so affected, and may authorise the destruction by the local authority, subject to payment by way of compensation of the value of the thing destroyed, of bees, or any other thing, which, in the opinion of the local authority, may spread a pest or disease affecting bees, or is liable to become infected by any such pest or disease, such value to be determined in manner pre- scribed by the order.

3. — (1) An order under this Act may impose fines recoverable on summary conviction for offences against the order, not exceeding ten pounds for any one offence. (2) An order under this Act may direct or authorise the local authority or anv committee thereof to which the powers of the authority under this Act may have been delegated, to carry into effect and enforce the order within the district of the local authority, and if a local authority or committee, when so required by any such order, fails to carry into effect the order or any provisions thereof, the Board shall have all such powers of executing and enforcing the order, or procuring the execution and enforcement thereof, and of recovering expenses incurred, as are conferred on the Board by section thirty-four of the Diseases of Animals Act, 1894, with respect to an order made under that Act. (3) In any proceedings under this Act, no proof shall be required of the appointment or handwriting of an inspector or other officer of the Board or of the clerk or an inspector or other officer of a local authority. 4. — (1) The local authorities under the Diseases of Animals Act, 1894, shall be the local authorities for the purposes of this Act. and any expenses incurred by a local authority under this Act shall be defrayed as expenses incurred under that Act.

(2) Every local authority shall appoint so many inspectors and other officers as the local authority think necessary for the execution and enforcement of orders under this Act, and shall assign to those inspectors and officers such duties and salaries or allowances, and may delegate to any of them such authorities and discretion as to the local authority may seem fit. and may at any time revoke any appointment so made. (3) Every local authority and their inspectors and offices shall send and give to the Board such notices, reports, returns, and information as the Board require. 5. — (1) An inspector of the Board or of the local authority may at any time, accompanied if he thinks fit by an expert adviser, enter any building or place wherein he has reasonable ground for supposing that there are or have recently been bees affected by any pest or disease, or that any order under this Act has not been or is not being complied with, and to examine any bees on such premises and anything thereon used for or in connection with bees : Provided that the powers of an inspector of a local authority shall not extend outside the district of the local authority. (2) If any person without lawful authority or excuse

(proof whereof shall lie on him) refuses to any inspector or other officer acting in the execution of this Act or of an order under this Act admission to any building or place which the inspector or officer is entitled to enter or examine, or obstructs or impedes him in so entering or examining, or other- wise in any respect obstructs or impedes an inspector or other officer in the execution of his duty, or assists in any such obstructing or impeding, he shall be guilty of an offence against this Act and shall be liable on summary conviction to a fine not exceeding ten pounds. 6. — This Act shall apply to Scotland in like manner as it applies to England and Wales, subject, however, to this modification, namely, that the powers conferred on the Board of Agriculture and Fisheries shall in Scotland be exercisable by the Board of Agriculture for Scotland, and that for the purposes of sub-section (2) of section three of this Act the Board of Agriculture for Scotland shall have the like powers as are conferred on the Board of Agriculture and Fisheries by section thirty-four of the Diseases of Animals Act, 1894, with respect to orders under that Act.7. — This Act may be cited as the Bee Disease Act 1912' (9)

On 16 October the order for the second reading of the Bill was read, which Mr. Runcimann called for that night due to the urgency of the situation. Mr. James Hope begged the Motion that the debate be adjourned that night, seconded by Mr. Ernest Jardine, but after objections from others in the House, Mr. Hope withdrew his Motion to adjourn the debate. After much discussion late into the night, however, it seemed that there was still work needed to get the Bill to a second reading.

The following year there were increasing concerns in the Commons about the worsening threat of Isle of Wight disease, while in May there were renewed hopes that the Bill would be passed during the current session of Parliament.

By 2 June 1913 Mr. Runcimann had to inform the House that a deputation of beekeepers with large apiaries, who made a living from beekeeping, had raised objections to the Bill, fearing abuses of the powers the Act would give inspectors, and that the Board of Agriculture recognised that it had to proceed with great caution regarding the proposed legislation. By July there were fears expressed in the House of Commons that if the Bill were not passed the fertilisation of fruit trees the following year would be greatly compromised. Among the deputation that had managed to stall the Bill's progress through the House was William Woodley of Beedon.

# MINDING THE BEES

## Chapter 3
### Queens of Foreign Varieties

'ITALIAN QUEENS DIRECT FROM ITALY –
Address, E. PENNA, Bologna, Italy.' [1]

*British Bee Journal 1910* [1]

Winter evenings, after a day's hard labouring, were probably the time when cottagers and farmers in the past maintained their spare skeps, ready for making increase from their stocks the following season. Skeps that had overwintered and produced honey that season or had been destroyed if they were too weak to overwinter, were brought in for storage and to be repaired and set up again for the next season's beekeeping. They would often wear at the lip, which was in contact with the base or stand and took the hive's weight, especially if the skeppist tipped up the hive to inspect the combs and the stock, resting it on a particular point that took all the strain.

Winter is the time I repaint equipment, build new hives and read about beekeeping. I also watch Youtube videos on the craft. I find and watch an old black and white film of a beekeeper on a cold winter night scraping the old covering of mud and cow dung from a skep and inspecting the woven ropes of straw beneath. If the lip is broken or uneven, which might create a gap at the base of the skep, the skeppist weaves in a new length of straw, twisting it into the rope. A piece of bramble, split and soaked, that was cut in winter, is inserted through a hole made by an awl and is woven around the new piece of lip-work and stretched into the woven rope of straw below. Where the bramble stitches are broken, they are cut out and a new, shorter length of bramble binding is inserted. Lastly, the inside of the skep is scraped, to remove old comb and propolis. A couple of pieces of old comb, a few inches long, and a finger's width are cemented into the top, inner surface of the hive. This comb is to encourage the swarm to build out new comb, which they will attach to these starter strips. Next, six thin skewers of wood, or *spiles*, are pushed through the outside of the skep, through the centre, until they pierce the opposite wall. Arranged in pairs, at right angles to the starter comb, the spiles create a framework through the middle of the skep around which the combs will be built. They give the combs stability, stopping them breaking off or leaning when the skep is tipped on its side or moved to another location. Repairs and cleaning completed, the skep is covered with fresh cow dung and left to dry out. I have a sudden overwhelming desire to make a skep. One day…

It's now the end of March and the beginning of Holy Week. At the Abbey we keep a quiet day at the start of this week, a day we usually have before Christmas and Easter before we get very busy. The forecast today is for sunshine and eighteen degrees Celsius, which would be ideal for opening the stocks and making the first spring inspection. There's a beekeeping saying that if the flowering currant *(Ribes ribes)* is in flower you can make your first inspection. After feeding the poultry I glance in the corner of the potager where we have a flowering currant and it's in full flower. It's cold though. By 10.00 am it's still cloudy and only ten degrees. Wind whistles along the cloister. Purple-brown flower buds are forming on the wisteria opposite the refectory. Each is the size of a queen cell, with a similar pitted appearance.

Wood pigeons spitfire across the meadow. Grey wagtails undulate overhead in short bursts, like toy planes, their engines cutting out intermittently – stalling, they fall…and rise again with renewed pulses of power. I notice the first green lacewing (*Chryoperla carnea*) of the year on the wall of the monastery North Block; pristine and lime-green as new spring growth, wings intricate and delicate as French lace. There are actually eighteen species of lacewing in the British Isles, the gold-eyed green species being the most common, and twenty-nine species of brown lacewing.

Beekeeping is governed by the weather - like farming. You can't book it in your diary, especially at this time of year. Some years you can't open the hives until mid April. I've done inspections as early as the end of February in the past when spring has been early. This year spring isn't early. March has been mostly dry and cold, but with sunny spells. The daffodils are in full flower and the tulips are opening. One of the Call ducks has been sitting on a clutch of eggs for nearly a week, and today I notice a second duck has started a nest and already has three eggs. Call ducks are seasonal layers, closely related to the wild Mallard, and their nesting is an indication of spring. Some years they don't even begin laying until late April and won't nest until early May, so this year they're strangely early, given the late spring.

I prepare my equipment and charge my smoker for the first time this season, so it's ready to light. I watch the sky all morning from indoors. Unless the sun breaks through, it won't be warm enough to open the hives without chilling the brood, and I don't want to risk knocking back the early spring build-up by killing off vital brood. By midday there's bright blue sky and sunshine, yellow daffodils brilliant in the garden, and cock pheasants strolling the monastery garden, polished as copper.

My monastic forebears at Douai Abbey didn't own a smoker. They used their tobacco allowance instead. In fact, when beekeeping began here in the summer of 1933 it was embraced with all the comical incompetence and crude improvisation of an episode of *Dad's*

*Army*. An old edition of *The Douai Magazine* narrates how our first swarm arrived in a June heatwave, having been promised in 1932 to Fr. Michael Young, a monk of the community, by one of our priests in a mission parish in South Wales. An old hive was requisitioned, minus its frames. The nervous monks deferred transferral of the bees until dusk, as the package was so ill-tempered, mistakenly believing that bees would not sting at night. They eventually approached their task at 9.00 pm, protective bathing suits on their heads and cigarettes smouldering. The operation proved to be rather hazardous, according to the article which ends on a cautionary note: 'Bees are wonderful little animals, but when roused...' (2)

The bees today are generally more cooperative. Inspection takes about an hour because three survival stocks are in nucleus hives, or nucs, which I transfer, frame by frame, into a full-sized hive. The stock in a National brood box is looking vigorous, with plenty of bees. I spot the queen, marked clearly with her blue dot on the thorax, reminding me she is last year's queen. There's not much brood though, indicating that she hasn't really started laying yet. This surprises me as the stock has been looking as though it might be my breeder hive this year. Now I'm not so sure. Why, I'm wondering, has the queen not started laying? On the other hand, I've seen good queens start slowly before, and catch up well.

There are three nucleus colonies, one overwintered in a wooden nuc hive, and this stock looks extremely strong. I locate the queen and see a couple of frames full of sealed brood. This looks like the kind of queen from which to breed. She's ticking all the boxes; laying early, a winter survivor, with a strong, docile colony, and building up early. The other two nucs are looking reasonably strong, with some brood, but I don't see either queen, despite moving all the frames into full-sized hives. Of course, it's possible to miss the queen, especially if she's hiding under a cluster of workers, or she can run around the back of the frame just as you're turning it, or even jump off the frame and run down between the frames into the box. I decide not to panic, but to close up and wait a week before checking again.

It's a bit of a disappointment to be left with so few stocks, especially if two more nucs have now become queenless. On the other hand, the more brutal natural selection is, the more reassuring it is that the remaining survivor stocks are the ones from which to make increase. That's beekeeping, and there's nothing new in losing stocks over the winter, or finding weak stocks in the spring. All beekeepers deal with these varying fortunes from season to season. In the past the old country skeppist would lose stocks too, and sometimes found that a poor season gave them no honey either. We do well, however, to realise that skeppists could also be smart; William Woodley tells in one of his columns of a lady skeppist neighbour who only ever kept and overwintered her middle-sized stocks, and she hardly had any swarming next season.

I reassure myself that even Brother Adam had his misfortunes while developing the Buckfast bee, despite his experience, dedication and expertise. Indeed that whole enterprise was initiated when Buckfast lost most of their stocks in 1915 in the last major wave of Isle of Wight disease. Then in the spring of 1932 he had secured the use of an isolated mating yard at Sherbeton, on Dartmoor, arranged with the Devon Beekeepers' Association. He and Father Benedict had transported two hundred mating nucs to the area and were resting, on arrival, with a cup of tea when their van caught fire while still loaded with all the nucs. Brother Adam lost a whole year's work that day. He must have felt sick with the blow. On another occasion in the Ligurian Alps he had some Ligurian queens packaged on the kitchen table overnight, ready for shipping. In the morning he found them covered with black ants, which had killed all the queens.

My misfortunes are slight by comparison. One spring, when I had fewer stocks, I had to rebuild the apiary from a single hive without any spare nucleus hives, which taught me a lesson always to make nucs as an insurance policy. At least this year my decision always to take nucs into the winter has paid off. It's left me, if nothing else, with at least one very vigorous stock and a good queen, which is a very secure foundation on which to build my resistant survivor stock. That's a positive, I remind myself. Hopefully I can also boost the apiary with captured swarms, if my swarm lures work.

The blackthorn hedge at the end of the meadow is flowering. *Native blackthorn (Prunas spinosa)* is sometimes confused with the later flowering hawthorn (*Crataegus monogyra*). Whereas hawthorn produces leaves before blossom, blackthorn flowers before coming into leaf, usually towards late March. Remains of blackthorn have been found in Iron Age excavations 3,400 years old near Glastonbury by excavators Arthur Bulleid and Harold St. George Gray. Folk names include *pear haw, pear hawthorn, snag, wishing thorn, faery tree* and *Mother of the woods*. According to ancient Irish mythology, moon fairies live in blackthorn, coming out in a full moon to worship the moon goddess. This is the time to cut the wood. Throughout Britain blackthorn has had a long connection with witchcraft. It is believed in folklore that the staffs and wands of wizards and witches are made of its wood, though in modern times the less superstitious rambler will often carry a blackthorn walking stick.

In British folklore witches were thought to use a blackthorn *stang* in cursing rituals, and to pierce dolls with the thorns. Probably for these reasons witches, as well as heretics, were traditionally burned on blackthorn pyres. In some parts heavy sloe crops signified winter sickness. Hence the rustic saying, *many slones, many groans.* T. F. Thiselton-Dyer's *The Folk-Lore of Plants* 1889, records the country worker saying: 'When the sloe tree is as white as a sheet, sow your barley whether it be dry or wet.'(3)

My frustration at the moment is largely because I want to work outdoors again after the winter. I dare say there are many modern hobbyist beekeepers who not only enjoy working with their hives, but enjoy the outdoors and nature as well. Many of us are no different from Isaac Walton and the angler, for whom catching the fish is the icing on the cake. For them and for beekeepers much of the pleasure of the craft is in the connection it gives us with the natural world. We do well to remind ourselves that the craft of apiculture can't be reduced entirely to economic concerns or goals.

That said, I have to avoid the unfair temptation to regard William Woodley's industrialisation of the honeybee as one of the moments in history that beekeeping began to be driven by purely commercial concerns. The truth is, in the thousands of years humans have kept bees there has always been an economic motive. Even before the economic necessity of the cottager's bee garden, people kept hives for one main reason – its products of wax and honey. It is a luxury today that people have the time for interests and hobbies, and so we add these reasons to the list of motives for our beekeeping. In the past it was simpler, and way harder; cottagers, and their ancestors, going back to Medieval times, Saxon times and earlier, never kept bees simply because it was an enjoyable or interesting past-time, but mainly because humans like to eat honey and because life was hard and so people did everything they could to make ends meet. Contrastingly, in the articles of the *British Bee Journal* about the readers' apiaries at the turn of the twentieth century there are many of examples even of the early hobbyist beekeepers with frame hives who were perfectly content to takes a surplus of only thirty pounds of honey per hive, or whose costs exceeded their income from selling honey. Many of them stated that their motivation for beekeeping went far beyond any commercial gain. It was because they were hobbyists.

Brother Adam concluded his book *Breeding The Honeybee* in 1982 with this summary: '...our only hope of achieving real progress in the improvement of the genetic composition of the honeybee, open to us, is by way of cross breeding and the synthesization of the wealth of economically viable traits.' (4) This led to his pursuit of the Buckfast bee which was the result of, '...combination breeding, that is, bringing together in one the commercially important characteristics of the various races.' (5)

I can't argue with the fact that a primary goal of beekeeping has always usually been economic. Even the hobbyist today hopes to sell some of their produce, especially to cover the costs of a hobby that can be quite expensive. Indeed here at Douai Abbey, although our apiary is not a commercial operation, part of the plan each year is to sell honey that is surplus to the monastic community's requirements, which helps to cover the running costs and the expenditure on new equipment such as frames and wax foundation.

William Woodley's bee farm at Stanmore and Worlds End, on the edge of the Berkshire Downs, was an economic enterprise that sought to maximise profits by industrialising the honeybee through the use of the newly-invented frame hive and the large scale of his operation. He grew his farm to 200 hives, selling swarms of bees and honey, including his prize-winning cut comb honey that he presented to Queen Victoria. William Woodley is notable, however, for only keeping the native British bee.

What interests me is that, according to Brother Adam's book *In Search of the Best Strains of Bees*, non-native races that might have had more economic value had already been available for decades '...on July 19th 1859, the first consignment of Italian queens reached this country.' (6) By 1900 even the Cyprian bee had reached Britain, though the jury seems to have been out on its suitability. J.P. Jackson wrote in 1881 of Mr Edouard Cori's breeding experiments in Bohemia with different races since 1864, and that Cori regarded the Cyprian as the best race. In 1881 another report appeared in the *British Bee Journal* that Mr. D.A. Jones had brought to Britain the first shipment of 150 Cyprian queens from Cyprus and the Holy Land, available for sale at thirty shillings. Some letters in the *British Bee Journal* asked for advice about the bee from those who had experience of it, and there was even a letter from someone living in Nicosia (possibly a breeder), who outlined the geography of the island and explained how long the post took to reach Cyprus. In April 1900 W.G. Atkins, of Deal, Kent, gave his own opinion that, 'The queens are exceedingly prolific breeders while the workers are very active and easy to handle.' (7) C. A. P. from Co. Kerry wrote that his Cyprians, Italians and Carniolans had all overwintered well. On the other hand, in May, John Walton informed readers that the late Mr. C.N. Abbot and the late Mr. Walter Marshall had both told him that all Cyprians (bees) were a '...vile lot'.(8)

Given that this was before William Woodley's career as a beekeeper, I am interested and surprised that he didn't keep Italian bees, but favoured the native Black bee. Had Mr. Woodley been strictly interested in maximising his profits, surely wouldn't he have obtained and made use of Italian queens for what was once the largest commercial bee farm in Britain? It was well-known, even then, that the Italian bee was the best race when it came to honey production, which was presumably the reason why the first ones were brought into the country. Despite this, not everyone thought the Italian bee was as good as its reputation, such as W. C. Mollett, Stonecoal, W.V. who compared his experience of Blacks and Italians in 1915, concluding, 'Taking everything into consideration I find that the Italians are often over-rated. I can get just as much money from the Blacks as from the Italians.' (9)

Brother Adam, on the other hand, favoured the Italian, especially the Ligurian over all other races, though he identified four separate varieties (the dark-coloured variety from the Ligurian Alps, the light-coloured one from Bologna, a very light-coloured variety from North and South America and New Zealand, and the Golden Aurea). Indeed, *Ligustica* was the basis of the Buckfast bee following success with his earliest crosses between Italians and survivor stocks of the native Black bee after Buckfast's apiary was almost destroyed by Isle of Wight disease in 1915. It was Brother Adam's opinion that: 'In the dark, leather-coloured Ligustica we have a unique combination of factors of economic and breeding value, thanks to which she has found a welcome in every part of the world. When properly handled she is second to none in answering the needs of the commercial and amateur beekeeper...' (10) After his experience of Isle of Wight disease, which he always maintained was caused by

Acarine, it was *Ligustica's* resistance to this problem that first won her his attention. 'Ligustica has over a period of more than 60 years proved to be one of the most resistant to Acarine,' (11) he wrote.

Mr Woodley clearly did not share this interest or enthusiasm for the Italian bee, though in 1909 F.W. Sladen published a report in the *British Bee Journal* on his efforts at Ripple Court Apiary in Dover to raise a British-bred cross from the Old English native bee and a Golden Italian strain developed in North America, using Mendel's discovery of the laws of heredity. The Journal cited , '...systematic work that is being done by him for the improvement of bees suitable to this country.' (12) In the June edition of 1910, the journal informed readers that: 'Mr Sladen has been breeding bees by selection for honey-gathering since 1892, and has succeeded in producing certain strains which he calls ''British Golden'', and which are noted for their good qualities. He also supplies queens of foreign varieties...' (13)

Sladen's first explanation of his selective breeding actually appeared in the journal 17 January 1907. Sladen had realised, with others who had experimented with the different races, that in the US the Italian did better as a honey-producer, while in Britain the Black was more successful. To this end he set about breeding a specifically British bee in the British climate, based upon the Black bee. In 1901-2 he crossed his Black bees with a good honey-producing American strain called American Golden Italians, which were distinguished by having a good proportion of yellow on their bodies. The second and third generation progeny of his breeding work produced very prolific queens, but he noticed that the trait for colour and prolific honey production were inherited independently. He had Black bees that were excellent producers, golden bees that were excellent producers; Black bees that were poor producers and golden bees that were poor producers. How, he wondered, could he bring together the two traits so that he could identify his most prolific producers by their distinctive yellow colour? The next stage was to eliminate all light-coloured bees from his selective breeding except those that were excellent honey-producers, from which he reared both queens and drones. In successive crosses his queens crossed with light-coloured drones became lighter. Any queens mated with poor honey-producers or local bees would produce darker progeny and queens which could be eliminated from the breeding. But as soon as lighter queens began laying, the light colour of their workers and drones verified that these queens had been crossed with good honey-producers.

Sladen's work was helped in 1905 – 6 when he came across a paper published in 1902 by Professor Bateson and Miss E. R. Saunders with their own observations of Mendel's laws of heredity through practical selective breeding experiments. Among their work they had explored the inheritance in chickens of different comb types (simple, walnut, rose and pea). With help from their paper, and his practical observations from experience, Sladen had laid the foundations for later breeders like Brother Adam, establishing certain principles of bee breeding. For example, he pointed out the challenges presented by parthenogenesis and that there are actually no F1 drones in a first generation, and the challenge that traits can express as linked together rather than separate. His ground-breaking achievement, however, was that he had successfully combined the best characteristics of the British Black bee with the best characteristics of the Italian bee in a way that they could be selectively bred and identified by colour inheritance. By 1906 he had produced the first commercial strain of honeybee, bred in Britain, called the British Golden.

There was also a series of articles in the December 1909 edition of the journal, including one on *Colour Inheritance in Bees and Mendel's Law*, which probably also accounts for the Literature section in the January 1910 edition publicising *Mendel's Principles of Heredity*. His article *Breeding The British Golden Bee In Ripple Court Apiary*, from December 1909, makes explicit the economic ambition of his venture with the British Golden: '...with the object of obtaining and improving a variety that will gather more honey than ordinary bees in the British Climate...' (14)

Sladen's report was mentioned by Brother Adam in his Preface to *Breeding The Honeybee* (though he dated its appearance in the *British Bee Journal* as 1910), referring to him as the first to have attempted to breed the honeybee, using Mendel's discoveries. One of the clever ways Sladen made use of the principles of heredity was to ensure that father drones were of the selected parentage by maintaining the golden body colour of progeny. This distinguished his selected bees from those mated with English Black drones. He published this, alongside the report, in a table of the matings of all his golden-coloured queens during 1908. The findings also suggested to him that a higher percentage of golden progeny resulted from queens mated between August and September than from those mated from June to July and that lower temperatures (in the 60s rather than the 70s) also resulted in more golden progeny. This pointed to the possible importance not only of Mendel's principles of heredity but that restricted-weather mating might also play its part. Sladen's findings included the observation that most of his matings occurred within two hundred yards of his apiary.

As a regular contributor to the journal in his *Notes By The Way*, Woodley would have been aware of this report. Doubtless, he'd also have been aware before this of the availability of the *Ligustica* bee that had by then been used by some British beekeepers for over half a century. Presumably, he would also have known that there were other bee farms in Britain and that there was an increasing interest in improving the honeybee through selective breeding, applying the new science of heredity and genetics. By 1910 William Woodley was already middle-aged and perhaps felt he was too long in the tooth for such new-fangled ideas, or his ambitions might simply have been more modest. On the other hand, he had been adventurous enough to embrace the new technology of the frame hive, section racks for producing section combs and the motor car (buying his Benz Volo model in 1906) so he wasn't shy of technology and innovation. I can only surmise that he was happy to work with the native Black bee just as his great aunt had done at her cottage in Stanmore, where, as a boy he had caught his first swarm. Maybe it was simply a case of *if it aint broke, don't fix it*. The native Olde English bee had served him well and he was making a decent living. Moreover, the cost of replacing all his queens every year with the British Golden or the other varieties Sladen sold would have increased his overheads substantially, and perhaps he didn't have the kind of margins to allow that. Good quality breeder queens remain expensive today ( I've seen a video of an American beekeeper introducing a $1,000 Caucasian breeder-queen to a hive. ) Whatever the reasons, Mr. Woodley could not have been interested in making his bees more profitable through the improvements of the strain developed by Sladen which would be advanced in the later work of Brother Adam of Buckfast.

Brother Adam's appraisal of the native Black bee was based on personal experience of working with them at Buckfast Abbey before Isle of Wight disease destroyed most of the Abbey's stocks. His observations of its strengths and weaknesses might explain why Mr. Woodley saw no reason to change. Brother Adam informed his readers, 'This dark brown bee was the possessor of quite an extraordinary assembly of most valuable economic qualities...' (15) He pointed out in particular her '...unusual longevity, wing power and industry...' (16) and that: '...she was a bee that for practical purposes did not require feeding and proved self-supporting. These traits, extreme thrift, ability to fend for herself, longevity, hardiness, wing power and industry, which were such a marked feature of the English bee, can hardly be found together in the same concentration in any other race.' (17) These strengths were despite the fact that '...she gathered only a third of the crop made by a prolific Italian colony of that time.' (18) Not everyone agreed with Brother Adam's assessment of the native bee's commercial value. W.B. Webster gave his own opinion of the Black bee in 1889, 'As a matter of fact, I should obtain a larger profit on the sale of one English, although the returns upon the two foreigners would be larger.' (19)

The same case can not be made for the Italian, despite her many merits. The Italian had her shortcomings, which Brother Adam admitted in *In Search Of The Best Strains of Bees*. For example, she often required heavy supplementary feeding and frequently turned heavy crops of honey into excess brood, while she was poorly adapted to the British climate. Brother Adam especially prized the English bee, however, for '...her incomparable cappings and her capacity for building comb...', (20) the latter which she often built out even in poor flows, so that, '...in regard to these two traits there is no other strain or race that can match her.' (21)

As Mr. Woodley of Beedon specialised in section honey (which he also took to the show bench where he won prizes), these two traits were of vital importance to him. Section honey production relies on the ability of the stock to build new comb, even in a poor year, and to cap it with the pure white cappings so esteemed on the show bench and necessary for sale to customers. Brother Adam, who had first-hand experience of the native Black bee wrote: 'The Old-English bee was the most zealous comb builder. Not only did she build with an extraordinary speed but also to a perfection which has hardly ever been equalled by any other race.' (22) He drew particular attention to a characteristic that would have benefited Mr. Woodley's production of section honey: 'The Old -English bee offered an unparalleled example of the most perfect and artistic cappings. No other race can show the same form of cappings; they were pure white, raised and dome-shaped, and the outline of each cell was clearly delineated.' (23)

Sladen's reports and articles on breeding the British Golden are interesting, historically, but also for his early attempts to fix and improve certain traits through line-breeding and crossing the honeybee. Interestingly, of all the traits he was trying to improve in his British Golden, an *Editorial of the British Bee Journal* (December 1909) identified one that was deemed more important and desirable than all the others: 'There is, however, one character, the quality called "constitutional vigour", which no breeder can afford to ignore, and which I regard as of the greatest importance in bees.' (24)

This Editorial's opinion was apposite, given the emergence of Isle of Wight disease at that time. It was constitutional vigour that was the real holy grail at that time when so many people's bees were dying, and which has come to the forefront in apiculture again since the emergence of *Varroa destructor*, and which seems to be the key to developing resistant bees.

There are a number of problems I have with the way selective breeding of the honeybee focuses on the selection of traits, as Brother Adam approached it. Firstly, particular genes don't actually equate to single traits. Brother Adam was aware of this. The idea of traits is really a human invention. Traits in bees and humans don't separate out as simply as the word suggests. A particular trait, such as docility, will be attached to other traits, and some of these might not be particularly desirable. For example, a breeder might want to select the trait for low production of propolis, but that trait might be linked to the trait for resistance to a particular disease. While expressing the desired trait to the degree desired, the trait for disease-resistance might be inadvertently switched off or weakened. Secondly, gene expression also depends on epigenetic factors (environmental, diet and conditions, for example). This means the breeder can't separate the traits from the bee's environment. It also means you can't engineer a strain of bee for every particular environment and situation. In short, this all means that the breeder can't actually forecast with any accuracy the effect of engineering a trait on every other trait in the bee. This is probably why Brother Adam did not leave a step by step precise blueprint for how he developed his strain. It seems to have been developed in a much more intuitive way, responding to situations and variables as they occurred. An analogy for this approach to selective breeding of the honeybee might be to think of squeezing a balloon; you don't quite know where the next bulge will appear, and if you push on that bulge another will pop up somewhere else. Bee breeding is forever squeezing the balloon of a pool's genetics without being able to predict where the balloon will next bulge in an undesirable direction.

During the years of foul brood and Isle of Wight disease few were looking in the direction of foreign bees as a possible source of the problem, but occasionally an interesting letter appeared in the *British Bee Journal* hypothesising how non-native bees might have been responsible for the increase in bee diseases and the mysterious Isle of Wight disease. In July 1915, for example, T.T. Taylor had been reading an article in *Gleanings in Bee Culture* by Mr. W.L. Roberts of Lavalle, Wisconsin entitled *Which Is The Dominant Race?* - referring to Black bees and Italians. Mr. Taylor quoted from the article: ' " Circumstantial evidence is added in the fact, admitted by all, that the so-called hybrids or cross-bred bees, after the first cross, show a breaking-up of all good qualities, lose their stamina, and become constitutional weaklings, and, of course, become much more subject to disease than pure bloods of any race." ' (25) He made the point that any native bee of a country or region should be strong enough to cope with any disease or vagaries of the climate to which it is exposed. He suggested that perhaps the cross-breeding and production of so many mongrel bees was contributing to the degeneration of the native Black bee's constitutional vigour (to use Sladen's terminology): ' It seems to me that the off-spring of first crosses of different varieties, as among domestic animals, generally may often be vigorous, but that further crosses will produce degenerates which are constitutionally weak. Thus, perhaps, the virulence of "Isle of Wight" disease is due to the extent to which our bees are crossed or mongrelised.' (26)

Controversially, Taylor was suggesting that imported queens and bees and their cross-breeding with natives could have been the root cause of the problems of foul brood and Isle of Wight disease, or at least a contributory factor, by weakening the mongrel progeny from random cross-breeding, because, '...we can never completely control their matings'. (27)

Taylor's solution was to call for the end of all importations of foreign bees and to try to fix a pure British bee that would be equal to all foreign bees. If this hypothesis proved correct, Taylor argued, then Parliament should be petitioned to legislate against importing bees, to help eliminate defective mongrels, '...while leaving Nature to do her part in producing the dominant type.' (28)

Of course Taylor's suggestions would never have been taken seriously because in the short term such measures would have probably further reduced the bee population in Britain, already decimated by Isle of Wight disease. The elimination of defective bees and the loss of imported bees would initially have impacted the fertilisation of fruit trees, which was already a concern for the Board of Agriculture. In Fact, the *British Bee Journal* reported on 1 May 1913 that Mr. Runcimann, President of the Board of Agriculture, had spoken at Pershore on 22 April at the Fruit Growers and Market Gardeners Association meeting. In his speech he referred to the Bee Diseases Bill and spoke of the harm Isle of Wight disease was doing to the bee industry and to agriculture. He told the meeting that he hoped to get a second reading of the Bill through the House of Commons the following week.

Mr. Runcimann was questioned in the Commons on 17 July 1913 by Mr. Bathurst who warned of defective fertilisation of fruit trees in Scotland and the South in 1914 if the Bill were not passed into law, and asked whether, '...he can and will by administrative Order take steps to check the ravages of this disease?' (29) The President of the Board of Agriculture replied, 'I am advised that the Board have no power, failing the passage of the Bill into law, to take action of the kind suggested by the hon. Gentleman.' (30)

What's striking about T.T. Taylor's letter, however, is that it could have been written today, a hundred years later; not about Isle of Wight disease, but regarding the same concern that imported bees and foreign queens continue to carry the risk of introducing new pests and diseases, and are compromising native bees through cross-breeding and the production of mongrel stock. In particular, the idea of fixing a pure native type that would be the equal of any foreign race, looks now to have been positively visionary in 1915. And it remains so, a century later. The times remain the same.

Brother Adam, however, was beginning to go further than Sladen and other bee-breeders with his experiments to develop the Buckfast strain. He dismissed some beekeepers' concerns at the time that foreign races were fundamentally unsuited to Britain's climate and environmental conditions. He believed that only the workers had to contend with unfavourable conditions, but that environmental conditions did not affect the queen and drones who were mostly within the hive or only flew in fine summer weather. He maintained that in their short lives, '...there is no question of any quality acquired and transmitted' (31) and that this, '...precludes any possibility of a permanent genetic adaptation to a particular environment.' (32) Any adaptation, he argued, is only acquired after a very long time. At that time, however, Brother Adam would not have known about an area of genetics known as epigenetics, a relatively recent area of science that has demonstrated the effect of stress and environmental factors over and above an organism's genetics; although the DNA is unchanged by these factors, certain gene loci can be triggered to switch on or off, and not always with positive results.

There is a recent example of how quickly such changes in an organism can happen, first published in *Nature* magazine 29 May 2014: on the two Hawaiian islands of Oahu and Kauai male crickets began attracting an introduced deadly parasitic fly *Ormia ochracea* with their chirping. The fly's larvae burrow into the cricket, growing inside them and killing the host. To protect themselves, large numbers of male crickets on the island of Kauai stopped chirping. In only twenty generations a mutation happened in the shape of the male crickets' wings, leaving them incapable of chirping. By 2005 male crickets on Oahu, over a hundred kilometres away, had also stopped chirping, evolving a slightly different mutation in their wing structures from the males on Kauai. Half the Oahu males today are unable to chirp.

This is not only a beautiful example of convergent evolution, but it also proves that a species can mutate more rapidly than was previously thought. In evolutionary terms this mutation has happened in a flash. The implications for our honeybees is that they might also be capable of relatively fast adaptation if put under natural selection for survival. There is already evidence that this is happening with populations such as those in the Arnot forest in NY state. It also suggests that adaptation does not necessarily involve only one strategy, but that species can adopt a variety of ways to adapt to a new situation. If we want to select strains of bee that are resistant to varroa, it would probably be a mistake only to select for one trait such as uncapping the brood which some breeders have attempted in recent years. It's more likely that we ought to allow the bees to select a variety of traits.

There is evidence also for epigenetic changes that can happen just as fast. For example, rats exposed to a particular fungicide have been found to manifest epigenetic changes in the first generation of male offspring which were passed on for four generations. A further striking example of epigenetic modification is the Agouti mouse (Agouti referring to a dark, wild-type fur colour). Two genetically identical Agouti mice can appear very different – one small and dark, the other yellow and very obese. In normal Agouti mice the Agouti genes are turned off, but at a specific gene loci a switching on can occur, affecting hair colour and weight. Mice whose gene is 'on' appear yellow and obese, but are also more likely to develop diabetes and cancer. These epigenetic modifications can also be inherited by their progeny.

It seems plausible to suggest that in the decades leading up to foul brood and Isle of Wight disease epigenetic modifications were probably happening to imported foreign bees and cross-bred strains. These non-native bees were exposed to environmental stresses which could have switched on or off certain genes as these bees were subjected to pressure of selection for survival. The question is, if genetic modifications occurred, did they also predispose those bees to a whole new set of problems, such as new diseases like foul brood? Did such possible changes also make these bees more vulnerable to Isle of Wight disease when it emerged?

Brother Adam set out to breed a perfect bee, and the more I have thought about this and its consequences the more it has made me question the underlying assumptions and ideas behind it; and the more I worry about continued efforts to breed the perfect bee, especially how to do it more precisely. Even if beekeepers could reach a consensus about what constitutes the perfect bee, maintaining a stable, closed population would be unsustainable. Bee breeding, because of the complexity of bee genetics and the need to maintain genetic diversity, can never be a fixed point of arrival at a desired outcome; it will always be an ongoing process because bee breeders have to constantly balance the traits they want to fix against traits they might lose, or undesirable traits that can piggy-back on an otherwise good characteristic. To maintain genetic diversity new blood has to be introduced quite regularly, to widen the gene pool, but this always risks introducing undesirable traits such as susceptibility to disease.

In fact, one of the most important characteristics of any strain is that they have sufficient genetic diversity that certain family lines can produce workers with specific specialisms. These specialisms give bee colonies their plasticity to respond to certain situations (such as varroa).

Let's suppose we achieve the perfect bee: sooner or later we will need to introduce new blood, to maintain enough genetic diversity to keep producing family lines with their own unique specialisms. But the minute you push on this balloon of your perfect bee's genetics you destabilise it, and a bulge appears in a new direction. Perhaps in that genetic bulge you lose a desired trait, by introducing another one and taking your strain in a different direction. Bee genetics and bee biology won't allow the bee breeder to ever reach a stable outcome. On this basis you can never attain the perfect bee. It is a forever shifting goal.

# MINDING THE BEES

## Chapter 4
### A World of New Possibilities

'BRICE'S RELIABLE QUEENS.
Well-known strain, one quality, one price.
Mated, tested Queen. 10s,6d, 15s, ready shortly.
Virgins 3s each.
HENRY N. BRICE, Dale Park-road, Upper Norwood.' (1)

***British Bee Journal 1896***

Traditionally, the beekeeping season begins very gently in March, as the bees begin to wake up. In March the skep beekeepers of the past would remove the flap of thin wood fixed across the skep entrance to stop the bees being attracted out by the brightness, if snow fell. We have no records to suggest that cottage beekeepers like Owen Thomas fed their bees at any time of the year, but March would be a time to feed if the weather was bad, the bees could not forage and if little was in flower.  A bowl of sugar syrup was placed under each skep, if feeding was required, a handful of straw placed into the bowl for the bees to stand on while drinking. (From late Autumn to March syrup must not be given because stored syrup can ferment and give the stocks dysentery.) From March, however, the winter fondant is replaced with thin syrup which mimics a nectar flow. This was also intended to stimulate the queen into laying, because increased brood production means an early expansion of the hive population. Whether you look at things from the perspective of the bees swarming, or from the beekeeper's perspective of honey production, early expansion of the brood nest is important. Early swarming gives stocks more time to establish a new home and lay up stores for the winter. Early expansion also means the stocks have the maximum number of foragers to coincide with the main nectar flow in June and July, which means more honey.

I've stopped feeding syrup in spring, mostly because it's a bother to make, and it's very heavy to lift and carry to the apiary. Instead, if I need to feed at this time of year, I feed fondant, which is just bakers' soft cake fondant. At this time of year I also feed pollen candy (fondant with pollen added), which supplements the bees' protein intake from the pollen they bring in from foraging, to feed the brood.

This year, though cold, March has been drier and brighter than for the last few years, and the stocks have been able to do plenty of foraging this month. They've had snowdrops, hellebores and hazel, and now there are crocuses and flowering cherry. I stand under one of the trees after feeding the Call ducks, look up towards the squeaky-metal calls of a Great tit, the bees humming among marshmallow-pink blossom. One or two tits appear nearer to eye level, busy working the flowers against blue sky in late afternoon sunshine. Bee farmers at this time of year move skeps or hives from their *home stands* to *out stands* among orchards and flowering crops. Commercial stocks are hired by farmers, especially in large production areas like California, with farmers migrating hundreds or thousands of miles across states to transport tens of thousands of hives to the vast monoculture fruit and nut plantations. During the spring their stocks expand while bringing back pollen and nectar, before being packed up and moved back to the home apiary when the pollination work is complete.

There's growing evidence that pollen from monoculture crops and sugar syrup are not the best foods for honeybees. In fact they might be contributing to the weakening of stocks and their increasing susceptibility to diseases. Granulated sugar, even when made into syrup, is always a poor substitute for the particular balance of natural sugars obtained from nectar, while research shows that bees forage for a variety of pollens which give them the nutrition they need to remain healthy. Pollens from monoculture crops are equivalent to humans eating a diet exclusively of one thing. Just as we would suffer from a deficit of certain vitamins, minerals and trace elements if we ate only yoghurt or chips every day, it looks as though bees also suffer from a diet of only one type of pollen.

After feeding the chickens and collecting eggs on a Sunday afternoon, I walk a few yards beyond the chicken enclosure to the hives. In my head I can hear Francisco Torregas' *Recuerdos de la Alhambra*, the piece of guitar music played at the start of the old Southern television programmes *Old Country* and *Out of Country*. I watch the bees for a few minutes coming and going at the entrances of the surviving hives. Today they are not in full throttle, but there's steady traffic back and forth. 12 degrees Celsius and sunny, but still too cold to open the hives. It would risk chilling the brood. In previous years I've inspected hives at the end of February in tee-shirt weather. It proved to be a useful exercise, as one hive had gone queenless.

This time of year, there are no drones yet, so queens replaced by the colony over the winter will be infertile, laying only drone eggs. A hive full of drones at this time of the year tells you the queen is a drone- layer. It's doom for the hive because she can never lay the fertile eggs that produce workers or queens, so she will never be superseded by the colony either. The first year I had a drone-layer I requeened them with a queen from a spare nucleus. It was a quick operation, but it had to be so as not to chill the brood that I moved with the queen. The easiest way to introduce a new queen to a colony is with her nucleus, because she is protected by her own nurse bees.

I learnt how to do it after reading how D. M. Macdonald worked it out for himself, who shared his own experience of fixing fertile workers (or a drone-layer) 19 January 1911 in the *British Bee Journal*. He advised taking a couple of frames from the affected hive, creating a gap. He then took two decent frames from a good hive, with plenty of brood, adhering nurse bees and the queen. He leaned the frames together against the hive and lightly smoked them, driving the bees with the queen all into the gap between the two frames. These frames were then introduced into the gap in the affected hive. The nurse bees would protect their queen and attract other nurse bees to the brood (nurse bees being more accepting of a new queen) until the queen was surrounded and protected by nurse bees and could not be attacked by older bees.

The queenless hive won't attack a queen under these circumstances, especially if she is a mature, mated and laying queen. I actually took out five frames from the centre of the queen less hive, shaking off the bees, looked for the drone-layer (who happened to be dead anyway), and simply slotted in the five frames of the nucleus, with the queen and all her bees and brood. With her entire nucleus to protect her the queen would, I surmised, be absolutely fine, and she was. Job done. Thank you D. M. Macdonald. Not something, I remind myself, that you can do with skeps. Neither something I can do this year without losing another nucleus, as it would reduce me to three stocks. Even if I take two frames, as Macdonald did, this weakens a valuable nucleus and gives it the problem to solve of requeenng itself at a less than ideal time of the year. If two stocks become queenless though, and I need two nucs to requeen them, I'll be down to two stocks. My nerves are on the edge this spring. Thankfully, older stocks are bringing in pollen, a good indication that they probably do have a laying queen and brood - for now anyway.

On YouTube I find *Old Country* and listen to that ponderous, calming title music played on a classical guitar. Jack Hargreaves walks into view, with a white pack pony at his side slung with his accoutrements - probably fishing tackle. You couldn't make the beginning of a television programme like that now, I'm thinking. It has to be exploding graphics, exploding music, and a speed to make you dizzy, with hand-held camera shots that look as though the cameraman is drunk, ducking and diving all over the place until you feel you're on a boat in a storm. I don't watch that kind of television. It's not just that I can't stomach it - but I refuse to. As Jack and his white pony cross a bridge, I glance below at the comments. Someone has written to the effect that they wish life was still like the late world of Jack Hargreaves

YouTube takes me to several episodes of *Old Country* and *Out of Country* that Jack Hargreaves began presenting in 1960. There he is, sitting in his shed, where I watched him when I was a boy, showing us his fishing rods, speaking completely naturally, mistakes and all, without script or auto-cue. You couldn't make programmes like that any more either, unless they're reality shows - one person just sitting in his shed, philosophising about the old ways or sharing anecdotes and memories, like an old grandfather. Television today is for young faces. Jack usually introduced what he was going to take us to see, showing us a roach rod or his ladybird edition of Isaac Walton's *Compleat Angler* before taking us chub-fishing; or telling us about the different kinds of painting - circus, barge and traveller, before showing us the cart-painter painting his new waggon. He also showed us thatchers working on a cottage first thatched in Elizabethan times, pointing out that the long-haired boy helping his father probably had the same hairstyle as the boy who helped his father first thatch the cottage. Jack made fascinating connections like that. But he wasn't only showing us old ways and crafts; he was showing us a type of person – the too easily overlooked local heroes of our cultural heritage. Jack wasn't only about environmentalism; he was about people, and that's a dimension of extreme environmentalism (and even beekeeping) too easily forgotten. I happen to think it was a decisive factor in William Woodley's opposition to Bee Disease legislation too.

Jack related an anecdote about a woman who lived in the town who once told him she wanted to visit the countryside but wondered what she could do when she got there. Jack's advice was as pedestrian as the slow, plodding start of the programme and the faltering delivery from his old shed. He told her to come and stand at the gate to a field and look over it at everything she could see. Then he told her to come back at another time and do the same thing. This time he told her to ask what she could see and how it had changed. Then, he told her, ask why it's there.

I'm watching, mesmerised. It's like reading *Winnie the Pooh again* later in life and suddenly realising that it's really wasted on children. There's a whole philosophy of life in Jack's shed or walking beside his pony or sitting by a pond, watching the tip of his fishing rod. It's something to do with connections, with wholeness. And that advice about simply observing is the foundation of all good beekeeping; you ask yourself what you are seeing and how it has changed.

That comment gets me thinking too about how someone in lockdown 2020 wished life was still like the one Jack Hargreaves shared with his viewers. Isn't it interesting that during the lockdowns of the Covid pandemic people reverted to that old country? In the biggest crisis since World War II, we started growing our own vegetables again in the back garden and making bread. People bought backyard chickens, to produce their own eggs. We slowed down and began walking and cycling again and opened our ears and eyes to the natural world. I don't think it was necessity alone that took us back to the wartime spirit of resourcefulness. There was enough food that none of us had to dig for Britain. I suggest it was for the same reason Walton's *The Compleat Angler* remains the most reprinted book in English after the *Bible* and Shakespeare's works. Walton not only gives today's angler the benefit of his wisdom, sharing information such as the best way to prepare bait for carp; he also makes connections that convey an entire experience anglers aspire to share. In that world, the angler and his rod become one with a stretch of water and the whole natural world. Walton's angling spoke in the dialect of happiness that we all understand.

Jack Hargreaves did the same with his programmes. Lockdown 2020 reconnected us with the basics of life, such as our food and our well-being, and where these are rooted. Bread, we've rediscovered, isn't just an instant commodity from a supermarket shelf. Bread is a basic of life, that can be deeply rooted in our health and happiness, because making bread, like gardening, growing vegetables, picking fruit and foraging, or home baking, keeping hens and coarse fishing, is deeply satisfying. These activities earth us. We are spiritually sick and disorientated in our modern culture. We are searching for healing and freedom. Gardening, making bread, fishing and keeping bees all help to heal us and begin to locate us spiritually. Walton's little book wasn't just a manual, and nor were Jack Hargreaves' programmes. There was something contemplative about them that reached in and touched the contemplative in our humanity. Jack's advice to the town woman, about looking over the gate of a field and paying attention to what you see, and asking why, is deeply contemplative. It's what the modern post-human world most fears, because it brings us face to face with solitude, the door of the heart that opens us to a sense of something that transcends ourselves. We realise that we are exiles from the *old country*, which is who we really are. We realise that there is something unique about humanity and that we matter. We realise that we are exiles from ourselves. We are disorientated. Someone else has written under an episode of *Out of Town how they miss Jack and the country we used to live in.* It is the voice of exile and disorientation-cultural and spiritual.

One of Jack Hargreaves' programmes especially captivates me. It takes me back to the cottager world of Owen Thomas and the Industrial Revolution's effect on the Dorset button- makers, a world that coexisted alongside the Beedon bee farm of William Woodley. The programme also shows the connexions that existed between a local industry, the lives of cottagers, and the local materials they used. It's precisely that interconnected world that we've lost in modern times and that we seem to want to restore.

In Twyford, between the seventeenth century until the Industrial Revolution, a huge cottage industry had existed manufacturing the *Dorset button*. From seven hundred button-makers in the 1700s, there were seven thousand in the Dorset area by the end of the century. While men went out to work, women and children worked at home making a range of button designs including the *Dorset knob*, a raised button used on a dress owned by Queen Victoria. Lengths of wire were coiled and cut, then soldered together, to form a ring. A small piece of linen was stretched and sewn over the ring, making the standard button for simple garments such as underclothes. A more ornate button called a *Dorset Cartwheel* was embroidered across the ring, to be used on ladies' dresses. Lastly, the *Dorset Hightop* and *Knob* were raised buttons made by building linen soaked in flour paste over a disc of rams-horn, then placed in a mould. The domed buttons, looking like miniature skeps, were used by the gentry to button long waistcoats or dresses.

In 1841 Humphrey Jefferies invented a button-making machine. According to Victor Houart's *Buttons: a Collectors Guide*, it was this invention that John Ashton patented and showed at the Great Exhibition in 1851. The effect on the button-making cottager economy was very damaging. There had been earlier legislation to protect the industry by making competition in the manufacture of cloth buttons illegal, but the button-makers could not be protected any longer from the Industrial Revolution. In the next decade, as many different out-worker industries died (including button-making), sixteen thousand Dorset families emigrated to the USA and Australia. Half of Twyford's cottages were left abandoned and fell down.

The tale of the Dorset button's demise throws light on the demise of the skep in English bee keeping. Think of all the out-workers in different cottage industries who were forced out of business in a similar way by the Industrial Revolution in every part of Britain. Consider also the mechanisation of agriculture, the falling demand for agricultural labourers and the range of related crafts and jobs that employed them. When you consider the huge number of these cottage labourers who would have once kept bees in skeps, it becomes clear that the skep didn't only decline in popularity because labourers couldn't afford frame hives; it declined because the Industrial Revolution was killing off skep beekeeping along with the traditional way of life  in which it was rooted. There were fewer skeps because there were fewer cottage beekeepers. Those who could afford to keep bees were those who had stable jobs and a stable and reliable income - people like William Woodley, grocer and watchmaker of Beedon Parish.

As well as an interesting window on social and economic history, the Dorset button story is of cultural importance too. It represents the old country's interconnectedness of locality, produce and local economy. The cottagers of Twyford were using linen, thread and sheep horn from sheep that would have been part of the local economy, the raw materials perhaps raised on farms by their husbands, fathers, uncles and brothers. It was a whole way of life that connected families and localities in a unique way.

When Isaac Walton used the word *complete* in the title of his best seller in 1653, he meant it in the original sense of the word; *skilled and highly accomplished*, the consummate angler. Today we might interpret another meaning that Walton alludes to but probably didn't intend when he used the word: *complete* in the way we speak of *being complete*, that is happy and deeply contented with life, as in *you've made my life complete* or *you complete me* and *I'm complete*. And that meaning corresponds closely to Walton's other title *or the contemplative man's recreation*. In other words, it is a book about one way of achieving happiness through  inner stillness. Today we might we call it *mindfulness* – maintaining, moment by moment, an awareness of our thoughts, feelings, bodies and our surroundings – or *being present*. Milton, the angler, was happy fishing because he was practising what we called mindfulness.

I imagine button-making is a mindful activity, just as angling is, because it catches you up in the stillness of the present. Beekeeping does the same, so those cottager skeppists of the past, like Owen Thomas, were also complete in a way that many of us are not in modern life. They were accomplished beekeepers and happy people who lived in the present in a way that we're trying to rediscover. Lockdown 2020 slowed us all down and made us more mindful. Free from the active schedule of a frenetic world that was suddenly suspended, we had time to reconnect with the present. When we rediscovered gardening and the joy of collecting fresh eggs and home baking, we were discovering much more than becoming accomplished or skilled at a craft or some kind of nostalgia for the past; we were rediscovering ourselves and what is essential and unique about being human. We were rediscovering the old country of ourselves – our deepest longing for happiness.

The modern science of apiculture is from a different world, with another language and mindset entirely. One of the characteristics of Brother Adam's writing is that he was scientific, not just with his necessary use of technical and scientific terms, but in his precise use of language. But there is something for me a little cold and clinical in his writing, as though he wrote in a lab coat. But his language does convey his goals in beekeeping, which were unashamedly to produce a bee that was commercially the most economic strain. His Preface to *In Search Of The Best Strains Of Bees* establishes this clearly: 'The task of modern bee breeding is to ascertain which of the races has the greatest value for breeding purposes, to collect them together, to test them, and by cross-breeding combine the best of the characteristics to form new types.'(2) His conclusion to *Breeding the Honeybee* similarly articulates this aim, 'Apparently our only hope of achieving real progress in the improvement of the genetic composition of the honeybee, open to us, is by way of cross breeding...' (3)

Embedded in this language, however, are the principles and assumptions of modern beekeeping that were the foundations of Brother Adam's development of the Buckfast strain, and they are principles and assumptions that many others today don't share. Firstly, he often used the adjective *perfect*, and secondly, he assumed that the honeybee could be domesticated. He explained, quite rightly, that: '...Nature never breeds for the perfection of the factors we desire for our commercial needs. Nature's aim is almost exclusively for the preservation and multiplication of a type. True to this goal of hers, she breeds within certain limits, to bring about the best possible adaptation to prevailing conditions.' (4) Nature, his writing informs us, never breeds an ideal bee, to satisfy our own commercial expectations, but through ruthless natural selection she breeds for survival.

The first point to make here is that the definition of a perfect or ideal bee, quite apart from whether that is possible or even desirable, is going to be a subjective matter established by the beekeeper and their particular needs. For example, on the internet I came across one beekeeper in the US who had decided to change to Causasian bees because they are prolific, docile and produce propolis to an extreme, which science now recognises benefits the bees' health. One beekeeper might be working in a hot Mediterranean country, while another might be keeping bees in North Wales. The different conditions in which their bees are kept will also be important factors influencing their own definition of a perfect bee. The beekeeper in North Wales might ask for a bee that is winter-hardy and able to produce a crop of honey in a rainy season or a poor nectar flow; the beekeeper in the other country might want a bee that keeps producing a lot of brood for most of the year, to take advantage of more than one nectar flow, and a strain that manifests a low propensity for swarming.

Brother Adam identified certain primary qualities, such as fecundity, resistance to disease and a disinclination to swarm; secondary qualities such as wing-power, hardiness and spring development; and management qualities such as good temper, calm behaviour and a disinclination to propolise. These and other qualities were identified as generally of most use to many beekeepers, which is fair enough. On the other hand, a criticism might be made about this selection, that Brother Adam was deselecting some of the traits that honeybees most need, to maintain health, such as their use of propolis. We know, for example, that propolis contains natural antibacterials, and that wild bees make a propolis envelope inside the nest that might play a key role in helping them fight off or manage diseases.

This example alone illustrates that not only is a selection of ideal traits essentially subjective, but it might not even be desirable. Brother Adam's goal was to produce a commercially economic bee, but that isn't only about honey production; it's also about the time involved by a beekeeper because time is also a commodity that costs money. These concerns, however, must be balanced by an understanding of the best conditions in which to maintain vitality, and vitality isn't only a genetic trait but is also dependent upon the conditions and the environment in which bees are living, together with the skill and experience of the beekeeper.

Elsewhere Brother Adam repeatedly used language such as, '...breeding material' (5); '...perfect' (6); '...economic value'(7); 'genetic possibilities' (8); 'breeding potentials'(9); which is more the language of the scientist and bee breeder than the average hobbyist. Bees are living organisms, not just *material*, and as such they need to be cared for by the beekeeper. Interestingly, Brother Adam wrote in *Breeding The Honeybee*: 'Bee breeding is concerned exclusively with the breeding of the honeybee and in no way with the care and maintenance of bees.' (10) Surely, the two are inextricably linked?

The assumption that the honeybee may be domesticated is a moot point, but I tend to the opinion that bees can't be domesticated because their breeding can't be controlled in the way we breed domesticated cows or chickens. In all other plant and animal husbandry we can put the parents in a field or cage and can control the mating. Consequently we also know the parentage. This doesn't work with bees. Firstly, the male does not play a dominant role in handing on its genetics. In bees, drones are haploid and each one breeds only once in its life, one of multiple matings between a queen and other drones; so that an individual drone's contribution to the gene pool is smaller than that of a bull which might mate with many cows over many years. Drones also fly to particular areas to mate, beyond the control of the beekeeper.

This production of haploid drones by parthenogenesis in queens (the laying of an ovum without fertilization) results in drones that have only 16 chromosomes (rather than 32 for workers and queens) from their mother, so that drones have no father or sons. This means there can be no mating of a father and daughter, a mother and son or a brother and sister in honeybees. A drone dies after mating, so he could not be inbred with any of these combinations. These kinds of crossings (inbreeding) are required in other forms of selective breeding, to fix and strengthen certain hereditary characteristics, but they are not possible in bee breeding. An added complication of parthenogenesis is that all the spermatozoa produced by a drone ( having 16 chromosomes from a single parent – a mother) are all genetically identical. To further complicate things, bee breeders are not concerned with breeding isolated individuals but with a whole society – a super-organism of many families of half and full sisters by between ten and fifteen different fathers.

Inherent in bee breeding is the susceptibility of the species to problems caused by the inbreeding that is the indispensable foundation of all selective breeding. As already mentioned in the previous chapter, the first problem with this approach is that bee biology does not naturally permit crossings between mother and son, son or daughter or brother and sister, while polyandry is a strategy designed to avoid inbreeding. In selective breeding of other domesticated species, crossing F1 individuals among themselves results in a segregation of the characteristics in the F2 generation, from which arise new combinations of genes that can then be passed on in line-breeding. In bees the segregation of F1 individuals happens only with females, while in drones it does not appear until the F2. Bee biology and parthenogenesis of the queen therefore makes inbreeding along the lines used with other species impossible, the nearest relations available in bees being half-brothers to half-sisters or drones with aunts. Moreover, the sperm of each drone (due to parthenogenesis) are genetically identical, creating greater stability in bee heredity than in other species. A consequence of this with inbred bees is a loss of stamina.

Given that Mendel's principles of heredity and the science of genetics and bee biology were still in their infancy when bee breeding began, and that bee breeding was at an experimental stage at the turn of the century, I wonder how many early strains ran into problems caused by excessive inbreeding of closed populations? How many queens might have arrived in British apiaries from strains that were already from closed populations with a narrowed gene pool and already lacking stamina? These strains then flooded areas of the countryside with alien drones which crossed with native bees in wild and managed colonies. Who knows what effect this began to have on British bees and their ability to cope with diseases? I wonder if what happened, as foul brood and Isle of Wight disease spread, was explained by Brother Adam when he wrote about, '...loss of vitality' (11) caused by inbreeding: 'It is an insidious and illusive deficiency which always shows up mainly in unfavourable climatic conditions against which a weakened constitution has no resistance. Nature then takes control and weeds out the unfit.' (12)

Another fundamental issue I have with selective bee breeding is the question about whether or not bees can actually be domesticated. The definition of a domestication is to breed a species whose characteristics are selected for human use, which requires humans to do the work of selection over a period of time. Domesticated animals that return to the wild are called feral, but arguably honeybees can never be feral because they are essentially always wild. We haven't actually changed or adapted them to the same extent that we have adapted the goldfish or the dog or the chicken. Moreover, due to the particular difficulties involved in controlled breeding of the bee, the only way it can be done is in isolated apiaries flooded with the particular drones desired to mate with queens. This was an advantage that Brother Adam had from living in such close proximity to Dartmoor. Even there he ran into problems with the mating apiary, with some years resulting in very low numbers of successful matings.

It's understandable, however, that Mendel's discoveries about heredity would open up a whole new world of possibilities in the area of plant and animal breeding that would capture the imagination and the ambitions of the beekeeper. Domestication of many species had been happening in an organic and slow way for centuries, developing many varieties of dogs, for example, or chickens that were suited to the different needs of humans. But Mendel had made possible a much more systematic, planned and precise approach to the development of different varieties, achievable in a much shorter period of time. Desired traits could be developed with greater understanding and accuracy in a more scientific way. Indeed it was the advent of genetic engineering, which would advance to the more controversial genetic engineering of Genetically Modified (transgenic) crops and animals, an issue over which many argue today.

Before Mendel's discoveries were known, selective bee breeding had been attempted. Indeed, Mendel was a beekeeper, but had little success with bee breeding due to the lack of knowledge at that time about bee biology and breeding. The Swiss Dr. Ulrich Kramer also made some attempts in 1898, but modern beekeeping was in its infancy and not enough was known about how a bee colony worked. Most of the early work, interestingly, was done in the German-speaking countries, starting with the Nigra strain, developed for thirty years in Switzerland.

Brother Adam was perhaps most influenced by the book *BIENENZEUCHTUNGSKUNDE* by Professor Dr. L. Armbruster (1886 - 1973), a German zoologist, which appeared in 1919. Armbruster's conclusions were based on a false premise, however, because multiple matings were not known about at the time he was writing. According to *Wikipedia*, a copy of the book came to Brother Adam in 1920 via Black Forest watch dealers travelling to England. Brother Adam wrote that, 'This book revealed to me a world of new possibilities.' (13)

The timing was right too. In 1919 Brother Adam took over Buckfast's apiary, at the end of the final wave of Isle of Wight disease which had destroyed most of Buckfast's colonies. Brother Adam noticed that the Abbey's survivor bees were crosses between the native Black bee and Italians. In his opinion, the disease had been caused by acarine, so he assumed that these bees were resistant to acarine. Using the ideas from Armbruster's book, Brother Adam set out to develop a honeybee based on this initial cross and improved by other combinations from breeding experiments and crosses between many varieties over the next fifty years of bee breeding.

It might be that the particular cross with which Brother Adam was working was resistant to Isle of Wight disease (which he believed was acarine). As it was obvious to him that the Black bee was susceptible to whatever caused the plague of Isle of Wight, Brother Adam went on to assume that Italians (the other half of his cross) were also resistant to acarine. Interestingly, however, there is an article from February 1915 in the *British Bee Journal* that contradicts Brother Adam's belief (and that of many others) that Italian bees were resistant to the problem of Isle of Wight disease. In his *Random Jottings* Charles H. Heap of Reading wrote: 'There has been abundant proof in this country that the Italian bee is just as liable to be attacked by "Isle of Wight" disease as the native black. I have seen many cases myself. Last autumn the only cases of the disease I found in one neighbourhood of a northern district were in two apiaries into which Italian bees had been introduced.' (14)

Brother Adam and others have proved that, despite the challenges with breeding bees that are unlike the challenges met in other areas of plant and animal breeding, considerable success can be achieved, at least in the short term. Bees like the Buckfast strain can be bred for docility, non-swarming and high production of honey. This does not necessarily mean, however, that the bee has been domesticated, or that this kind of selective breeding is desirable in the long run.

The logical consequence of this trajectory in the development of the honeybee could be a GM bee, which has certain parallels with GM crops, in so far as there are possible unknown environmental consequences. GM crops involve the modification of genetic codes in plants and the introduction of specific genes from another species. Some GM crops release toxins that can harm pollinators and other insects such as butterflies. GM crop planting can also affect biodiversity, while there are fears about cross-contamination with other plants. What happens if a transgenic fish, such as a salmon, gets into the ecosystem and breeds with wild fish? At least one variety of transgenic tropical fish is reported to have escaped into the wild in Brazil and now poses an environmental threat with its potential to become invasive.

Selective breeding actually interests me, and genetics is fascinating, which is why I do marvel at the work of Brother Adam and other bee breeders. I developed an interest in goldfish when I was a small boy, and goldfish breeding is extremely interesting. After school my mother would take me to the goldfish pond in the park next to the school. I loved watching those large, fat golden fish gliding through the black water. I'm sure that was where I caught my interest in ponds as well. By the time I was in secondary school, less than a decade later, the golden fish had all but gone, replaced by bronze-coloured fish. I couldn't understand where the big, bright goldfish had all gone.

As a teenager I got my first fish and built a small garden pond one Saturday afternoon. More fish followed and they began to breed. I would take out the eggs and raise some of them in an aquarium, and I began to notice a number of things. First, goldfish don't start off gold, but bronze, developing their ornamental colours after a few months. A small percentage revert to type in every generation and never develop the mutated colour. Next,I began to notice variations in their colours and patterns and even their fin length. There were even differences in scales, with some being very metallic and other being matt. In some varieties, called *Shubunkins*, the black pigment under the skin manifested as a lovely blue. There were short-tailed examples of these multi-coloured fish, called London shubunkins, and long-tailed ones called Bristol shubunkins. When I read about the history of goldfish I discovered a world of selective breeding from the wild *Carassius auratus* in China. The original wild form was bronze, but people noticed occasional mutations with yellow or gold scales and fins. They began to breed these fish and fixed the colour trait. After many centuries the goldfish had been developed by the Chinese and Japanese into 125 different breeds, in a variety of colours, scale types, body forms and fin types. Some fish had flowing fins like veils or double caudal fins. Others had telescopic eyes or bubble sacks under their eyes. Some had excrescences like raspberries on their heads or pom-poms on their nostrils. There were even some breeds, like the lionhead, which had no dorsal fin.

What happened to the goldfish in that pond next to my old primary school suddenly made sense to me, and it's a good way to explain why I am so interested in locally adapted bees. The golden fish were caught by herons, or died in some other way, because they stood out. Nature does not select a golden *Carassius auratus* because nature always selects for survival, and golden fish advertise themselves as an easy meal. Under pressure of natural selection for survival the small percentage in each spawning that reverted to type and retained the natural colour ended up surviving, because they were better adapted to camouflage in the dark water. Over the years, without human intervention, the fish population in the pond had regressed to the original wild colour because wild-coloured carp tend to avoid predators.

In the same way nature selects traits for survival in honeybee colonies that escape from managed apiaries and return to the wild, or in native bees that have always been wild. Nature edits out any traits that do not help the bee adapt for survival. This is how locally adapted bees develop. Just as the wild traits came to the fore in the goldfish in that pond, the traits for hardiness and survival in the wild come to the fore in colonies of bees that survive in the wild. But there has to be a kind of regression, or back-engineering of the bee by nature, to strip it down to the essentials for survival. Locally adapted bees are often very dark, as Italian or Buckfast yellows and oranges are edited out along with other traits that can't survive in the wild here in Britain.

I've bred all kinds of animals – guinea pigs, rabbits, mice, finches, goldfish and various poultry, and have always been fascinated by the different mutations we have produced by selective breeding. We've produced large chickens for the table, bantam ducks like the Call duck as hunting decoys, and poultry that lay eggs in various shades of blue, brown and white. It was egg colour that first interested me in my own selective breeding of chickens when I came to Douai Abbey and looked after our poultry.

I had large white eggs from leghorns, chocolate-brown eggs from Marans and Barnevelders, and small blue eggs from araucanas and cream legbars, but I wanted to add a green egg to the boxes of eggs I was selling. There are hybrids called *olive-eggers*, but I couldn't find any, so I had to breed my own *olive-eggers*. Eventually I succeeded, but the eggs were small and either too blue or too brown, so I had to make further crosses between birds carrying the genes for brown and blue eggs of the right shade and birds that laid a larger egg. I got there after a few years and those mossy-green eggs were very striking too.

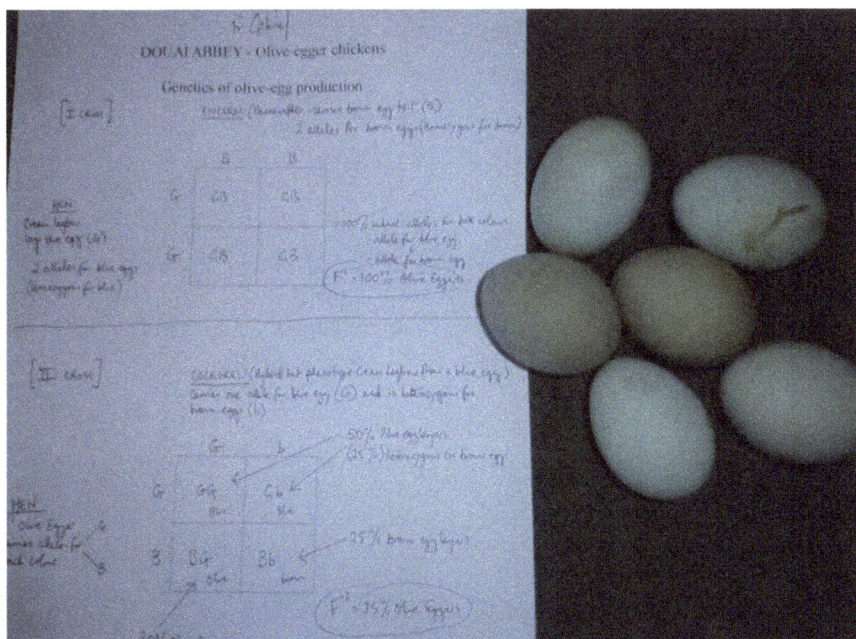

*Douai Abbey olive eggs*

It's not then that I am against selective breeding of animals and plants, but I am against the particular way it has been done with honeybees because their biology and genetics are much more complicated and they are not, in my view, domesticated in the way we define a domestic duck or pig. I'm also uneasy about the history of selective breeding of bees because it has all been predicated upon the idea that we can domesticate the bee and that we can perfect it. This obsession with perfecting the bee , if we're not careful, could lead us down the path towards transgenic bees in the decades ahead, and that worries me.

Just as there are concerns that transgenic crops or fish could affect other plants and animals in an ecosystem, there are undeniable consequences from flooding an area with either pure-bred or hybrid drones. Drones are flying gametes, equivalent to the pollen of plants. Just as there are concerns that GM pollen might disperse and cross with other plants, the flying 'pollen' of drones can also mix in an uncontrolled way with local bee populations, creating random crossings of mongrel bees and interfering with the selection of genes best-suited to a particular environment.

This raises the question that we have to ask increasingly about modern science – that just because we have the technology to do something, does it follow that we should always use that technology? I would respond with a 'no'. If the answer to this question is 'no' and that the bee can't (or shouldn't ) be domesticated, this creates a controversy – especially from a monk in a sister house of the same Benedictine Congregation as Buckfast Abbey, who asks:

*was Brother Adam's genetic experiment with the honeybee based, like Armbruster's work, on a faulty premise?*

More seriously perhaps, in the light of modern knowledge of bee biology, genetics and epigenetics, do we regard his experiment as having been another example of what William Woodley referred to as misapplied modern methods; or, accepting his premise that the honeybee can and should be domesticated, will we inevitably want to look to transgenics as a more efficient and faster way to achieve the same outcome?

Although not an immediate concern, the idea of a transgenic bee resonates today with the concerns William Woodley had about government interference in apiaries like his own and the freedom of the skeppist and their traditional ways. In our own time concerns about genetically modified organisms (GMOs) is not only about the ethics of engineering different species and the possible consequences for the food chain, the environment, human health and the freedom of the consumer to choose; it seems to me that there is also a greater anxiety about the morality of allowing most of the world's food supply to be controlled by only a few agrochemical companies. Would a GM bee be owned by a company like Monsanto?

There may be benefits to GM foods, but there is always the worry that in the wrong hands something good has the potential to do more harm than good, or that something can go wrong. There have been cases of GM crops that have had adverse effects. A positive use of the science is exemplified by the Hawaiian papaya. The papaya had been susceptible to crop damage and losses due to ring-spot virus, but in 1998 a transgenic papaya was introduced called the Rainbow papaya which was created to resist the virus. There is also the example of the Innate potato, a blight-resistant potato. Then there is the Sudden Death mosquito, gene edited so that its offspring drop dead, lowering the mosquito population. On the other hand there have been failures of the technology, such as the bt Cotton crop, engineered for India. It promised to decrease the use of pesticides because pests attacking the plant are delivered the pesticide. But these pests became resistant and required increased use of pesticides, which in turn increased costs and decreased the yield per acre. In this case GM crops didn't end well.

What about animals, such as goats modified with a spider gene who can produce milk from which silk can be extracted? A little more alarmingly, scientists have also produced a singing mouse whose vocalising trait is heritable, and cats that glow in the dark. In Canada a genetically engineered Atlantic salmon has been created and patented by a company. After two decades the American FDA finally approved it for the market in 2015. It reaches market size in only two years, with less food than non-GM salmon, and is by then double the size of its non-GM counterpart. Yet there is no law that retailers have to declare to consumers that they are being sold a GM salmon.

Although there is not a huge amount of evidence as yet that GM foods are harmful, there are fears that they might be and that perhaps we have too little data about their long term effects. For example, in 1996 a GM insect-resistant maize was created which produces proteins that are toxic to caterpillars. There are fears that such crops could cause allergic reactions or have toxic effects in the rest of the food chain. On the other hand, there has been research into the effects of bt Maize on nurse bees and larvae, none of which showed up any problems in terms of the weight, functioning and death rates of the bees and larvae.

It seems to me that one of the biggest concerns with these GM organisms, however, is that they are monopolised by just a few big companies, and that GM foods are being used to bolster industrial agriculture, which is driven by financial profit. One multinational organisation in particular is the world's largest seed company and could be said to own the world's food supply, but its history is morally questionable. It started off as a chemical company producing chemicals for war, which were used in the Vietnam war in the 1970s. It then switched to waging war in agriculture and produced a famous weed-killer called Glyphosate, which not only kills weeds, but beneficial soil bacteria, beneficial plants and even people. A successful court case brought against the company in 2019 by a couple who sued after the chemical gave them a type of cancer, was followed by 125,000 more law suits from people who claimed the chemical was carcinogenic. Despite this, the company is still free to sell its famous weed-killer, which contributes to its multi-billion dollar profits each year.

The same company then produced GM seed for crops that were resistant to Glyphosate so that plants could be planted more densely, increasing the yield per acre. But the company still makes billions a year from the chemical, while also selling neonicotinoides that kill insect and are harmful to honeybees, and which is known to be a contributing factor in bee Colony Collapse Disorder. How can such a company morally ride two horses – promoting widespread use of pesticides and weedkiller while also producing GM crops that are resistant to them? This company and a few other multinational organisations own 60% of the global seed market from their patented seeds, and require farmers to buy new seed every year. If a farmer's seed even blows onto another field the companies will prosecute. Farmers become locked into business with these multinationals.

This might seem a long way from the reality of ground-level hobbyist beekeeping and William Woodley's resistance to Bee Disease legislation a century ago, but it's not as far removed as we might think. There is an argument today that we need to move away from the monopoly of big agriculture and return to smaller, more intimate food systems. Smaller farmers have gone the same way as the small skeppists in the First Industrial Revolution. But we still have the equivalent of the small skeppist among our hobbyist beekeepers and a GM bee might pose a real threat to them.

What happens if one of these multinationals decides to engineer and patent the honeybee, to commercialise a strain that would be resistant to pesticides and varroa mites while maximising crop yield? You can guarantee they would replicate with the honeybee what they have already done with GM seeds. Beekeepers would probably have to buy new queens every year, there would be a monostrain of bee, and because it would be impossible to make it sterile (as can be done with Atlantic salmon) it would eventually breed out every other honeybee, becoming the dominant bee. The small hobbyist beekeeper would eventually be owned by the multinationals. Most would have to give up, unable or unwilling to become locked into one of these big company monopolies. Would you sacrifice your freedom for a utopian vision of the perfect bee?

William Woodley was protecting the beekeeper's freedom against what he perceived as a kind of terror. If, however, in principle, most beekeepers agree that engineering the honeybee is a worthy pursuit, towards combining the best traits for honey production, it isn't too hard to make a case for GM bees which would seem to make life easier for the hobbyist as well as the commercial beekeeper. What's wrong (they might argue) with a fast and precise route to accomplish what Brother Adam took a life-time to achieve? What's wrong with a bee that resists disease and parasites, has high immunity to many viruses and that produces a lot of honey?

On the other hand, could it open the door to a new terror, far beyond anything William Woodley foresaw? My concern is that unless we move away from the premise that we can perfect and commercialise the honeybee, we could open the door to a terror that won't necessarily affect our health, but would outrage the fundamental freedoms upon which hobbyist beekeeping is established.

These suggestions might seem outrageous and unlikely, and there is an argument that transgenic bees would be an absurd idea because it would be plain stupid; but the history of GM organisms in the hands of multinationals has already been outrageous, as far as I can see, with some equally stupid results. At the moment a transgenic honeybee doesn't seem likely in the immediate future, but it is possible to imagine, and big business agriculture and aquaculture have already demonstrated that there will always be someone ready to exploit commercial possibilities of transgenics. Moreover, it's possible in today's world to wake up in the morning and imagine whatever absurdities you like. Just look at the number of conspiracy theories there are. Once upon a time it was just a little speculation that Elvis was still alive and worked in the local chip shop. Now there is serious speculation that aliens from outer space are about to arrive on earth. And there are even stranger ideas closer to home. Stranger things have happened than we could have imagined even a decade or two ago, so is it too outrageous to imagine a transgenic bee when the technology already exists? If history proves me wrong, all well and good. Personally, I'm not confident that it will.

# MINDING THE BEES

## Chapter 5
### The Spirit of Progress

'HEALTHY STOCKS IN SKEPS.
1901 Queens. 9s each. Two 17s 6d. I am overstocked.
Hybrid Cyprian Queens 4s each.
SPEARMAN, Colesbourne, Cheltenham.'(1)

*British Bee Journal 1901*

I'm a worrier, especially since this year it would seem that T.S. Elliot's *The Wasteland* was right about April; *i*t certainly feels like the most cruel month at the moment. Since the stocks began to crash after Christmas, I've had to work quite hard at trying not to be over anxious about it. I could do a mite count, but what's the point? Firstly, if the mite load has reached a dangerous level in any of the stocks, I've no intention of using any treatments anyway. Secondly, even if the mite load is high, this doesn't necessarily mean the bees are in danger. If they are developing resistance they should be more tolerant of a high number of mites.

What's the worst that could happen? Worst case scenario - I lose all the stocks. That's why I've decided I need another string to my bow with my treatment-free strategy, in the form of wild swarms, so that if I lose all the stocks I will still be able to start again. If I can lure wild swarms, not only can I replace lost stocks, but there's a chance that some swarms might already be resistant to varroa. I can honestly admit that the moment I hit upon the idea, I ceased worrying about the crashing stocks. If I lose them all, I decide, that's just natural selection. It just means none of them had what it takes.

I calculate by April that if the bees have made it this far, they'll probably be all right going into this new season, but I need to have a plan for reducing the inevitable build-up of varroa. As a result, I feel much more 'in the moment'. I'm not worrying about what might happen; and that's as it should be, not only as a beekeeper but as a monk and as a human being. Why should beekeeping not be like Isaac Walton sitting mindfully by a river, enjoying his fishing and the natural world?

Sitting mindfully by the duck shed, I notice Garlic mustard *(Alliaria petiolata)*, also known as Garlicwort, Poor-man's mustard or Jack-in-the-hedge, growing nearby, in the shade of an oak. It's a wild edible weed that can take over forest floors, but it's nutritious, containing vitamins A, C and E and some vitamin B. It was also once used for medicinal purposes to treat gangrene and ulcers. The roots taste like horseradish – apparently.

Looking up at the crumpled new leaves of the oak, I see tassels of male catkins that emerge with their first leaves - lime-green and two or three inches long.  The male flowers of oaks hang down in catkin-curtains while the female flowers are small and red, located on short stalks called peduncles. It's from these that the acorns form. Oaks are monoecious, meaning they have both male *(staminate)* and female *(pistillate)* flowers on the same tree.

Cuckoo flowers are also beginning to appear in the apiary – the colour of those Parma violet sweets I enjoyed in my childhood. Parma violets were created in 1946 by the Derbyshire company *Swizzels Matlow*. They are based on a similar confectionery from India eaten after a spicy meal. You can also make Parma violet gin, by crushing the sweets to a dust and combining them with gin. Parma violets, the flower *(Viola alba)* first appeared in Italy in the sixteenth century.

Increasingly I realise that since I started beekeeping, I've been dominated every season by worry - probably in common with most modern beekeepers. The worry starts in April just at the point when you stop worrying about getting the bees through March. From April you're usually walking a fine line if you want to produce a honey crop. You want a steady expansion of the brood nest, but not so fast that the stocks get crowded early and swarm before the main nectar flow. You want to keep them expanding until you have the maximum population of workers to hit the nectar flow in June and July. Meantime you're watching for swarm preparations (preventing them if possible) and responding if you see queen cells appear. It can be a nail-biting time.

I've released myself this year from wanting a large honey harvest. Firstly, the stocks are low and I need to split them, to make increase and to make nucleus stocks to go through next winter. Based on this winter's losses, I'm going to need at least ten nucs if I want half of them to survive. Secondly, in order to make a large surplus of honey I'm going to have to build very large colonies, but last year those super-colonies were the ones that crashed first in the winter. More brood simply leads to more varroa. Instead, I must run smaller colonies and expect a smaller harvest from those stocks, much as skeppists would have done in the past. Of course, I don't want swarming because I can't afford to lose more of my precious stocks, especially as they are now survivor stocks from which I want to make increase. This means I must make artificial swarms. Not only will I be in control of swarming, but it will create a brood break in the split stocks that have to rear a new queen. Brood breaks are nature's way of reducing the mite load.

It's ironic then that whereas I spent my early years of beekeeping trying to avert swarming so that I could build up large colonies, this year I'm going to be reliant on artificial swarming and on smaller colonies. I've sacrificed hopes of a large honey harvest for the survival of the stocks. I've calculated that survivor stocks are going to prove a lot more valuable to me than honey at the moment.

Taking that pressure off myself feels good. I'm sure the old cottagers in past centuries worried about certain things, and that their lives were not all about old roses growing around the cottage door. They lived in the present much more successfully than modern people do, I'm sure, and they didn't spend all day worrying about their bees. Life was hard enough without that added burden. Minding the bees meant expecting them to swarm, just as they expected the dawn chorus to begin in February and the cuckoo to arrive in April. For all their hard lives, they were more attuned to the natural world and the rhythm of the seasons than we are. They rose when the sun came up and worked while the light lasted. They stopped for holy days. They dug their vegetable gardens by the lamp of a full moon after a long day's work. They were earthed to their piece of acreage and the cycle of life and death in ways that were natural to them and make our modern lives seem completely unnatural and disorientated in almost every way. Probably the reason I appreciate this old way is that it's very monastic. In every way my life and routines are rooted in cycles and seasons, in light and darkness.

I'm noticing, year on year, that it gets harder to hear the cuckoo because there are fewer of them around. The old cottagers would have been tuned to hearing the cuckoo, as they were alert to the appearance of different kinds of blossom and flowers:

*Cuckoo, cherry tree*

*Good bird, tell me*

*How many year before I die?*

According to this Buckinghamshire children's rhyme, the number of calls the cuckoo gives after someone recites this rhyme indicates how long they have left to live. No cuckoo yet though! And that silence might be something of a relief, if there were any truth in rustic superstitions because when I do hear the cuckoo these days I seldom hear more than a few calls. The earliest I've heard the first cuckoo is 6 April, but some years I hardly hear one at all more than a couple of times.

Other folklore associates the cuckoo with the cherry tree. It is said that the bird needs to eat three good meals of cherry before it can stop cuckooing. I recall one year, several years ago, when we heard cuckoos incessantly all day, all around the Abbey for some weeks during May, but I have no idea about the cherries that year. It's an odd folklore though, as cherries wouldn't be available this early for a cuckoo. These and many other folklore traditions and sayings ran in the veins of the old country cottagers and their beekeeping. For example, counting the cherry pits from the eaten cherries on your plate is accompanied by the invocation, *This year, next year, sometime never.* Counting the pits as you recited was believed to indicate when you would marry, according to where you stopped in the recitation when the pits ran out. (Not applicable to monks!)

In Kent another superstition held that if you visit a cherry orchard and don't rub your shoes with cherry leaves you will die of suffocation from a cherry pit. In North-east Scotland it is known as the *witches' tree* that should never be cut. Despite this taboo, cherry is fine wood for making furniture, gun-stocks, smoking pipes, spoons and bowls, and musical instruments, polishing up nicely to a reddish wood.

Britain's native cherry trees are thought to originate from Central Asia, the name resembling the French *cerise*, and the Spanish *cereza* which each evolved from the Greek place name *Cerasus*, now in Northern Turkey, where European cherries were first grown. Stones from cherries have been found by archaeologists on Bronze Age settlements, which means they've been in Britain for at least two thousand years.

At the back of our kitchen and monks' refectory we have a single wild cherry that is so out of the way that it is overlooked every year. Last year I tried its sweet cherries, and very lovely they were too! *Prunus avium*, the wild sweet cherry, is another ancestor of our cultivated varieties. The seemingly misnamed Bird cherry *Prunas padus* (whereas paradoxically *avium* means *of the birds*) has bitter fruit. *Prunas padus* has an acrid smell and in the past people put it at the door of the house to ward off plague. Its bark makes a fabric dye in various shades of brown, the roots making a purple dye. The leaves are less obliging, unless you want to make hydrogen cyanide, for which reason chewing them is best avoided! On the other hand, resin from its bark can be used as a cherry gum or a treatment for colds and coughs, or for gall and kidney stones, if mixed with wine. Sour cherries have natural antibacterial, and anti-inflammatory qualities.

Cherry blossom is significantly less abundant than it was in the past, though the fortunes of British cherry farming are on the rise again. Sixty years ago, there were forty thousand acres of cherry orchards, reduced now to one thousand. Production hit an all time low in 2000 but has been growing again with something of a revival. For cherry farmers life now really is a bowl of cherries.

The main pollinator of commercial cherries is the honeybee, but there is a pollination deficit from natural pollinator species and hobbyist beekeepers. It's vital for the orchards to be correctly pollinated, as supermarkets require a standard-sized fruit in the right quantity. If the trees are under pollinated the fruits are fewer but larger; if they are over pollinated the yield is higher but with smaller fruits. Orchard farmers of cherry and all other fruit therefore require the right density of bees per acre to achieve the pollination levels required for economic success. Bee farmers supply a vital pollination service to these crops, hired for a period during which the beekeeper earns part of their income, and the bees are fed. Apparently hiring bees for this service produces a 20% increase in crop yield.

There's a controversial side to this kind of beekeeping, however. In the US bee farmers work on a truly industrial scale, with tens of thousands of hives transported on lorries large distances to California where cherries and almonds are among many orchard crops grown in the equally industrial monoculture of the orchards. Not only do the hives have to migrate long distances with the bees packed up, but for many weeks they feed on an exclusive diet of one type of pollen, which is not what bees do naturally through the seasons. There is still much research needed on the effect of this over-dependence on one kind of pollen and the relationship it might have with *Colony Collapse Disorder* in the US that has blighted so many commercial beekeepers for so many years.

Am I in the middle of a kind of 'mid-hive crisis' as a beekeeper I wonder? In his charming book *The Bee Master of Warrilow* Tickner Edwardes identified three kinds of beekeeper, and there's a bit of them all in my beekeeping, I think. First there are the old skeppists: 'Men can still be found who keep their bees much in the same way as bees were kept in the time of Columella or Virgil; and are content with as little profit.' (2) But he identified such beekeepers as, '...the man who obstinately shuts his eyes to all that is good and true in modern bee science.' (3) Then there are the, '...ultra-modern pushing young apiculturists of today.'(4)

Among the treatment-free subspecies of beekeeper are those who describe themselves as natural beekeepers, and these are a broad church. If treatment-free beekeepers are the Protestants, natural beekeepers would surely be the nonconformists. Another group could be described as fundamentalists, and they are the log-hive keepers who want to return bees to their ancestral home in the ecosystem - namely trees. There was a time in Mediaeval Europe when bee forests existed specifically for bee colonies living in the wild. Any given forest had perhaps twenty nests occupied at a time in tree cavities created by beekeepers and owned by nobles. In Eastern Europe this developed into log hives that were upright sections of hollow log, often carved, free-standing on the ground, with a base and a roof. In Britain it's interesting that the skep developed, as opposed to the log hive, probably because it was brought by the Saxons who were accomplished at agriculture. They were already using the woven bushel *skeppa* to measure out grain produce, while skep hives would also seem to have been a characteristic of their more organised agricultural systems.

I'm not against the log-hive beekeepers of today, who are re-wilding bees. It's a complete reversal of the process of farming bees that goes right back to the Egyptians. But it's not for me as a beekeeper. The buzzword today is *diversity*, and perhaps diversity among beekeepers is good for honeybees in the long run. What the re-wilders want to do is move away from farming bees to a consideration of honeybees as a vital part of the environment. Through re-wilding they hope to build up wild survivor colonies that adapt to local conditions and become more resistant to diseases. One of the arguments is that keeping bees in more natural conditions results in fewer winter losses compared to losses in managed frame hives. Another is that wild bees can repopulate areas where bees mate *(Drone Congregation Areas)* with the huge number of drones required to ensure genetic diversity and a higher mating success for queens. I'm all for these aims. As a beekeeper, I want my queens properly mated with strong drones carrying good genes for varroa tolerance and a general resistance to diseases. I read recently about some recent research from the University of British Columbia that queen bees with viral infections have smaller ovaries than healthy queens. Log and tree beekeepers are doing me a big favour then, especially if I want to lure swarms.

But I also want to be hands-on with beekeeping as a craft. I want to know my stocks and to be able to monitor what's happening in their hives. I enjoy raising queens too, while I can't say I'd enjoy beekeeping quite as much if I were not interested at all in a modest honey harvest. Even natural beekeepers who keep stocks in frame-hives will often frown upon the idea of opening the hive more than once or twice a year because they claim the bees are then less stressed. From some beekeepers I've watched who crush bees and bang boxes and frames together with no regard for the bees, I can appreciate that view. On the other hand, I've worked carefully and gently with stocks and they've hardly seemed aware of my presence. Ultimately, for me the higher the stock is up a tree, the further away I am from the beekeeping craft, and I enjoy the craft.

In defence of the tree and log beekeepers, many of us hardly appreciate the importance of the relationship between bees and trees, especially early in the spring. The spring build-up of stocks is probably more dependent on trees than the sparse early flowers, particularly as early sources of pollen for feeding brood. In fact, trees give the largest pollen and nectar yield per acre when compared with flowers. There might be early snowdrops and crocuses, but bees are more likely to forage more successfully on early flowering trees such as alder, birch and hazel. Alder yields fifteen hundred pounds of pollen per acre. Native willow yields the same, later in April. In some species two mature eighty-foot trees have the same potential for nectar and pollen as a whole acre of flowers. Pollen especially is important to the spring build-up, as it's needed to feed brood. Unlike nectar that workers pass to another bee to store, workers store pollen themselves, as though it's so valuable they don't even have time to share the job. It's a case of dump and go. Elm used to be an early source of pollen before *Dutch Elm disease*. As a boy I lived in a road called Elmsleigh avenue, that had huge English elms planted its entire length. During the Dutch Elm disease of the mid 1970s they all died and were felled along that road almost overnight. Think of all the Elm streets, Elm lanes, Elmsleigh roads and High Elms lanes that now remind us of the dangers of monoculture planting, not to mention the danger of a single pest, pathogen or virus like the Dutch Elm beetle. And don't talk to me about importing them.

Skep hive stocks relied on early flowering trees such as hazel, alder and birch. Bee farmers would move their stocks from home stands to out stands in March, to take advantage of the early availability of pollen from these trees. The skeps would be tipped up and covered with a piece of cheesecloth or muslin, to stop the bees coming out. The entrance would be stuffed with moss or grass and the skeps stacked on their side on a cart or lorry. Installed in the out stands, which were usually near farmers' orchards, the stocks were then well positioned for the nectar flow from the next flowering of trees in April and May – fruit trees such as cherry, apple and plum. The big take-away from this is to plant trees, if you want to help honeybees and only have a small area in which to do something. One large tree in a back garden will yield more pollen and nectar than a small area of lawn left to grow wild flowers.

The long range weather forecast is uncertain, due to a number of variables, so who knows what the weather will do? It's frustrating. Maybe I'm missing something too, but I don't understand how, on the one hand, meteorologists can be unsure about the long range weather forecast that is based on computer models; but we're told to follow the science and trust in other computer models that claim to be accurate forecasts of climate change years and decades from now? It's a mystery to me.

A preoccupation with the climate and the weather this spring has got me thinking again about the importance of locally adapted bees and environmental factors. Brother Adam pointed out, interestingly, that it isn't always the case that bees introduced from different places fare badly, or that bees that have existed for a long time in a particular environment are necessarily the most successful bee, by beekeeping criteria,  for that region: '...another race of bees brought in from a totally different environment can produce results which on average far surpass those of the native bee. In fact the introduction of different races and types of bees has many real advantages. For instance the Cyprian bee is able to winter in England much better than could the old native bee; we have never lost a colony of Cyprian bees due to climate even in the wettest of winters and coldest of springs.' (5)

Bees, he explained, adapt themselves to their environment over long periods rather than become acclimatised, though he also pointed out that issues of acclimatisation don't really pertain to queens and drones anyway who spend much of their lives in the warmth and comfort of the hive rather than outdoors in the wider environment. It is really only the workers who have to deal with the challenges of a particular environment and its conditions.

On the other hand, his assessment of the Italian bee *(Ligustica)* was that: 'The pure Italian does very well in favourable honey-flow conditions as found in North America and other parts of the world, but in conditions such as ours she often proves a dismal failure.' (6) In particular he drew attention to, '...our changeable climate'. In the British climate, Brother Adam explained, spring doesn't arrive suddenly with a burst of blossom, as it does on the Continent, but here spring is slower and more gradual in its arrival, with cold spells and changeable conditions, much as we have had this year, '...which imposes a severe strain on the vitality of the bees and can exhaust them prematurely.' (7) Another drawback related to our conditions rather than our climate is the Italian's tendency to drift, because of its poor sense of direction. This means that Italian bees will end up in and out of neighbouring hives. If those hives carry pests and diseases, such as varroa, Italians will be adept at picking up and spreading them through its tendency for drifting between hives.

His appraisal of the Carniolan *(Carnica)* highlights that she is a good bee for places where there is an early flow from fruit blossom and dandelion, but that the Carniolan can't take advantage of a late flow from heather, and might even miss a main flow with her tendency for swarming after an explosive spring build-up. Brother Adam added that: '..this bee reacts with great impetuosity to spells of unfavourable weather and will in times of dearth reduce or stop breeding altogether...' (8) and that, '... the Carniolan is not suited to our conditions.' (9)

Between 1950 and 1962 Brother Adam travelled extensively to search for the best strains of honeybee and to find them in their native regions. He made some interesting observations about the plight of these distinctive ecotypes and geographical races because of random hybridisation, which raises questions about continuing hybridisation today. In 1950 he observed the native pure French bee which at that time was nearing extinction, and lamented: 'In spite of these rather serious defects, it would be an irreparable misfortune and loss if the native bee of France succumbed to the current trend of indiscriminate hybridisation.' (10)

He identified a similar problem with the Greek bee *(Cecropia)*, partly due to indiscriminate cross-breeding and partly because of the large scale transportation of colonies from different parts of Greece to northern areas for its honey flow. He warned that this cross-breeding and use of mongrel stock was a great danger to the pure races of honeybee throughout the world and that it would one day become difficult to find genuinely pure specimens of the various geographical races. This was important because cross-breeding and combination breeding that produced the Buckfast bee and might produce other hybrids with economic value required fixed characteristics from breeding close relatives within a particular strain or race before crossing could be done with other races. To this end Brother Adam recognised that: 'To preserve and promote these breeding possibilities it is essential to establish reservations to maintain these different races. This maintenance of the races with their original hereditary wealth and individuality is a pre-requisite for any progress in breeding the honeybee.' (11)

His emphasis on this necessary protection, however, seemed based simply on the idea that these regional ecotypes and geographical races are just a resource of genetic material required if future crossings are to continue to develop the honeybee commercially. However, by identifying the issue of indiscriminate hybridisation and the threat to pure races, he alluded to the real heart of the problem, as I see it, which is the attempted domestication of the honeybee. The problem is that its genetic engineering requires isolated apiaries, to avoid indiscriminate mating, but at the same time those who are not bee breeders but who queen their hives with his engineered queens end up flooding ecosystems with drones that contribute to further indiscriminate hybridisation of local wild stock as well as the hobbyist's stocks. For me, that model doesn't work in the long run. It affects local bees at a genetic level and interferes with the selection of adaptive genes, which are important because, as Brother Adam pointed out again and again as a sort of premise for his project: 'Nature's aim in breeding is limited exclusively to the preservation and dissemination of a species and her sole means of doing this is a ruthless selection...The one aim was the survival of the most adaptable and the fittest.'(12)

From his journeys in search of the best strains Brother Adam identified that Nature does this even within a particularly small region, creating different ecotypes and strains that are all delicately tuned to the conditions of a particular environment. For example, he identified that there were different strains of the Greek bee (*Cecropia*) that is a subvariety of the Carniolan, just as there are many subvarieties of the Carniolan, such as the Banat bee, those in the Carpathians and the Pester Plateau of Serbia, and others in the Montenegrin Alps. The French bee too was identified as more varied than had previously been known, 'Although all these local strains have basic characteristics in common, they show essential differences in the way they emphasise one or other of these characteristics.' (13)

I must add that I see this aspect of Brother Adam's work as far more valuable than the economic potential of any particular crossings he produced, as little had been known before this work about the different races of bees and the genetic relationships between individual races. Brother Adam observed their particular morphological and physiological features and variations and contributed much to the knowledge we now have of the existence of the great variety of races and local strains. Whereas he regarded this knowledge as indispensable for his cross-breeding, I would argue that its greatest value lies in helping us to understand the importance of the different environments and local conditions for each race or strain of bee to which it is delicately tuned for survival.

Brother Adam's work also alerts us today to the dangers of selective breeding of the bee, with the example of the Carniolan, of which he observed: '...in the recognised strains of Carniolan of today we look in vain for the one-time vitality she manifested. The efforts to produce uniformity in the external markings together with the accompanying inbreeding has undoubtedly resulted in a weakening of her vitality.' (14)

With the Greek bee he recognised a similar problem with the effects of hybridising: 'My fears on this score were already raised in 1952 when I saw the  accumulation of tens of thousands of colonies brought there from all parts of the land. This is leading a to a steady decline anyway of the variety as once found in Macedonia.' (15)

In the context of this valuable research by Brother Adam, it is alarming that one in fourteen hives in Britain at the time of writing is currently queened by an imported queen. I looked at just one well-known bee supplier's website, to see what varieties of queen they had on offer. What did I find?

First, there were Carpathian queens, advertised as originating from Romania where they are open-mated. Apart from the fact that they are imported, they are open-mated (mated with any available drones) so there is no guarantee that the workers will be pure Carpathian. Next, they advertised Carniolans, which are actually described as F1 mated, which means they are not pure-bred, as F1 will have lost 50% of the Carniolan genetics. These are raised by European partners, so they are also imported stock. Next there were *Ligustica* from Greece and Romania, which were described as open-mated. Once again, this carries no guarantee that they are 100% mated with *Ligustica*, and these are further examples of imported queens. I also found F1 Buckfast-cross mated queens from Greece, which are hybrids, for starters, but also *open-mated*. The description of the latter was honest enough to point out that hybrid second-generation crossings can differ greatly, depending on the drones with which they mate. In my view whoever buys these queens is playing lucky-dip. Their Buckfast mated queens are bred from Buckfast queens imported from a Danish breeder, so are in fact F1 Buckfasts and not the pure Buckfast strain at all. These F1 queens are produced and mated in the UK and there is no indication that the mating station is isolated, so they could be mated with anything. Again, it's lucky-dip what you get from the workers these queens produce, as there will be variations among the different patrilines, depending on the type of drones with which the queens mated. Lastly, I found the European Dark bee for sale (*Amm*). The description informed me that they are F1 queens, which are open-mated and from Greece. These are not going to guarantee pure *Amm* stocks and are another import.

Looking online at beekeeper forums, it's clear that there are some beekeepers who see the designation Buckfast or *Ligustica*, for example, and think they are getting a bee produced by Buckfast Abbey, or at least in the UK, and that it will be pure-bred. The F1 description can be misleading because anyone not understanding genetics and bee breeding might assume that F1 simply means first generation, but if the parent was open-mated or naturally-mated these are usually euphemisms for *indiscriminately mated*. If such bees are open-mated but in an isolated apiary, they might well be mated with Buckfast drones from the next generation flooded into the area, which would be pure, carrying the genetics of a mother and the drone's grandparents. In all probability, however, the F1 label is a way to avoid admitting that these queens are not pure-bred, or they would simply be called by the race or the strain's name instead of F1. In essence what you have to remember is that when you buy a mated queen you're also buying the colony she will produce. She might well be genetically a pure Buckfast (or whatever race or strain) but her worker progeny will not be, if naturally or open-mated – they'll be 50% the mother's genetics and 50% whatever the drones were who mated with her.

Allow me to clarify what's going on here by way of an analogy. Imagine I have a pedigree Siamese cat and a pedigree tabby and I cross-breed them. My first cross will be hybrids, but I will have some idea what I might call them. The F1 kittens will be a known cross, but to call them either *F1 Tabbies* or *F1 Siamese* would be misleading because they are neither Siamese nor tabbies any more. They are, in fact, an *F1 Siamese / Tabby* cross. Now imagine I allow my Siamese pedigree queen out every day to be open-mated. In the local area there might be Toms of every description – mongrels, perhaps a Maine Coon, a white British short hair, a ginger and a ginger and white, but I have no idea of the full genetic diversity that exists. My Siamese queen comes in pregnant with kittens which she delivers some time later. I have no idea which Tom is the father. When I come to sell my kittens I decide I want to make them sound as special and attractive as possible for the market, especially if I want to make a good sale on each kitten. I decide then to call my kittens *F1 Siamese* in my advert. Of course, it will be obvious to any cat owner that I couldn't get away with that dishonest description because Siamese cats conform to certain distinguishing characteristics. Somebody will quite correctly tell me that my kittens are not *F1 Siamese*, just by looking at them. My name has attempted to mislead. They are more correctly called *F1 hybrids*.

When it comes to bees, you can't always distinguish the race or variety by its appearance in the same way that you can with different breeds of cat. Secondly, the queen bee mates with a number of drones, unlike the queen cat who usually mates with only one Tom. It follows that if I have a Carniolan queen that is then open-mated, it's analogous with putting my Siamese cat out one night and having her mated with ten different Toms, all of which father at least one of her kittens. My first generation of bee workers and queens will be indiscriminate mongrels. Calling them F1 Siamese is simply misleading, even if it's not entirely dishonest or illegal.

It's playing into the oldest trick in marketing, which is that people are attracted by a label and a story. The name Buckfast has the same attraction as a designer label and there's the additional story of the monk and the bee. Of course, breeders can get away with calling their queens F1 Buckfasts because they are indeed F1 queens and they are indeed pure Buckfast. In fact, they also produce pure Buckfast drones, as there are no F1 drones in the first generation due to parthenogenesis in the queen; the first generation of drones having the genetics of their Buckfast grandparents. But the designation leads some to think they are getting an F1 Buckfast *colony* as well when the queen has often been open-mated, which means the colony you get from her will not be pure Buckfast. True, they might well perform nearly as well as pure Buckfasts – F1 hybrids usually closely resemble their parents. But the fact remains, an F1 Buckfast queen that is open-mated will probably not give you a pure Buckfast colony. It's another example of why we need to always think of raising colonies and not just raising queens.

If beekeepers near the Douai apiary have imported queens from suppliers such as the example I've given, bee biology means I can get away with calling them *F1 (variety)* because in the F1 generation there are actually no F1 drones. F1 queens produce drones of pure descent, because of parthenogenesis in the queen, which means that an F1 Buckfast will at least produce pure Buckfast drones; a Carpathian or *Ligustica* queen, even if open-mated, will produce pure Carpathian or *Ligustica* drones. To put it another way, there are no F1 drones in a first cross. Only the F1 workers and queens will be the hybrids.

Furthermore, the pure-bred drones now flooding the Drone Congregation Areas where my queens are mated will be introducing pure-bred ancestry from a particular race. (That's how I can get away with calling a Buckfast queen's open-mated daughter an F1 Buckfast.) The problem is that these drones from other beekeepers in my area will be mating with my queens and introducing genetics into my next generations from pure races, and that messes up my goal of raising a locally-adapted bee. Some of these crosses can even be problematic to handle.

A common difficulty is that first crosses in particular can produce a poor temperament. In fact, Brother Adam assured us, from his vast experience of many crossings, that some crosses invariably guarantee this problem, '...the cross Buckfast-Anatolian gives a bee equally unwilling to swarm, very productive but very bad-tempered.' (16) Moreover, reciprocal crosses don't in most cases produce the same result; gentle strains crossed with gentle strains don't necessarily produce gentle strains, but can produce an aggressive bee. One example is that Carniolan queens mated with drones of other races often produce bad-tempered bees, but not so if you use Carniolan drones. Brother Adam wrote, 'As far as temper is concerned all crosses between races have a bad reputation.' (17)

This tendency for indiscriminate and random matings to produce aggressive bees was known by Brother Adam and even as far back as William Woodley, and has been called F2 aggression. Quite apart from interfering with the adaptive genes selected by local bees, F2 aggression is a further interference in the beekeeper's selection of the most gentle stocks with which to work.

In the UK at the time of writing there are roughly 25,000 hobbyist beekeepers who belong to the BBKA. If they each have a minimum of two hives, this means there could be at least 3,500 queens imported by these beekeepers in a given year, and that's a conservative estimate. Considering the fact that many beekeepers have a few or more hives, we might expect to at least double this number, which brings us to 7,000. It would also be a conservative estimate that at least several thousand queens were imported for UK hives every year up to Brexit and the pandemic, though these figures would have been affected by our new circumstances. This does not account for bee farmers or unregistered beekeepers. In fact, according to the Beebase website, voluntarily-registered colonies in 2020 increased to 224,000, which is ten times the number belonging to the BBKA. Of course these numbers would also need to be adjusted to take into account the fact that on average queens are replaced every two years and not annually, unless a queen fails to survive the winter or fails in another way. The calculation of drone populations would also probably need to take account of those beekeepers who cut out drone brood, particularly to monitor or reduce the mite load.

Based on the BBKA figure, let's assume that drone numbers per colony at their maximum are about 24% of worker brood, and that an average-sized hive has a population of 50,000 bees. This means each hive, at optimum production, can produce up to about 12,000 drones a season. If there are several thousand imported queens each producing perhaps a total of 12,000 drones or more in a season, that adds up to a cautious estimate of 84 million drones flooding the UK's Drone Congregation Areas with foreign genetics every year. If we do this maths with the Beebase record of colonies, we have to multiply 84 million drones by 224,000, which is deeply alarming. A high percentage of these drones will carry foreign genetics and will be mating with hobbyists' bees and wild populations, and will necessarily interfere with the selection of adaptive genes in locally adapted bees.

Now let's take a closer look at those characteristics, based on that website. Firstly, Carpathian queens are a subspecies of the Carniolan. They are described as good-tempered and have good winter-hardiness, with an early spring build up. Brother Adam claimed that they could be prone to bad-temper. But the temperature variations in Romania are more extreme than in Britain, with very cold winters and very hot summers. *Ligustica* come from Greece and Romania, and the variety (there are different varieties of Italian bee) is not specified. The climate differences between the country of origin and UK are significant. *Ligustica* will consume more food due to excessive brood rearing and having a loose winter cluster which loses heat. Carniolans are described as prolific and gentle, and well-suited to our climate. They are bred in Europe, which could mean anywhere, but Brother Adam thought they were unsuited to our conditions. *Amm* queens come from Greece, which has a very different climate. Those Buckfast imports that are open-mated F1 queens will bring in an undetermined mixture of genetics, having mated with many drones, which will not all be Buckfast. Brother Adam did not believe the Carniolan or Italian were suited to British conditions, while it seems strange that the Buckfast now also comes from abroad, from places such as Romania. An additional concern with all these bees, regardless of claims that they might be suited to conditions in Britain, is that they increase the risk of introducing new pests and diseases such as Small Hive Beetle (SHB).

Imported queens currently arrive from about twenty-five different countries, among which are Greece, Hawaii, Czech Republic, Malta, New Zealand, Denmark, Poland and Ireland. It isn't only that these queens might introduce new disease and pests, but that they might be susceptible to diseases already here to which local bees are resistant or immune. An example of this might be the emerging problem of Chronic Bee Paralysis Virus (CBPV). In a study published in *Nature Communications* and reported in *The Times* in 2020 scientists believed they might have identified a link between the virus and imported queens. Cases of CBPV have increased exponentially since 2007, and were found to be 1.98 times more likely in professional beekeepers' hives from 2014 onwards. It has been suggested that either the foreign bees are susceptible to the disease picked up in Britain or they are responsible for introducing it. Moreover, the pattern of the disease's emergence and burning out over a couple of years, in waves, seems to suggest a link with the replacement of queens every two years.

There are more accurate figures available of imports from the EU before Brexit, that are evidence for the joint concerns of exotic races and strains that are not adapted for survival in Britain and the threatened introduction of new diseases and pests with them. According to the *National Bee Statistics*, published by BIBBA's website, the largest importer of package bees, nucs and queens into the UK between 2014 and Brexit in 2020 each year was Italy. In 2019 Italy exported 3,304 queens to Britain, while Greece exported 4,928 queens here, Italy's exports increasing by 2020 to 9,701. The total number of queens exported here in 2019 from all exporting EU countries was 20,081, rising to 21,405 in 2020. Before Brexit then the number of queens we imported was actually three times higher than my conservative estimate of that figure now. More alarmingly, despite all the EU's regulations and paperwork for these exports, Small Hive Beetle still managed to enter Italy where it is now an increasing problem. It seems obvious that it is only a matter of time before SHB arrives in Britain, just as varroa arrived and other diseases, unless we stop them arriving, by banning their importation.

A 2014 Europe-wide study has confirmed that locally adapted bees consistently perform better than imports of exotic strains. Imports have evolved in different ecogeographical conditions which differ greatly from those in Britain. When these bees don't survive our winters (or perhaps pick up fatal diseases over here, such as CBPV) a cycle of demand is produced which benefits the commercial activity of bee breeding and exporting, but does nothing to advance the cause of our own bee population and its survival. The dilution of adaptive genes in local bees simply weakens our own stocks. Further evidence that exotic imported queens have unsuitable genetics for our conditions may be inferred from the observable fact that bees living in the wild in Britain don't resemble imported races and strains because those characteristics are deselected by natural selection for survival. Nature is telling us, by the darkening of locally adapted bees, which bees are best suited to Britain.

What do you get if you cross a bra with a beekeeper?  You get the modern all-in-one bee-suit. The modern bee-suit is a hybrid and is actually made from the same materials as bras and swimwear, the prototype run up by a beekeeper and his wife  back in 1968, who then founded the B.J. Sherriff beekeeping equipment company.

It all began with Edgar Sherriff, who bought the Langridge Ltd. corset factory in the 1890s, founded to manufacture crinolines and stockings. His son became the director of the Bristol company, which even made parachutes during World War II, and panty girdle for *WRNs* and *WAAFs*.

Brian (born 1928), the third generation of the family, took over the business in 1946, expanding from seventy machinists in 1959, producing five hundred dozen bras a week, to one hundred machinists a year later. In 1962 they diversified into swimwear as well. From the mid -1960s, however, the company began to run into competition from cheap foreign manufacturing, and the receivers were called in during the 1970s. Brian and his wife had also established the *South Cornwall Honey Farm* with four hundred hives in the 1960s.

In 1968 Brian was stung badly on the back of neck while beekeeping and complained that his traditional protective clothing hadn't worked. His wife, Pat, went into the factory and put together a new kind of bee suit from the same boning, net and fabric used to make bras and the gussets in swimwear. The invention gained much support (pardon the pun) from bee shows and exhibitions, and the orders took off. Brian Sherriff went quickly from bras to bees, and from bust to boom, which is as unlikely a diversification as you could make.

A first spring inspection is usually quick. You don't want to expose brood to chilly winds for too long or keep the hives open, even if it's between fourteen and eighteen degrees. Brood must be kept warm, and as soon as you open the hive the temperature in the brood chamber falls. I'm looking as fast as possible in each stock with a particular check-list: is there a queen, and is she laying? How well is she laying? Is she laying worker-brood or is she a drone-layer? How much brood is there, and are there any abnormalities that might indicate disease? Does the colony look healthy, with plenty of bees, or is there a small, weak population? Do they have space, especially if the colony is large? Are there plenty of stores of nectar and pollen?

At this stage, if all looks well, I can close the hive and allow the bees to get on with the spring build-up.

Skeppists were not able to make such detailed inspections. By turning the hives upside-down they could see how many seams of bees were between the combs and how far down the combs extended. A triangle of comb could be cut out from the central combs down as far as the brood nest, to check for eggs and brood; this was all that told the beekeeper a queen was present and laying.

Perhaps in April I might see the first drones begin to appear. Beekeepers tend to overlook drones, or even think they are rather useless, and that's a great shame because drones are not only fascinating but are extremely important – probably also in ways that we still don't understand. Not least of all they are 50% of the genetics of the honeybee. Until recently I paid little attention to them, but once you discover them it completely changes your perspective of the hive and everything you're doing.

Drones are produced when the colony can afford them because they are expensive to produce and maintain. They are therefore an indicator of a stock's wealth and its preparations for the swarm season. In autumn, before the cluster settles down for the winter, all surviving drones become redundant because a drone is actually little more biologically than a flying gamete, and no mating takes place in the winter. It follows then that no drones are taken through the winter by the colony; in the autumn they are ruthlessly sacked and expelled from the hive.

Drone biology, as with all bee biology, is quite extraordinary. They are raised in enlarged cells, usually on the edge of comb, and the capped cell of the larva has a slight dome, unlike capped worker brood which is flat. The male larva also takes a few days longer to develop in the cell, which suits the varroa mite's larval development. Hence some beekeepers will allow bees to produce sacrificial drone comb, which is cut out as part of an integrated pest management strategy to reduce the mite population.

Many beekeepers have a commercial view of drones. Drones, they observe, need a lot of protein, which requires plenty of pollen. As they mature, they eat the honey and require a lot of energy to make perhaps just a few mating flights a day. If a colony is allowed, it will produce several thousand or more drones in a season, compared with only a few queens. None of these drones will mate with queens from the same colony because drones fly further afield to mate while queens stay close to home, to avoid inbreeding. You can see why many beekeepers consider drones a bit of a waste of time and have no conscience about cutting out drone comb; the result is fewer mites and a reduction of the costs to the hive in valuable resources. It also leaves more comb available for worker brood, to maximize the worker population and to increase the honey crop.

Our traditional skep beekeeper at this time of year would cut out the first signs of drone brood with a long-handled, curved blade called a *drone knife*, but later in the season drones were left to increase in number. At this time of year, the skep beekeeper wanted the bees to draw the combs down to the bottom of the skep, and removing the drone comb encouraged them to do this. It also prevented the early production of swarm cells, as colonies time their swarming with the peak production of drones. The traditional skep beekeeper therefore was not reducing the overall population of drones to a detrimental degree. Modern beekeepers, on the other hand, who cut out or put in drone comb frames every week for *varroa* control, are significantly reducing the population of drones.

Why does this matter? Because our drones will mate with queens from someone else's apiary, and their drones are going to mate with our queens. If most, or all, the beekeepers near me are decimating their drone populations, they are also reducing the total number of drones visiting Drone Congregation Areas (DCAs) throughout the summer, where our queens are mated.

Beekeepers in many places are concerned that queens are not getting properly mated, probably because there are fewer drones around. This also affects the genetic diversity and the mating competition between the strongest drones. That is not good for bees, especially if the situation continues for years, or decades. I might decide for purely selfish reasons that varroa control is my main objective in sacrificing most of my drones, but I would be making problems for my neighbours' bees and vice versa, not to mention bees living in the wild. Quite simply, drones are undervalued, and they are as precious as queens. Once I realised this, I stopped thinking about *raising queens* every year and began to think of *raising colonies*, which include drones.

In defence of the re-wilders and the natural beekeepers, they're allowing our Drone Congregation Areas to build up the vast number of drones it takes to run a viable and healthy mating location where a queen can be mated with maybe as many as fifteen drones from many different colonies. These natural beekeepers are doing me a favour. The chances are that some of my neighbours, who belong to the local BBKA and are doing courses and exams in integrated pest management (including cutting out drone brood), are working against each other's best interests and mine by working against nature.

You might think that the jump William Woodley of Beedon parish made from watchmaker to bee farmer in the nineteenth century a strange one. Then again, a bee colony is as precise, delicate, beautiful, intricate and balanced as a watch mechanism, and just as easily broken. Just as you wouldn't have jolted or dropped a watch mechanism in Victorian times, you can't metaphorically drop or jolt a hive of honeybees without damaging their delicate mechanism. We still have an incomplete understanding of the intricacies of that system operating in our hives. I read recently that researchers have only just discovered that bees blow pheromones from the queen to each other by intentional fanning; they pass it around the hive like an email.

Furthermore, we have no real idea how drones who have no knowledge of *Drone Congregation Areas* manage to find the same area each year, often several miles away. It's a mystery how the colony and the queen know the correct ratio of drone comb to worker comb to build, or how they monitor and regulate the entire economy of the hive. Drones are part of this delicate and beautiful mystery, but it's almost standard practice now to cut out drone brood in the name of an Integrated Pest Management. We have no idea what effect that has on a particular colony or the extent to which it is damaging DCAs and the successful mating of our queens.

Tickner Edwardes identified a third kind of beekeeper, exemplified by the old bee Master of Warrilow: 'Born and bred amongst the hives and steeped from their earliest use in the love of their skeppist forefathers, these interesting folk seem, nevertheless, imbued to the core with the very spirit of progress. While retaining an unlimited affection for all the quaint old methods in beekeeping, they maintain themselves, unostentatiously, but very thoroughly, abreast of the times.' (18) I like this third class of beekeeper. Like the bee Master of Warrilow, this beekeeper seems to me a thoughtful, intelligent kind of person with that quality of, '... wholehearted reverence' (19) Mr. Edwards identified, not only for past customs, but also for honeybees. Above all, they have a healthy recognition that progress is not a bad thing. With Mr. Edwards, I believe that: 'To say, however, that modern ideas of progress in bee farming must inevitably rob the pursuit of all its old-world poetry and picturesqueness would be to represent the case in an unnecessarily bad light.'(20)

It's true that not everything about the old ways of beekeeping was informed by good science or was good practice at all. Equally, not everything done by modern beekeepers of frame hives is bad either. Sometimes we're doing the same thing but simply in a more controversial way. One example is equalising the bees during the spring build-up. Colonies don't expand or even get going at the same rate; some explode into action and need weakening, to stop them swarming; others are lethargic and slow to get going, and risk not being strong enough to make a decent harvest. In the past skeppists would weaken one hive or strengthen another by substituting a weak for a strong hive through swapping their positions. When you move the hive like this, older foraging bees will continue to return to the same position as their hive. If you put a weak hive there, the foragers from the strong stock originally in that position will now return to the weak stock, making it stronger. Conversely, you can use the same method to bleed out excess forager bees from a colony that's building too quickly.

Modern beekeepers often have a more sophisticated approach to equalising. They will remove frames of brood and even nurse bees from strong hives and donate them to weaker hives. Some natural beekeepers abhor this approach because moving frames and bees between hives carries the risk of cross-contamination of hives, if the donor hive has a disease or old comb that's full of pathogens or parasites or a build-up of chemicals. Another technique that's possible with frame hives is checkerboarding in order to encourage the queen to lay in a larger area, which will expand the brood nest. The nest is roughly rugby ball-shaped across the frames of comb;

wider in the middle of the nest and thinner at the outer edges. If you remove alternate combs and reposition them, disrupting the even rugby ball of the nest, you will create gaps that the bees will want to make regular, to restore the rugby ball shape. The queen lays in those disrupted gaps, expanding the area of the nest. As one natural beekeeper explained, that's like opening someone's house and rearranging the furniture. From my point of view, however, I no longer want the largest brood nest I can achieve because with treatment-free beekeeping more brood means more varroa. More brood also increases the chances your bees are going to want to swarm. Even from the wisdom of the old bee Master of Warrilow: 'Success is all a question of numbers. The more worker bees there are when the honeyflow begins, the greater will be the honey harvest. The whole art of the beekeeper consists in maintaining a steady increase in population from the first moment the queen begins to lay in January until the end of May…' (21)

Brother Adam maintained that, 'The ability to equalise colonies effectively may be regarded as the hall-mark of an accomplished beekeeper.' (22) Yet he was, surprisingly, not that far away from respecting the customs of the old beekeepers. Despite his industrial-scale set up and scientific methods, he wrote: 'The spreading of brood, the removal of pollen-clogged combs to hasten the spring build-up, stimulative feeding, every unnecessary examination and disturbance are strictly banned and have no place in our management.' (23)

When Brother Adam took over Buckfast Abbey's apiary from Brother Columban in 1919, the mysterious Isle of Wight disease had reached epidemic proportions, almost completely wiping out the native British Black bee (*Amm*). Though many beekeepers at the time contradicted him, and still do today, Brother Adam had to respond to the situation of British beekeeping at the time. Beekeepers had to depend on queens imported from abroad – the most common being the Italian (*Am ligustica*). Brother Adam explained: 'Bee-keeping had perforce to re-orientate and adjust herself to new conditions. The old-time ideas and tenets, which before the demise of the old native bee had a measure of justification, no longer held good.'(24)

His breeding programme set out to cross the Italian Ligurian (which had survived the Isle of Wight phenomenon) with survivor stocks of the native Black bee, producing the foundation of the Buckfast bee. Through decades of selective breeding and crossing with other subspecies, to fix the best traits in his super bee, Brother Adam produced a stock that was prolific, disease-resistant, docile, hard-working and productive. There is no doubt that the Buckfast strain is an extraordinary and superior achievement. I started with Buckfast queens and my first honey crop was double or even three times what the bees now produce. Even at the time, Brother Adam admitted, there was widespread opposition to the use of non-native stocks: 'It was widely held at that time that a non-native bee, as well as foreign methods of beekeeping, were quite unsuitable for the conditions prevailing in the British Isles. These views, held with the utmost conviction, were largely based on a misapprehension.'(25)

Faced with an entirely new set of problems every bit as threatening to bees and beekeeping, we in turn must reorientate and adjust to our own new conditions, just as Brother Adam did. He could not have foreseen them. Ironically, in the light of these new conditions, many of his ideas and tenets have now, in their turn, become old-fashioned and no longer hold good. Some of them, dare I suggest, are based on misapprehensions of his own.

Brother Adam might wring his hands at my objections, I have little doubt. He would be dismissive of anyone championing the merits of local, native stock, as much as he dismissed the attributes of the native Black bee: 'Their undesirable traits far outweigh their good points.' (26) My defence is that there are different definitions of what constitute the most desirable traits and there are different definitions of what 'progress' means. If, like Brother Adam, the sole aim in beekeeping is the commercial production of the best possible yield of honey, then you want a bee that produces the largest colony possible, and local mongrel stock won't reliably deliver that goal. The disadvantages of that objective are that you'll need a bee that is pure-bred, such as the Buckfast, Italian or Carniolan, which will mean buying-in mated queens every year or two to maintain pure stock.

There are other beekeepers (and I think I'm one of them) who don't want the maximum amount of honey at any cost. We also enjoy the craft of beekeeping, which includes swarm prevention and managing problems. For us the problem of varroa is now part of the craft that has always been about understanding how bees work and about working with nature. As Brother Adam admitted: 'The choice will in every case mean a balancing of one set of advantages against another set of disadvantages, and an eventual adjustment to the particular idiosyncrasies of the bee favoured.'(27)

The most important consideration, which brother Adam never mentioned, because it was not an issue in his day, is our responsibility towards our environment and our responsibility towards each other. The need for an understanding of bee ecology is inextricably linked with the moral imperative of a new human ecology too. No beekeeper, in my opinion, can remain morally detached from the environmental conditions that affect their bees, or from other beekeepers who share their craft and the consequences for each other's beekeeping. Everything I do or don't do with my bees will have some effect, though probably unnoticed, on every other beekeeper around me. Our queens and drones are flying to the same areas for mating, our bee genetics become shared among our stocks, as do our diseases and pests.

The problem is that modern beekeeping is not so much a broad church as a confederation of different denominations that don't have a common hymn book. Around a core set of tenets and ideas which are broadly accepted, there are other varied beliefs and practises. Arguably, the root cause of this diversity is the choice of bee – and that choice is down to the beekeeper.

If I were keeping bees in the town, as more and more people are doing these days, my choice of bee might be influenced by a different set of criteria. Urban beekeepers don't want swarmy bees that can upset the neighbours or lead to litigation. Swarmy local stocks or Carniolans are not going to be their first choice. The Buckfast, with its reluctance to swarm, would be an ideal bee for an urban beekeeper. Another beekeeper might want to introduce children to beekeeping and is looking for docility as the most important trait. For them the Buckfast or Carniolan are going to be reliable for their docility. Another beekeeper might have a special interest in varroa resistance, and the progress made in this area by the Russian bee. Someone else might favour Italian bees for no other reason than it's what they've always worked with, or it was the bee their grandfather had when he introduced them to the craft. These different beekeepers are the equivalent in gardening of the dahlia grower, the sweet pea enthusiast and the auricula fanatic. One man's meat is another man's poison, and what one person sees in a David Austin rose someone else will see in wildflowers. Beekeeping, however, is not quite as simple as gardening.

The one issue that unites both hobbies in immediate controversy is the environment. Think of the call for peat-free compost, growing the grass long and choosing flowers for pollinators. But there are still obsessives who pour chemicals into their lawns and manicure them to within a millimetre of their life until they are more like bowling-greens than gardens.

It isn't mere sentimentality and nostalgia that makes us look back to a gentler era of beekeeping: 'The old bee-gardens lay on the verge of the wood. Seen from a distance it looked like a great white china bowl brimming over with roses; but a nearer view changed the porcelain to a snowy barrier of hawthorn, and the roses became blossoming apple-boughs, stretching up into May sunshine, where all the bees in the world seemed to have forgathered, filling the air with their rich wild chant.' (28) Mind you, the old country folk, including William Woodley, weren't sentimental about nature, and could be as red in tooth and claw as Mother Nature herself. In February 1908 Mr. Woodley wrote that he had been shooting bullfinches for taking the buds from his fruit trees and that they were the worst budders. Later in the year he also admitted to not allowing swallows to nest on the house because they swooped over the apiary and ate the bees. (He did not indicate his methods for discouraging them). He also wrote that sparrows were worse for this particular crime. There are frequent letters throughout the 1900 editions of the *British Bee Journal* on the subject of *Tom-Tits* as well and whether or not they eat bees. Mr. Woodley wrote in February's *Notes By The Way* that he had trapped and killed a number of Tom-Tits after he caught them hanging around his hives in the snow that month. *The Bee-Master of Warrilow* even begins its first page with the subject of blue-tits around the winter apiary, 'They are the great pest of the bee-keeper in winter time...all the blue-caps for miles round trek to the bee-gardens.' (29)

We look back longingly, (and not with mere sentimentality) as we have done since Covid-19, to the *Old Country* with Jack Hargreaves, and Isaac Walton with his simple roach rod, or Mr. Middleton digging for victory during World War II, in a black and white Pathe film, for one reason - life then seemed much simpler; or, as Jack Hargreaves explained in his 1987 biography

*Out of Town – A Life Relived on Television*, recalling letters received from his many viewers: most people who wrote to him, from all generations, thought that something had gone wrong with the world. For these people the country life Jack lived as he grew up had become a dream.

First responders at a crisis are trained to assess a situation, and they have a check-list to do it. Firstly, they ask, does the person know WHO they are? Secondly, WHERE they are? Thirdly, WHEN they are? Lastly, does the person know HOW they got there? During Isle of Wight disease there was probably a lot of confusion among beekeepers about who they were; were they traditional skeppists or modern beekeepers? Many, it seems to me, were somewhere in between. It's only now that we can see them in their historical context, so that it's clearer to us than it was to them why they responded as they did. Probably many of them didn't fully understand where they were in the development of beekeeping, or HOW they had arrived at the situation they were in. That's the definition of crisis – some would probably have scored a zero for every part of the assessment.

We're in many kinds of global, domestic and individual crises today, of which beekeeping is one. How do we score on those assessment questions? How many beekeepers know what kind of beekeeper (or who) they are or aspire to be? How many have a sense of where they are in relation to global beekeeping? Do we have a good sense of when we are beekeeping (meaning our historical context and the history that has shaped our beekeeping) and how we got into this situation?

The beekeeping crisis is a symptom of a far deeper existential crisis. Many people today are looking at the chaos of the world in which there doesn't seem to be any stable meaning in life any more, or any objective truth; which means many people don't know who they are; coupled with a sense of hopelessness and disorientation. We wonder where we are too. If life is meaningless and you can't do anything about it, and you also feel lost, and can't see a way out, that's a crisis! And most of all, we don't really understand how we got into this state in the first place. Many of us (beekeepers included) are questioning the spirit of progress and are asking in what ways we have really progressed at all.

When I made my first visit to William Woodley's grave in Beedon I hadn't driven a car since before the lockdowns. I was very aware as I sat in the car that day of the skills we take for granted when we carry out our lives in automatic. As I sat in the car, I had to remind myself of the routines of looking in the mirror, checking my blind-spots and reading the road ahead. Now I think of it, this is a good analogy for why I decided to write this book. It's been a rear-view mirror, a checking over my shoulder at the blind-spots beekeeping has developed over the last century; and it's a reading of the road ahead. Quite simply, I'm not beekeeping in automatic any more. I have glanced over my shoulder and can see the hazards coming up behind me from the past. I am also looking at the road ahead, and I am noticing all kinds of worrying situations that might develop which I would not call progress.

# MINDING THE BEES

## Chapter 6
### A Change o' Air

'WANTED STOCKS OF BEES,
on Standard Frames and in straw skeps.
POSTMASTER, Breachwood Green.' (1)


*British Bee Journal  1910*

Every start of a beekeeping season I  like to try something new. I always learn something new and unexpected too. Honeybees never do quite the same things every year, and different stocks behave differently and have different histories. Even so, it feels this year as though I'm starting beekeeping all over again. That perspective is enhanced by having to build up the apiary again from relatively few survivor stocks. Whereas in the past this might have depressed me, with the certainty that this year's honey harvest will be low, and might even be non-existent, I actually feel quite upbeat and excited. It's as though I'm on the edge of an adventure, especially with the prospect of catching wild swarms.

Starting again and having to rebuild from a few stocks is sometimes part of beekeeping. As with any animal husbandry, things can go wrong, and losses go with the territory. In the US, since 2006 some beekeepers have lost as many as 90% of their stocks to Colony Collapse Disorder (CCD), and even now average annual losses are at about 30%. And we're not talking about William Woodley of Beedon parish with his 200 hives; we're talking about commercial bee farmers losing tens of thousands of hives in one go - enough to put them out of business overnight.

William Woodley did have his own catastrophe though. Towards his final years he was hit, as were most beekeepers in Britain, by the mysterious Isle of Wight disease, which by the First World War had caused untold damage to beekeepers' stocks, in skeps and hives across Britain. To add insult to injury, Mr. Woodley was also blamed by many beekeepers for worsening the Isle of Wight problem by obstructing proposed legislation to make bee diseases notifiable. Despite this, Mr. Woodley's bee farm was hit hard by both the Isle of Wight phenomenon and criticism from other beekeepers so that in 1917, writing again in the *British Bee Journal*, he considered himself, '… a scourged member of the craft.'(2)

Attempts at legislation for powers to intervene in outbreaks of bee diseases began in 1896 with the *Bee Pest Prevention Bill*. It focused on the prevention of outbreaks of *foul brood*. At that stage, however, the Board of Agriculture wasn't keen, arguing that the organisational machinery needed to deal with foul brood was too costly compared to what seemed a relatively minor problem. It was 1919 by the time the Board of Agriculture finally recognised the full impact of bee diseases on pollination and food production. By then it wasn't only foul brood that was an issue, but the Isle of Wight mystery had wreaked havoc on bee stocks across Britain and in agriculture. As early as 1904 it was evident in the *British Bee Journal* that there was already a problem, with reports of fruit crops suffering through poor fertilisation of blossom from the lack of bees.

Another attempt at the *Bee Pest Prevention Bill* failed in 1905. In a vote by beekeepers, 229 were against, who between them owned over 7,000 hives. Those who voted 'for' were 421, owning about half the hives as the 'against' voters. This was evidence that most beekeepers who voted against the bill were those with larger apiaries, such as Mr. Woodley.

Woodley had written of his opposition to the Bill in the *British Bee Journal*, so the defeat of the Bill left him wide open to public criticism. J. Pearman of Derby wrote in 1916: 'I am satisfied we should not be in the position we are today had it not been for Mr. Woodley and a few of his friends; while true friends of the BBK were trying to get a bee diseases bill passed, a few of these gentlemen of knowledge were giving their time and money and visiting the Board of Agriculture, trying to defeat the bill.' (3) A year later, George E. H. Pratt of Sheinton, Salop, exemplified a continued resentment towards Mr. Woodley: 'Mr. Woodley and his clique are quite welcome to all the credit they claim for having inflicted what is in reality a serious injury upon the beekeeping industry...' (5)

The Bill, however, was intended specifically to focus on foul brood, not the problem of Isle of Wight disease (whatever that was!) To have intervened in the Isle of Wight phenomenon, the bill would not have been sufficient without an amendment, and even then, how effective would intervention by the law have been ? No one really knew what was causing colonies to collapse during these years, and if they had, no one knew of any cure. What would legislation have achieved – the systematic destruction of many stocks by BKA experts and the Board of Agriculture's officials that were being destroyed naturally anyway? Not to mention the destruction of otherwise healthy stocks that were deemed a threat, such as those in skeps.

William Woodley was probably in the wrong place at the wrong time when angry beekeepers needed a scapegoat; and he was probably not against the Bill simply to preserve his own interests as a bee farmer, but because he genuinely championed the small beekeeper, especially the skeppists like his beloved great aunt and his village neighbours. Evidence of this can be found in an account of a visit made by John Walton to Mr. Woodley's apiary at Worlds End on 24 September 1888, which was published in the *British Bee Journal* on 18 October that year. John accompanied William to nearby Hermitage to pick up some stocks of skeps. He also reported that, 'Mr. W said he had taken especial notice of something like forty stocks of skeps this season, outside his apiary.' (5) This suggests that Mr. Woodley was in touch with many local skeppists – probably local cottage labourers and poorer people in the area. Further evidence for this comes from his obituary in the *British Bee Journal* on 25 October 1923: '…and the poorer people looked to him as a friend and counsellor.' (6) We also have testimony from John Walton who described an encounter with a young man on their way to Worlds End, who was critical of beekeeping associations, '…saying that he, and others had told him the same, did not reap any benefit by them for the half-crown subscribed.' (7)

Mr. Woodley's opposition to the *Bee Pest Prevention Bill* was no doubt motivated to a large extent by his solidarity with rural labourers who were his neighbours and fellow skeppists. It's reasonable to assume that he never forgot the skeps in his great aunt's garden at Stanmore, or the first swarm he captured and hived in a skep. And there was also his sympathy for the labourer who could not afford the modern frame hive and its expensive equipment.

Perhaps the skep stocks Mr. Woodley and Mr. Walton collected from Hermitage were of some cottager who had died or who had moved away due to lack of work. Maybe the forty or so hives he took an interest in outside his own apiary belonged not only to fellow beekeepers but to people with whom he was on friendly terms. Clearly, he knew where all the local beekeepers were because in 1915 when he wrote to his landlord about giving up the tenancy of his out apiary due to losses from the Isle of Wight disease, he wrote, 'I don't think there is a stock alive within Beedon parish.' (8)

His friend, Charles H. Heap of Reading, who wrote William's obituary notice and also wrote for the British Bee Journal, had made a similar observation some years earlier, in 1912, in his 'Random Jottings': 'In the Windsor District of Berkshire very few bees remain, and in other parts of the County the losses are considerable.' (9)

This reality does not seem entirely consistent, however, with William Woodley's repeated defence of the cottager skeppist when there were hardly any left anyway in the wake of the disease for which Woodley had no adequate response except opposition to legislation. Whatever his views about the *Bee Pest Prevention Bill*, he was as susceptible as everyone else to the ravages of the *Isle of Wight disease*. In 1915 all his bees had died of the mysterious phenomenon at Stanmore, and nearly all his stocks at Worlds End. He made valiant attempts to restock, especially with hybrid Dutch and Italian bees which many claimed had proved more resistant to whatever was killing British bees. Despite such attempts, by the following winter 1915-16 (the same year Brother Adam's Buckfast apiary was badly hit) Mr. Woodley's stocks were decimated again. Interestingly, among the stocks he lost were his imported Dutch bees. He wrote: 'The winter of 1915-16 reduced me to a few stocks and as the spring advanced these developed symptoms of Isle of Wight disease.' (10)

As things stood, the only option Mr. Woodley had was to import more bees. By the end of the First World War the British Government had caught up with events as well, and was restocking from abroad. The nation's food production depended on it.

Conditions today are very different; our historical context is different too, and different situations require very different approaches. Even at the time of writing I have had an email from the Chair of the BBKA, expressing concern and pressure the association has put on the Government over a certain business that intends importing bees through Northern Ireland, from Southern Italy where Small Hive Beetle has become a problem. Whereas in 1918 foreign bees were the only means of restocking Britain, imported bees today are a possible means of introducing the biggest threat to honeybee biosecurity since varroa arrived in the 1990s.

William Woodley could not even have recourse to catching swarms; there weren't enough bees remaining to generate sufficient swarms. At least I have some stocks that have survived, and at least I have a good chance of catching some swarms this season. Like Mr. Woodley, I'm adapting to my situation in ways I could never have imagined. Since the day skep beekeepers disappeared, modern beekeeping has developed every possible method, manipulation and breeding programme to eliminate the problem of swarming, but this year, for the first time in my beekeeping, I'm actually depending on swarms, and I'm excited at the prospect of the swarm season, whereas it used to be a cause of continual headaches and anxiety.

I've never used swarm lures before, but I've done a lot of research about them. There seem to be a few important considerations when using them. The size of the box, or whatever container is used, has to be at least twenty-five litres (five gallons). If it's too large the bees won't be interested. It must smell like an old hive too, preferably with old wax and hive debris melted and painted inside the chamber. Many recommend that a frame of very old, black wax comb is irresistible to the bees. Next, the location is important, including its height from the ground and which direction it faces. Lastly, the lures must be set at the right time for swarming, which of course is a bit of a moveable feast from season to season, depending on the weather.

My bait hives are ready and I've been thinking for some months about where to place them. Although much research suggests that honeybees prefer their nests to be at least fifteen feet off the ground (and often more) I don't want to have to go up and down ladders on my own, especially carrying a bulky hive containing a couple of pounds of bees in addition to the weight of the box. There's plenty of evidence that bees will readily take to a hive that's fairly close to the ground, so I've looked for places that are several feet up, which seems a good compromise. Another consideration is the way bees use prominent landmarks – especially tree-lines. We have a lot of tree-lines at Douai Abbey, but also lots of outbuildings with accessible roofs where I can place hives. Lastly, these locations should ideally be part shaded and facing south to south east. All this aside, catching bees this way is not unlike fishing. You can have all the right equipment, and do all the right things, but the bees can decide not to cooperate and might do something quite unexpected. You might provide a lovely old hive, full of old comb and the delicious smell of wax, but a swarm might still ignore your offer and set up home in an electrics box on the side of a house or behind an air brick high up in a building.

My ideal locations at the moment are the tree line beyond the car park, the trees bordering the meadow, and the trees between the meadow and the field to the north, and some miscellaneous outbuildings between the 1918 workshop and the compound and potager, the poultry and the apiary. I plan to suspend the bucket hives in the trees and stand the wooden bait hives on the roofs of the buildings, in different positions.

The swarm season can start as early as April if there's an early spring and good weather, but that would be unusual. We start the month with a slow build-up of brood and not yet any sign of drones. Drones are a sign of a colony's prosperity and readiness to swarm, so when they first appear you know that a couple of weeks after emerging they will be ready to mate. That means if drones appear in the middle of April they are ready to mate at the start of May. All other factors being equal, such as the spring nectar flow and fine weather, swarming could start this year from early to mid May.

✺

In my experience swarming in most years begins towards the end of May here at Douai Abbey, with the majority of prime swarms issuing throughout June, with a few straggling casts in July, although one year I did catch a decent swarm in late July. On this basis I decide to wait until I see drones in the hives before setting my bait hives.

We know from Owen Thomas' records when his skep swarms 'rose' from 1757 to 1776. He records two in May, issuing on days 4 and 30 of the month. Most rose 2, 6, 7, 10, 13, 16, 17, 19, 22, 23, 30 June, showing that June was the main swarm time, with an even distribution of swarms throughout that period. In July there were four recorded swarms, on days 2, 4, 7, and 18 of the month. 3 August was the only swarm that month. On some of the swarm dates in June more than one swarm issued, increasing the number of swarms recorded in June to 23.

If we consider the fact that Owen Thomas used the old Julian calendar, which removes 11 days from September, we can adjust the number to compare them with what might be expected to happen today on our Gregorian calendar. This would give a total of nine swarms in May, nineteen in June and two in July, with none in August. By Owen's records then it's clear that swarming was mainly in May and June, which corresponds with what tends to happen today. Even allowing for differences many might attribute to climate change and global warming, my experience of swarming is much the same as Owen Thomas' over two centuries ago, with swarming tending to start at the end of May and going through June.

The old cottagers, if they worked at home, would probably have been on hand if their stocks swarmed, but they would also have had some ability to anticipate swarming by observing weather conditions. In other cases a member of the family, such as a child, might be given the job of *mindin' the bees*, as Willam Woodley was when he was a small child at his great aunt's in Stanmore. On his own bee farm later in life Mr. Woodley employed a swarm watcher. We know from his *Notes By The Way* in 1897 that during the winter he left the apiary in the charge of an octagenarian beekeeper. We also know that he employed a swarm catcher in the out apiary at Worlds End who hived swarms there in straw skeps and carried them the two miles back to the home apiary at Stanmore. He paid the swarm catcher ten to twelve shillings a week. I've read that, according to records, there was only one octogenarian in Stanmore in 1897, called Enoch Brind (1816 – 1897). It's highly likely that old Enoch was William Woodley's swarm catcher at that time.

On a skep bee farm in Lower Saxony, filmed in the 1970s, workers on the bee farm literally sat near the bees during swarming, ready to catch swarms as they rose. On the farm they employed the use of four-foot-long catching bags made of linen and gauze that were fixed over the hive entrance and attached at the other end to a long pole in the ground, the long stocking of the catching bag angled upwards towards the pole. As the bees issued they gathered and settled at the pole-end of the sock where they could be shaken down and tipped into an empty skep. With several hundred stocks, many of which swarmed at the same time, the swarm catchers could be on their feet catching and hiving swarms all day.

Such a system had operated for centuries, which through natural selection developed swarmy bees. When I read speculation from modern beekeepers that a characteristic of the old British Black bee was its reluctance to swarm, I have my doubts; it doesn't make any sense when you consider how skep beekeeping would only select bees with a strong tendency to swarm. Non-swarmy bees were no good to the old beekeepers because they couldn't make increase from them so easily – a necessary part of the system of managing skep hives. Non-swarmy bees would not be selected, and as they were discarded, would favour selection among apiary stocks for a strong tendency to swarm. We know from Mr. Woodley's columns too that his bees swarmed every year and that swarming was part of his management system.

Tickner Edwardes described a charming visit to a bee master of the old cottager-type during swarm time: 'Keeping watch over the comings and goings of his bees was always his favourite past-time, year in and year out, but it was in the later weeks of May that his interest in them culminated. He had always had swarms in May as far back as his memory could serve him. As a rule the bees gave sufficient warning of their intended migration some hours before their actual issue.' (11) The picture evoked of the old bee man is of someone completely in tune with this natural ritual of the beekeeping year. There is no sense of panic or inconvenience, but an atmosphere of quiet and solemn tradition that is positively enjoyed as one of the greatest pleasures of beekeeping: 'At these momentous times a quaint ceremonial was rigidly adhered to by the old bee-master. First he brought out a pitcher of home-brewed ale, from which all who were to assist in the swarm-taking were required to drink, as at a solemn rite. The dressing of the skep was his next care. A little of the beer was sprinkled over its interior, and then it was carefully scoured out with a handful of balms and lavender and mint.' (12)

Modern beekeeping has taken a less relaxed view of swarming. L.E. Snelgrove wrote the classic *Swarming: its control and Prevention* (1935), identifying that swarming '...has remained an insuperable difficulty for the majority of beekeepers.' (13) Brother Adam of Buckfast believed that, '...swarming must be regarded as one of the foremost obstacles to successful beekeeping.' (14) Despite this, it hasn't stopped beekeepers from keeping the swarmy Carniolan bee. Carniolans (*Apis carnica*) are the native subspecies of Slovenia, Southern Austria, Croatia, Bulgaria, Serbia, Herzegovinia, Hungary and Romania. *Carnies*, as they are usually called by English-speakers, are the second most popular race of bee kept by beekeepers after the Italian. Although renowned for swarming reliably but inconveniently, they have many other redeeming traits. Many regard it as the best bee, while even Brother Adam admitted that Carniolans were, '...the key that unlocks the hidden potential of other strains.' (15) A Buckfast monk who could recall the debate once told me that there was a long-running controversy about the direction taken by Brother Adam with the Buckfast bee, with many beekeepers arguing that a better bee to improve through selective breeding was the Carniolan. In fact, this has now happened in some countries where the Carniolan has even been bred to be less swarmy.

*Carnies* are often called *the grey bee* for their dusky-brown-grey colour. They are generally docile, so are easy to work with, and they produce a good harvest. They have a good defence against pests and diseases and overwinter successfully in smaller numbers than some other races. Although they have a slow start in spring, once they begin expansion it is generally explosive. On the other hand, because they match brood production to available forage, this can hold them back or halt them, leading to poor harvests in some years. They compensate for this to some degree by their ability to forage in the early morning and later in the evening than other races, and in cool, wet conditions. Carnies are not prone to robbing or drifting between different hives in the apiary and are not usually prone to brood diseases.

Clearly, many a modern beekeeper would disagree with Brother Adam when he wrote: 'Of all the qualities a strain may possess, there is probably none more important than a disinclination to swarm. A strain may possess every desired characteristic, but a highly developed swarming instinct will effectively neutralise the qualities of economic value.' (16) A different perspective shared by many modern beekeepers, however, in contrast with the view of beekeepers of the past such as Snelgrove and Brother Adam, is that we no longer have a purely economic approach to beekeeping. Whereas, '...the bugbear of modern beekeeping in the last century was swarming', (17) today it is varroa and the increasing threat from Asian hornets and the Small Hive Beetle. Snelgrove did admit in his day that the difficulty of swarm control was closely followed by bee diseases, but today this situation is reversed; even if beekeepers today keep bees unashamedly with the goal of honey production, most would admit, I think, that their greatest concern is actually to keep their colonies alive. Preventing colony collapse is far more of a priority than preventing swarming, simply because unless you control and prevent diseases and parasites such as varroa, you can't hope to harvest any honey anyway.

Part of the interest for me of catching swarms is that they could contain any number of races, including Carniolans, while natural selection in the wild will select some of the most desirable traits for survival; such as the ability to overwinter, and disease-resistance. At the same time natural selection won't necessarily produce the gentlest bees or the most productive. Nature selects for survival, not human economics or convenience. But that's why this year's swarming isn't something I regard any more as, '...a bad legacy left us by the old skeppists.' (18) Instead it's something I now anticipate with excitement, hope and curiosity.

There are many modern beekeepers (and I was one of them until a year or two ago) who regard swarm prevention as not only essential for honey production, but the true measure of the beekeeper's skill at the craft. I can understand that view because, having lost swarms when I started beekeeping, I know the satisfaction of finally managing to prevent or manage swarm preparations so that you don't lose your bees. The first year I successfully prevented swarming in every one of my six hives I felt enormously accomplished as a beekeeper.

The Bee-Master of Warrilow speaks for most modern beekeepers, however: 'To the modern bee-keeper...a swarm in May is little short of a disgrace. There is no clearer sign of bad beemanship nowadays than when a strong colony is allowed to weaken itself by swarming on the eve of the great honey-flow, just when strength and numbers are most needed.' (19) This year, however, I am closer to the mind of the old bee-man with his garden of skeps, 'When, as sometimes happened, the swarm went straight away out of sight over the meadows, or sailed off like a pirouetting grey cloud over the roof of the wood, the old bee-keeper never sought to reclaim it for the garden.' (20)

It's interesting to speculate about the possible consequences legislation might have had on swarms when Isle of Wight disease was it its worst. As with bird flu that has caused such problems for poultry keepers in recent years, it's important to keep wild stocks away from managed stocks during an epidemic. It's much easier to monitor managed hives for a problem than to know the state of a swarm's health if it turns up in your garden or in nearby buildings. During Isle of Wight disease it's likely, I think, that swarms would have been viewed automatically by many beekeepers as a possible threat to managed hives. They might have issued from a beekeeper's apiary a few miles away where the disease was present, making them a significant threat to other stocks if they landed nearby. But you wouldn't necessarily have known they were carrying the disease. Alternatively, they might have been wild swarms from colonies in good health, but there would be no way of knowing for certain. I can't help but wonder how many perfectly healthy wild swarms, including those from wild colonies, might have been destroyed in the name of biosecurity by beekeepers, experts and officials from the Board of Agriculture had the full force of the law enabled such measures. William Woodley was especially concerned that such people would head for the skeppists and destroy their stocks on the basis that they could not be inspected and were deemed a biosecurity threat simply because they were not in managed hives. No one during those years, including William Woodley, seemed aware, however, that wild bees also needed looking after and that they might have had a better chance of surviving the disease if they could keep out of its way. In fact, this is exactly what seemed to have happened, as we now know, native bees having reemerged in enclaves throughout Britain in recent years. Perhaps that is largely because swarms were not targeted as possible sources of transmission.

'' 'Tis gone to the shires fer change o' air,' the old bee-man would say..." allowing for the occasional swarm that escapes the beekeeper. (21) That said, I don't want to lose swarms from our bees this year – the swarms I want to encourage are from elsewhere. Our own bees, now reduced to a few survivor stocks, are too valuable to me to allow them a 'change o' air'. I'm going to need them for splitting into nucleus colonies to make increase of my stocks and to raise new queens, to continue the development of my survivor strain for next year. If I lose half of even one of these valuable stocks to swarming, that's potentially at least a few decent nucs that I can't raise to increase my chances of having more survivor stocks next spring.

No change o' air for the bees in the apiary then, if I can help it, but my generally more relaxed approach to the issue of swarming is certainly for me a breath of fresh air.

# MINDING THE BEES

## Chapter 7
### Doing Nothing

'WANTED stocks of
pure Carniolans or Italians,
1914 queens.
HUTCHINSON, Bryarwood, Kendal.' (1)

*British Bee Journal 1915*

The melting moon slides across the teflon-dark pond like a lump of lard. Easter morning we wake up to a hard frost with weak sunshine. Winter clenches the world again for a few hours before the sun thaws the day. By early afternoon glorious Watercolour washes of sunshine illuminate the undamaged blossom and the last stands of tremulous daffodils. The greengage has burst into white blossom on an area of lawn near the Abbey Church where a former beekeeper kept the bees decades before my time. I've been worried that a frost would knock back the new blossom.

Throughout the afternoon the bees work the white boughs of the greengage. In the apiary there's a steady passage of bees in and out of the hives. Many are carrying pollen – yellow pollen, though pollen comes in many colours, including red, green and pink. Pollen is usually taken as a good sign that the stocks have a laying queen because they need pollen to raise brood, but this isn't always the case. A colony can still be queenless and collecting pollen and nectar, as though the bees are desperately trying to continue with business as usual, as if that will put everything right.

In fading light after Compline, at the start of Holy Week, I see movement through the dead bracken and ferns between the apiary and the meadow. I walk into the apiary and startle a badger back into the line of hazels where some trees have fallen in recent storms.

Easter week the weather turns Arctic again. There are heavy frosts at night and snow flurries during the day, though there are also some sunny spells. The wind is sharp and cold as a razor-blade, stinging my face and hands as I feed the chickens or smash the ice on their water. Though the greengage and blackthorn are frothy with blossom, the bees have retreated again into winter mode. I watch a tangle of three kites hanging a little above the trees at the edge of the field across the lane from the Abbey gates. They rise and fall, twisting at the ends of invisible strings.

It's not a good situation for the stocks. If they are confined for too long and the weather doesn't warm up, they can develop dysentery. In a crowded winter cluster this can be disastrous – that's supposing the colony is large enough to generate enough heat to look after its brood and keep itself alive and that it's not dwindling. The latter is more likely with our stocks, as they had hardly started building up when I inspected them just over a week ago.

The Isle of Wight Disease, though it has never been fully explained or understood, seemed partly triggered into its third and final major wave by a similar spring in 1916. Easter that year was hot and people took to the seaside on Bank Holiday Monday, but then a cold spell followed that kept the bees in their hives for some weeks. When they emerged, as the weather finally warmed up, many appeared to be flightless and weakened, gathering in huge numbers beneath many hives across the country.

No one has ever adequately explained the Isle of Wight phenomenon, but it actually began appearing in the first several years of the twentieth century on the Isle of Wight, and it continued for at least a year after the end of World War I. Some said it was acarine, then known as tracheal mites; others blamed dysentery or nosema. Equally, there were many who dismissed these explanations, either through scientific investigation or simply because their own bees were not affected or seemed to recover. Even today it is a controversial topic, together with the claim that emerged afterwards - that the native British Black bee was wiped out entirely by Isle of Wight 'disease'. Brother Adam of Buckfast believed this, while others disagreed in articles and correspondence in the *British Bee Journal* at the time.

Whatever happened, in my opinion it was probably not all down to one single cause, as with Colony Collapse Disorder today. Even if a pathogen had been partly responsible, pathogens and parasites can have varying effects, depending on many other factors, such as the weather, the conditions in the hive, the beekeeper's care and experience and the condition of the colony in the autumn as they prepare for winter. During the war years there were certainly other factors to consider. Firstly, there were many beekeepers who were not yet experienced at keeping bees in the relatively new invention of the wooden frame hive, having recently made the change from skeps to modern hives. This and the number of beekeepers fighting in the Great War, and the lack of sugar available to feed bees, must have resulted in many stocks suffering neglect and becoming weakened even before the worst of the winter of 1915-16.

I'm not unduly worried about the Douai stocks being knocked back by winter conditions, as happened at Easter 1916, but it's other factors combined with this that concern me. For example, I don't know how compromised the stocks are from *varroa* and my treatment-free approach. *Varroa* will still be present, though in smaller numbers at the moment while there is little brood. What I can't tell is how weakened the stocks are from the various viruses transmitted by varroa. In any case there is nothing I can actually do, whatever is going on. I am at the mercy of nature.

A week after moving the stocks into full hives I really need to check them again for increasing brood production. It's been a good, dry week, with some sunshine – hopefully enough to stimulate the queens into laying more eggs. But I'm also concerned about two stocks in which I didn't find a queen. Did I just miss them when I transferred the stocks to hives, or were the colonies queenless? If there is no sign of brood or eggs next time I make an inspection I'll know that these stocks are hopelessly queenless. At this time of year it's not a situation I can remedy; I have no spare queens with which to requeen, and it's too early in the year for them to raise a new queen and for her to be mated. As soon as I can inspect them, I'll know for certain. If they are queenless, I can unite them with a queen-right stock, which will also boost that stock. Easter week, however, the weather is cold – on average below ten degrees for most of the week, which is too cold to carry out inspections. My curiosity and anxiety will have to wait – probably until mid April.

No two seasons are the same in beekeeping. You have to develop a feel for the craft – an intuition gained only by regular inspections and close observation every year. You have to learn to listen to the bees and not rely on everything you read in books or that every beekeeper will try to hand on to you. You have to balance the science and the art. Sometimes, as with ringing the bees, you can't always distinguish between the two.

One of the criticisms of the old cottager skeppists might have been an over-reliance on old knowledge that was simply handed down to them by tradition. The old bee man in *The Bee Master of Warrilow* believed, for example, that his oldest hive still contained the remnant of his original colony, which is obviously a ludicrous idea. It reminds me of the character of Trigger in the old comedy series *Only Fools and Horses* who claimed to have had the same broom for thirty years, explaining how many different heads and handles he had gone through to maintain his beloved broom. The old bee-man is portrayed, not always sympathetically, with a complete contempt for modern methods of beekeeping, while: 'In its place he had an exhaustless store of original bee-knowledge gathered throughout his sixty odd years of placid life among the bees. His were all old-fashioned hives of straw hackled and potsherded just as they must have been any time since Saxon Alfred burned the cakes.' (2)

Doing things the old ways because we've always done them that way is quaint, but it doesn't allow for progress. Without some progress in the old craft perhaps today we'd still be killing bees over sulphur pits at the end of the season to harvest their honey. Some of those old ways were also plain superstition, like the belief Thomas Hardy captured in the old Wessex of his novels, that milk goes up into the horns of cows when they stop producing milk. Similarly, the bee man believed that pursuing an escaped swarm would be, '...naught but ill-luck' (3) Then there was the old bee tradition of *tanging* the bees or, '...ringing the bees' (4) which to the bee man was, '...an exact science'.(5) The idea was that banging on metal objects could bring a swarm down, where it would settle within the beekeeper's reach, so he could take it and hive it. Tickner Edwardes narrates skeptically: 'Whether this ringing of the old-time skeppists had any real influence on the movements of a swarm has never been absolutely determined; but there was no doubt in this case of the bee-keeper's perfect faith in the process...' (6)

Tanging the bees was first recorded by the Roman poet Virgil in his *Georgics (Book IV)*. In an effort to assess this ancient custom for myself, I once tanged a swarm that left a nucleus box and went up out of reach into a tall beech tree at 2.00 pm one torrid June day on the edge of the apiary. Within minutes of my rhythmic banging of a hive tool against a ladder the swarm had fragmented from its pixelated cluster and reappeared, bearding the entrance of its original hive. Within minutes of returning the whole swarm had gone back into the nucleus hive. Whatever Tickner Edwardes believed, tanging the bees does work; it isn't just old superstition – it's part of the ancient wisdom of the craft. I have the evidence of my own experience.

It wasn't just superstition either that enabled the old cottagers to read the weather, before the advent of weather forecasts. Just as country folk could read the weather in the surrounding countryside – by the clouds, the behaviour of wildlife, and sounds, so too the old bee man exemplified a breed that could tell the weather simply by observing the bees:

'He never observed the skies for tokens of tomorrow's weather, as did his neighbours of the countryside. The bees were his weather-glass and thermometer in one. If they hived very early after noon, though the sun went down in clear gold and the summer night loomed like molten amethyst under the starshine, he would prophesy rain before morning. And sure enough you were wakened at dawn by a furious patter on the window, and the booming of the south-west wind in the pine-clad crest of the hill. But if the bees loitered afield far into the gusty crimson gloaming, and the loud darkness that followed seemed only to bring added intensity to the busy labour-note within the hives, no matter how the wind keened or the griddle of the black storm-cloud threatened, he would go on with his evening task of watering his garden, sure of a morrow of cloudless heat to come.' (7)

I've had the same experience as the bee man – of the bees responding to the changes in the weather, of which we are usually unaware. I recall one sultry June afternoon while I was inspecting the hives, when they seemed more agitated than usual, running on the combs and flying at me in squadrons of irascibility. Just as I closed the last hive, giving up in the face of their disagreeable behaviour, the sky darkened to cobalt and the sky shook with a low, lingering boom; the bees had known a storm was coming.

I'm feeling my way this year, as I have to do with all my beekeeping. It isn't an exact science because it's also an art. You have to develop a feel for beekeeping as well as good knowledge. And even the science in beekeeping is still young. As Brother Adam reminded us in 1982 when he wrote Breeding The Honeybee: 'Bee breeding by up-to-date methods has hardly begun. Dr. Ulrich Kramer, a Swiss, provided the initial impulse. However at that time Mendel's laws of genetics were hardly known and modern beekeeping was still in its infancy...' (8) In the 1880s controlled bee breeding was first attempted in rotating mating cages, followed in 1901 by early experiments involving confining them in large tents. It was not until 1926 that Dr. Lloyd Watson had the first partial success with instrumental insemination, a technique not advanced until 1944 by Dr. Harry Laidlaw.

At the time Isle of Wight disease raged, though genetics was in its infancy, science was only just discovering the secrets of bee biology and that the universally valid laws of heredity established by Mendel didn't apply in quite the same way to honeybees. It had been observed two hundred years earlier by Anton Janscha and Francois Huber that queens returned from mating several times with the tell-tale sign of mating that we understand today. It wasn't until 1944 that Dr. W. C. Roberts of Baton Rouge, Louisina, published his research on 110 queens more than half of which he had observed several times with what Brother Adam rather delicately called, '...the mating sign.' (9) Brother Adam's own experiments in 1947 confirmed that multiple matings were the norm in the honeybee. This was where Mendel had failed as a beekeeper, because he did not understand parthenogenesis in the queen, or multiple matings and that drones carry only half the chromosomes of the female bee. Coupled with an inability to control the drones, Mendel's attempts to explore heredity in the honeybee therefore came to nothing.

It's important to remind ourselves as beekeepers where we are in the history of beekeeping and how we got here, and that new discoveries are being made all the time that humble us in the face of such mysteries presented to us by bees. It's little wonder that during Isle of Wight disease everyone floundered when trying to account for its cause; viruses had not yet been seen because electron microscopes had not been invented. Immunology was not understood, which is why more people globally died from Spanish flu than had been killed in the Great War. If viruses were so little understood and could not be cured, is it any wonder a virus or viruses attacking honeybees (whose basic biology wasn't even fully understood) would rage uncontrollably once it got a foothold, in the face of such ignorance and helplessness? Our own Covid-19 pandemic has been a wake-up call that we don't understand everything and that we can still be reduced to helplessness by something as small as a virus.

Bee disease legislation back then, which William Woodley and others vehemently opposed, would not have solved a problem that science didn't understand. It would have needed to address Isle of Wight disease specifically, which would only have been effective had the cause or causes of the disease been understood. Even then, without a cure, what could legislation have accomplished? Legislation would have meant more experts with powers to visit affected apiaries (had the owners bothered to notify the authorities in the first place) and it might have been that they'd have burned all the affected colonies and equipment, then walked the disease on their boots to another area, a concern Mr. Woodley and others had expressed. We know today that avian flu and mad cow disease and other epidemics and pandemics can be transmitted in a variety of ways, such as on people's clothing or on boots and car tyres. In William Woodley's day, at the height of the problem, biosecurity methods were still in their infancy, as well as knowledge and understanding of honeybee biology and the variety of pests and diseases that can afflict bees.

Could it have been the case (as William Woodley believed) that bee disease legislation might actually have made Isle of Wight disease worse? If so, William Woodley can remind us today that, despite our instinct to be proactive in the face of a problem, sometimes the best course of action is actually to do nothing. Sometimes too we can't always be led by the science alone, or have complete faith in scientific models (as weather forecasts demonstrate) but we have to balance this with a good case that can be made for human intuition. We shouldn't knock such a notion. Two people, for example, enter into marriage, not on the basis of the best empirical evidence that they are suited and that it will work, but on the basis of discernment that can only in the final analysis be an intuitive act of faith.

Not that treatment-free beekeeping is by any means merely reactive or that it can be justifiably accused of 'doing nothing'. It is a proactive response in its own way. More importantly, Mr. Woodley believed that the proposed answer to Isle of Wight disease through legislation was pointless because it would not address the underlying problem of an incurable disease. I have come to the same view about chemical treatments against varroa. They are like so much modern medicine that treats the symptoms but does not always treat the underlying problem.

# MINDING THE BEES

## Chapter 8
### The Grave And The Cul de sac

‘WANTED NATURAL MAY SWARMS.
British Bees, 2s 6d. 1 lb given -
NICHOLSON, Langwathby, Cumberland.’ (1)

*British Bee Journal 1907*

No one knows how they do it: a drone that only lives a few weeks of the year leaves its hive when sexually mature, about twelve days after emerging from pupation in the comb. Then it flies, usually a few miles away to a mating location known as a Drone Congregation Area ( DCA). These areas are used for decades by generations of bees who possess some mysterious knowledge of them. As if that's not impressive enough, drones and queens from the same hive go to different Drone Congregation Areas, to avoid inbreeding.

Over two hundred years ago the naturalist, Rev. Gilbert White, heard a strange buzzing at Sheep Downs on Selborne Common, Hampshire: ‘There is a natural occurrence to be met upon the highest part of our down in hot summer days, which always amuses me much, without giving me any satisfaction with respect to the cause of it; and that is a loud audible humming of bees in the air, though not one insect is to be seen. This sound is to be heard distinctly the whole Common through, from the Moneydells, to Mr White's avenue-gate. Any person would suppose that a large swarm of bees was in motion, and playing about over his head. This noise was heard last week on June 28th.’ (2) This, we now know, was the earliest recorded Drone Congregation Area, in 1792. How do we know? Because it's still active today, which also makes it the oldest known DCA.

I have come across a blogger on the *Vita Bee Health Blog* site who mentions that Beowulf Cooper claimed to have found the Selborne Sheep Down's DCA in 1973, though he did not give its exact location. Another blogger apparently searched in 2012, though unsuccessfully. The blogger from 10 August 2015 claimed to have found the location, though their experience was less dramatic than Gilbert White's. The blogger reported some drones coming to their lure and having been hit on the head a few times by other drones. They reported the location to be in a small clearing just before Sheep Down, though in White's time there would have been fewer trees and the Down would probably have been more open and extensive. Perhaps there were also more drones around back then, which explains White's dramatic account of the noise, though the 2015 blogger admitted that they had discovered the location in August, while White's experience had been reported in June when there were probably more drones and more mating.

Drones are, I suspect, the most disregarded members of most beekeepers' stocks. Many beekeepers today hardly notice them, attempt to reduce their numbers to a minimum and probably view them as useless, at least as far as the beekeeper's immediate interests are concerned, which is often just honey production. In my own beekeeping, drones are a newly-discovered wonder of the world of honeybees. These male bees, with their stout, square-ended bodies and bulging eyes, have one goal in life, which is to mate. At least a dozen or more will be successful at mating with a queen, after which they drop to the ground and die, their mission complete. Until they are successful, each drone might make several trips a day to their chosen DCA, returning to the hive to refuel before setting out again.

Although many beekeepers think drones are a waste of time and are only interested in eating and mating, they are actually much more complex than this. It helps to begin by thinking of them as part of a super-organism in which every part, or individual, is intrinsically related to the whole. Just as a plum tree produces pollen in vast quantities, a bee colony produces drones in large numbers; drones being basically flying pollen, which means we have to start regarding drones as flying gametes. These gametes do not exist to fertilise an organism by inbreeding, but to spread out, as pollen does on the wind, where they can fertilise other colonies in the surrounding area. Because many beekeepers are only interested in their own hives they don't value the drones present among their stocks, but these males exist to benefit the genetic diversity of colonies elsewhere. Similarly, other beekeepers' drones exist to mate with our queens, securing the genetic diversity necessary to our own stocks. That's why every beekeeper's drones matter and should be of interest to everyone.

At a typical DCA. there will be between ten thousand and thirty thousand drones a day turning up, from as many as two or three hundred different local colonies over several square miles. This number of drones is vital to secure the right genetic diversity from the strongest, fittest, healthiest and fastest drones. Each queen will mate on average with fifteen to twenty drones in matings every fifteen seconds, often in more than one mating flight. Every sperm made by an individual drone is identical because drones develop from haploid eggs, so they only receive chromosomes from their mother. From an average of fifteen matings a queen will have fifteen family lines (patrilines) among her worker daughters, which is a very clever strategy bees have developed to ensure the genetic diversity to contain in a colony the variety of behaviours needed to cope with situations that might threaten their survival.

For example, multiple matings fix genes for specific specialist jobs in different patrilines of the same colony. If there is a heatwave, certain bees will have the gene for ventilating the hive. Others might have the gene for hygienic behaviour to remove dead or diseased brood. Problems might relate to health or work issues, but genes to cope with the multitude of possible variables a colony might encounter are not general to all the workers in the colony. Mating with only a few drones or with weak or poor quality drones will compromise the diversity of genes present in the few patrilines of the workers and make it less likely that the colony is able to adapt and survive. They might also contribute to a loss of the colony's vitality, also reducing its chances of survival.

Drones are highly specialised components of a complex super-organism, but their highly specialised evolution also leaves them vulnerable. Drones have no second set of chromosomes, so weak drones that have inherited poor genetics through lack of competitive polyandry can't hide behind a second set of chromosomes that might compensate another organism for deficiencies in its bloodline. Unlike most other animals, in which the male has a dominant effect by mating with many females, a single drone has a relatively small genetic impact. It's the queen who has the dominant influence genetically. As Brother Adam demonstrated in *Breeding The Honeybee*, 61.8% in the pedigree of bees are actually females because the genealogical tree of this species isn't based on the usual set of two parents, but the males' series of ancestors always begin with one parent; male ancestors, therefore, are always in a minority. Other genetic traits aren't carried by drones, but by queens, so a haploid drone can't be the way natural selection might filter out poor genes in other animals by simply allowing the weak organism to die off. Instead, beekeepers' interventions, and especially selective inbreeding, might actually perpetuate less desirable traits in drones that are then passed on, all for the sake of a particular set of selected characteristic, such as docility, that the breeder is trying to fix in their stocks.

In fact, this very process of selective breeding for traits, as in Brother Adam's work with the *Buckfast* strain, can be very damaging in the long run because the desired traits of the bee breeder's stocks can only be fixed in a closed population at the expense of limiting genetic diversity. A certain amount of inbreeding is also necessary, as in all selective breeding of animals and plants, but bees are biologically set up to avoid inbreeding and are very susceptible to problems when inbred. Moreover, the idea that there are outstanding colonies from which the breeder can make increase is based on a mistaken idea of bee genetics; in fact, colonies that are outstanding are ephemeral because of the necessity for constant genetic diversity through polyandry. Nor can the whole range of desirable characteristics be fixed without compromising long term genetic diversity. A colony might be outstanding for docility or honey production, but not for hygienic behaviour or winter-hardiness. Nature hasn't designed the honeybee to be that inflexible in its genetics because its the very plasticity of colonies through genetic diversity that has enabled their survival.

Dual *et al* found that drones damaged by *varroa* viruses are less likely to fly or breed successfully. Studies in 2002 show that one mite at a drone's larval stage could reduce its flying power and its sperm production by 24%. Two mites at a drone's larval stage could reduce its flying power and sperm production by 45%. Given the importance of a high number of drones needed to sustain the viability of a DCA throughout a breeding season, and the need for athletic, healthy drones to ensure competitive mating, its clear that varroa has much more of an impact on bee colonies than many beekeepers realise. It's probably fair to suggest that most beekeepers regard varroa as a problem inside the hive, for their workers and the honey yield, the drones being expendable and therefore useful for trapping varroa as part of an integrated pest management (IPM) approach. But as soon as we think beyond individual hives and consider the way bee genetic diversity is secured through DCAs, drones become worthy of our increased attention. Indeed, as soon as we stop thinking merely of raising good queens but of raising good colonies, the importance of drones comes to the fore.

Another overlooked issue with drones is that they range across an average five to seven mile radius, compared with the smaller three mile average radius ranged by workers. Drones are also more prone to what beekeepers call *drifting*, or moving between hives outside their original home, where they are generally welcomed. But this has potentially huge repercussions if there are high mite levels in a colony; drones leaving for DCAs and drifting into other colonies risk introducing *mite bombs* into other colonies that might have already been given treatments and are assumed to have reduced mite loads at safer levels.

As someone who wants to stop chemical treatments against varroa, I accept that I could be accused of posing a threat to fellow beekeepers through my roaming drones as well as infested workers shedding mites where other bees can pick them up. On the other hand, I'm deeply concerned about how modern beekeeping is damaging genetic diversity by removing drone brood, and the practice of keeping drone numbers low because drones are unfairly regarded as a waste of time and of the colony's valuable resources. The bee master voices the view of many mistaken modern beekeepers, 'It is only the drone that rests. He is very like some humans I know of his own sex – he lives an idle life, and leaves the work to the women kind.' (3)

I suspect one of the biggest reasons for the lack of interest in drones is that many beekeepers import queens rather than raise their own. I read a statistic that in 2018 one in fourteen hives in Britain was queened by an imported queen of that year, from sixteen different countries, but mainly Greece and Italy. That's a lot of queens! That has to have a negative impact on a species whose genetic success depends upon adaptation to its local environment. It does affect them, and it's called *outbreeding depression*. Quite simply, it means that imported stock and local bees interbreed, but their progeny are unable to fully exploit the local conditions because of inherited genetics that are in some ways unsuited to the new environment, while they also lack genetics that are adapted to the local environment which are deselected by outbreeding.

This was illustrated strikingly in 1966 by Louveaux who investigated the honeybees of Landes in Southern France. In that region he noticed two nectar flows – one in late May and another in September. He found that the Landes bees had two population peaks that exploited these two nectar flows. It only happened in Landes – nowhere else in France. Louveaux then tried an experiment, swapping colonies between Paris and Landes. He found that the Paris colonies in Landes only had one population surge which could only exploit one of Landes' nectar flows. The Landes colonies moved to Paris, however, continued to produce two population peaks, but the second happened just as they should have been preparing to settle down for the winter. Neither colony had surges in brood production or populations that were suited to their environment and so neither could fully exploit the nectar flow in the new environment in which they found themselves. Each was a ecotype that was finely tuned to its own unique locality.

I'm wondering about the implications of that statistic – one in fourteen queens being foreign imports in 2018. It means that a large proportion of drones in DCAs are going to be drones from a variety of races – probably Italian, Carniolan, Buckfast (mostly produced abroad) and a variety of other crosses that might include some Carpathian and Dark European bees. On the other hand, beekeepers who buy mated queens aren't that interested in drones and are probably the same beekeepers who cut out drone brood as part of their varroa management. Nevertheless, it means that some drones from these imported queens are reaching DCAs in Britain and that some of them could be mating with my queens, diluting my efforts to produce locally adapted genetics in my stocks.

It also has implications (and more alarming ones) for my own swarm catching, because some of these alien stocks will then throw swarms, depending on their inherent propensity to swarm. If there are many local beekeepers with Buckfast stocks, for example, these are less likely to issue swarms, but if Italians become crowded or a beekeeper is inexperienced these will swarm. If a high proportion are Carniolans, these will invariably swarm, especially with less experienced beekeepers who haven't yet mastered swarm management techniques. My guess then is that in an average DCA there are likely to be a high proportion of Carnies.

I'm hoping, however, that in rural West Berkshire where Douai Abbey is located, surrounded by farmland and woodland, there will also be a higher proportion of wild bees which are under pressure of natural selection. Such bees are more likely to have developed hygienic behaviours and are probably expressing traits that were found in their *Apis Mellifera mellifera* ancestry. There might be some swarms from imported stock, which are no use to me in the long run, but the chances are that I will catch some locally adapted bees. If some of these have survived for some years, natural selection will also have deselected most of the genetics that are unsuited to our conditions.

There is some scientific evidence, however, that bees can adapt to outbreeding depression. Analysis of selected colonies has shown that those which were originally Carniolan in areas where there are also native Black bees show a tendency to select the native traits over time and do not remain Carniolan – the native traits being by far the most dominant traits in the DNA of these colonies. This was first proposed by Beowulf Cooper in his book *Village Bees*. He observed that when *native* and non-native honeybees cross-bred, the traits of both parents remained, albeit hidden in the next generation. All these characteristics, he argued, will appear in some form, perhaps individually in some cases, in later generations. What beekeepers regard as mongrels therefore might actually have all the original characteristics of the native British bee in their genes. This was why he argued that native bees still existed after the Isle of Wight disease, their genes having been passed down through survivor colonies to locally adapted bees.

Although some gene inheritance will be random, Cooper argued that most of the characteristics that will be selected under pressure for survival will be the ones that make up the native Black bee, such as their dark colour, an ability to work and mate at low temperatures and in the rain, and resistance to certain diseases.

By the second week of April the temperature is struggling to reach thirteen degrees Celsius, but I decide to make a quick inspection of the hives, to see if they are producing brood in sufficient quantities to begin the spring build-up. The answer is no! There's little real progress from two weeks ago, though in the queen-right colonies there is more brood than there was when I last looked. This suggests to me something about the stocks' genetics. A late spring build-up is typical of Carnies, although once they really get going brood expansion is usually explosive. Given that one in fourteen hives have imported queens and that Carniolans are the second most favoured bee by British beekeepers, this points to probable matings between my queens and at least some Carniolan drones. In one hive the bees are also very defensive – another Carnie trait.

It's incredible to think how recently we have begun this interference with honeybees, to the point that we are now compromising their genetic diversity, which could ultimately turn out be more catastrophic than varroa. Up until the middle of the nineteenth century there were no imported honeybees in Britain. All bees were native *Apis Mellifera mellifera*. This situation had been undisturbed for thousands of years in a land that was originally mostly forest, with a small human population and with minimum destructive impact on the environment. Native bees probably swarmed every three years, and new colonies probably balanced out winter losses and other failures of survival from disease or weak colonies. Natural selection constantly eliminated weaker colonies, and a generally stable population was maintained of the strongest bees.

In my more radical moments I think that we probably need in Britain (in an ideal world) a kind of reset in beekeeping in which everyone stops importing and all chemical treatments are stopped. This would, in the first few years, bring about a seemingly catastrophic crash in honeybee numbers as natural selection removes the weakest colonies that today we keep alive artificially. After a few years Beowulf Cooper's survival under the pressure of natural selection would result in a very small population of survivor bees that would be forced through a genetic bottleneck. Emerging from this would be the very traits aiding survival that are present in the genes of *Amm* hidden in these so-called mongrels. The traits that make up native bees, that are adapted to survive the British climate and environment, would be reselected.

There's nothing new about the idea of such a reset. I've read a suggestion made in June 1921 in the *British Bee Journal* that called for all stocks in Britain to be cleared, with the help of the Board of Agriculture, while native Black bees might have then been raised in the Scilly Isles and used to restock the mainland once it had been disinfected. A huge operation would have been involved in such an undertaking, not to mention a huge financial cost. By then, however, the Board of Agriculture had realised that there was a crisis in food production, so support for the scheme was never given and Dutch and Italian stocks continued to be the only way to maintain a population of honeybees in Britain with which to safeguard food production.

Of course, that kind of reset could never happen today either. It's just not realistic. It is as unrealistic and probably morally suspect as those environmentalists who demand zero carbon emissions virtually overnight. Beekeepers would never reach such a consensus of opinion. There would be those who have always kept Italians or Carniolans who would regard this reset as an outrage upon freedom of choice. The main reason it won't happen, however, is because in the short term there would be a crisis in food security. Think of all the foods reliant on honeybee pollination that would not be pollinated. The fall in production would have a drastic effect on the economy, pushing up food prices because of our inability to produce the full range of foods in our diets, including fruits and vegetables. Even onions rely on bee pollination! This is the problem with extreme environmentalist positions: they are always at the expense of people – especially the poor. You can't save the planet at the expense of people's lives.

It won't happen, but such a reset would work, at least for bees, if not for people. It happened in the Arnot Forest region of New York State. In 1978 Professor Thomas Seeley of Cornell University, studying bees in the Arnot forest, identified and counted all the nests of wild honeybees. In 1992 varroa hit the US and in the Arnot forest most of the wild tree-dwelling colonies died off. Probably only a couple were eventually left, which were forced through a genetic bottleneck under pressure from natural selection. Out the other side, some years later, bees emerged with varroa tolerance. In 2011, when Seeley returned to the region, he found that there were the same number of colonies as there had been in 1978, but they all had varroa, and yet they were all surviving.

Since then other experiments have been done elsewhere, such as in Gotland, Sweden. In four years 94% of 150 colonies on the island left without beekeeper intervention crashed. For the first two years every swarm died out, but from the fourth year onwards swarms from survivor colonies didn't die out. At year four the same genetic bottleneck happened that had occurred in the Arnot forest; the pressure of natural selection created survivor stocks. Nature always looks for a way to survive.

Such a reset, if applied in Britain, would have the same result as at Gotland and the Arnott forest, but the initial cost of losing 94% of our bees over four years is simply too high a price to consider. The fact remains that the alternative comes with potentially as great a cost in the long run; our present beekeeping approach continues to impact on selection pressure to the detriment of every natural mechanism honeybees have developed for survival. Italian bees don't take winter brood breaks because they are adapted to a warmer climate where winters don't necessitate tight clustering to keep the colony warm. In Italy, even in winter, light foraging can continue, as nectar and pollen can be collected all through the year. Contrastingly, in Britain winter clustering and a brood break in summer are essential. Italians wintered here in large colonies and in loose clusters eat more stores and run out of food earlier in our long winters, requiring syrup to keep them alive in early spring. Their drones will then spread non-adapted alleles (traits) throughout Britain so that non-adapted genetics enter local gene pools. Then in a cold, wet summer the same bees have to be fed syrup again.

Carniolans, the second most popular race, are in other ways poorly adapted to Britain. In particular they have a facility to take brood breaks when there is no forage available. This means they often don't get started until warm weather arrives and allows them to forage. That's a very good trait in Eastern Europe where long, cold winters can bite well into the spring. In Britain the native Black bee evolved to begin brood production early, in order to be ready to swarm in May and June and to take advantage of a harvest even in a cool, wet summer. Faced with a cold, wet June, for example, Carnies will stop brood production completely, reducing their population and their honey production. As they switch on expensive brood production well into September, this means that in bad years they need syrup to boost their winter stores.

All this leads me to wonder why the genetic make up of the Douai bees isn't a complete mess? Why are our mongrel bees in Britain not completely dysfunctional? What's preventing a complete melt-down, it seems to me, is the increasing existence of wild bees that are under natural selection to reselect the hidden native traits in their DNA; together with beekeepers like me who raise their own queens and have locally adapted stock. This, and the fact that beekeepers of imported stock are probably cutting out drone brood or using it to trap varroa, means that many drones reaching DCAs are probably already locally adapted and their numbers give them a good chance of breeding. Although the Douai bees are expressing *Carniolan* traits, such as slow brood production in early spring, they aren't *Carnies* and their DNA probably by now contains many other traits from *Amm* that are not yet fully expressed but will be reselected in subsequent generations to favour survival in Britain.

It's interesting, I'm thinking, how William Woodley made the jump from watch-maker to bee farmer. When you think about it, there are striking similarities between horology and bee keeping. Both are essentially solitary occupations and both require the accumulation of years of knowledge and experience before you master them. Most striking of all, I'm beginning to realise, is that honeybees, like clocks and watches, are precision mechanisms, finely tuned for a specific function. By considering the complex and intricate ways various races of honeybee have evolved to survive in particular environments, I'm becoming more aware that locally adapted bees are as finely tuned as a clock to take advantage of local conditions. I can see exactly why William Woodley became as fascinated by bees as he was by clocks and watches. More disturbingly, the delicate mechanisms of both means that they are also too easily broken or damaged.

***

I've had a monastery car booked out all week for Saturday morning. All Easter I've had meetings and paperwork and homilies to write, and I'm trying to organise retreat speakers for the Abbey's retreat programme. I'm also producing a document to present to the community next week, and struggling with my homily for tomorrow. Monastic life is a lot busier than many people imagine. Sometimes I think it would be nice to be able to spend all my time in the apiary, as Brother Adam did, but these days monasteries have smaller numbers of monks and no one has the luxury of indulging one interest, while we all have to multi-task. It's also the first time I've attempted to go out in the car since the autumn because of two lockdowns. I've actually been out twice since the pandemic began.

As I'm driving the twenty minutes to Beedon, along backroads through Cold Ash, Hermitage and Chieveley, I'm wondering what William Woodley would have thought of a beekeeper making something of a pilgrimage to his grave almost a hundred years after his death. I notice plenty of blackthorn blossom in the hedgerows, but it still feels like a fridge outside – the kind of cold that you can feel lingering in your bones long after coming inside; the kind of cold that makes all the nerves in your teeth jump.

Driving through Chieveley where Mr. Woodley started work as a grocer, I come eventually to a sign for Worlds End. Somewhere here he had his out apiary. I decide that's a line of investigation for another day. It starts raining suddenly – hard, as I reach Stanmore and turn left, following a sign for the church. At first I see nothing, then just a glimpse of what looks like a small church spire protruding behind a hamlet. Left again, and I arrive at St. Nicholas' church.

In the *Domesday Book* Abingdon Abbey is entered as holding the whole of Beedon after King Edgar granted the Abbey some lands at Beedon, other land being left in 1015 by Eadway, a Prince of Wessex. In 1505-6 the manor was held by the Abbey, but in 1538, when the Abbot surrendered the Abbey estate to the king, the Abbey's overlordship ceased. The Grade One, listed church is small, with a nave just fifty-one feet long and twenty-two feet wide, dating almost wholly from 1220, with its original window flints and stone dressings.

I put up the hood of my Paddington Bear coat against the cold rain and stand at a little brown gate, looking ahead at the small porch. To the left of the church, as I'm looking at it, I can see only two gravestones. I open the gate and walk towards them, rain spitting in my face. Immediately the rest of the small graveyard opens up to view. As I draw closer to the nearest grave, to the left of the church I can see a familiar carved lozenge-shape in the centre of the headstone, which I recognise from photographs. I'd given myself a maximum of half an hour to find the grave, but I never expected it would be this easy. Sure enough, when I reach the grave I can see, carved in relief, in the lozenge-shaped recess, a queen bee, several inches long. Orange lichen has obscured some of her features and has hidden some of the text on the stone, but under the queen bee I can make out the name *William Woodley*. Above the carved bee, in clearer, black lettering, the name of his wife – *Annie*. I've found it!

Photographs taken, I stand for a while - as you do, in a graveyard. I'm thinking how this grave had already been here a decade when the first bees arrived at Douai Abbey. I'm thinking how quickly we fade into obscurity and how ephemeral was the fame Mr. Woodley enjoyed at the height of his powers:

'Beneath those rugged elms, that yew-tree's shade,

Where heaves the turf in many a mouldering heap,

Each in his narrow cell for ever laid,

The rude forefathers of the hamlet sleep.' (4)

*The Woodley grave*

All of a sudden it strikes me that the queen bee carved in its recessed lozenge is in her cell, as Mr.Woodley is, and it's a grave-cell too; she's lying in a kind of coffin, just like her bee master. It's suddenly rather touching to be there.

Perhaps the one lesson William Woodley's story has taught me is that the best beekeepers, like the best bees, are probably the most adaptable. Woodley must have been adaptable in the first place to move from grocer to watch-maker and then to bee farmer. Then he adapted from skep beekeeper to the modern frame hive technology. When disease almost destroyed his apiaries in 1915 he started again, but the following winter 1915-16 Isle of Wight disease hit his business a second time. Still adapting to the changing situation, in 1918 he began restocking again with hybrid Dutch and Italian bees that seemed to be more resistant to Isle of Wight disease. Adaptability is the ability to change when circumstances require it. That is never easy, especially for someone in their seventies who has done things a particular way for decades. William Woodley was willing to change in the end, but he had no choice by then. Who knows, but perhaps his doubts about modern methods in apiculture at the end of his career, and their possible contribution to the advance of the Isle of Wight scourge, might even have brought him back, full circle, to skeps.

We get some sense of the scale of the crisis from a letter Woodley wrote in March 1915 to his landlord who rented him the tenancy of his out apiary at Worlds End. The letter suggests very strongly that Woodley felt defeated: 'I am very sorry to give up the garden but all my bees have died with the Isle of Wight Disease at Stanmore and nearly all my stocks at Worlds End. I don't think there is a stock alive within Beedon parish.' (5)

What seems to have changed his mind and opened his thinking to new possibilities, is that a year later, following the winter of 1915-16, he was hit by Isle of Wight Disease a second time, 'The winter of 1915-16 reduced me to a few stocks and as the spring advanced these developed symptoms of 'Isle of Wight' disease.' (6)

The only way he and everyone else could resurrect their apiaries was by importing foreign queens and stocks. While this was an admirable adaptation in some ways, and a necessary one at that time (for economic reasons), it further opened the door to a century of cross-breeding and mongrel bees, of random out-crossing and the gradual depression of traits specifically adapted to Britain.

At the second Commons reading of the 1979 *Bees Bill* on 25 October 1978, Parliamentary Secretary to the Ministry of Agriculture, Fisheries and Food, Mr. Jerry Wiggins, addressed the House in an effort to stop varroa mites entering Britain via imported queens. He gave the statistic in that speech that there were 240,000 hives in Britain at that time, and that there were 6,000 imported queens, which means at that time one in forty hives were headed by foreign queens. Varroa has been here since 1993 and we now have one in fourteen hives headed by foreign queens. In 1978 when Mr. Wiggins addressed the Commons there was a maximum penalty under the 1954 *Act on Bees*, of £20 for illegally importing queens. Mr. Wiggins wanted it raised to £1,000. For a determined business person even £1,000 is a meagre figure when you consider that a queen will sell for £40-50. 18,849 queens at £40 (average cost) – you don't even need to do the Maths; there's good money in selling queens. That's why a beekeeper in the US will pay $1000 for a top quality breeder queen. So beekeepers, quite frankly, are to blame. Even those suppliers who operate within the law and have the required health and biosecurity paper-trail for their imports are not exonerated. If these imported bees are not yet spreading diseases and parasites, they're still spreading undesirable genetics. In my view it's also only a matter of time before they introduce new parasites and diseases. It happened with varroa; it likely happened with Isle of Wight disease. History has a way of repeating itself when we learn nothing from its lessons.

We've come a long way in a very short time and have ended up somewhere with the honeybee that Mr. Woodley could never have imagined a hundred years ago. But he also made a similar journey in only a few decades; from straw skeps to the frame hives he used in 1878, to the end of the native Black bee and the increased importation of foreign queens. In the *British Bee Journal*, on 11 November 1909 the writer of an article about him titled *Prominant Beekeepers* described him as a beekeeper who, '...relies entirely on the old native black bee, and believes in no other; and, while endeavouring to improve his strain, no foreign blood is allowed to mix with it.' (7)

One wonders today why the number of imported foreign queens in Britain has been permitted to rise to one in fourteen hives, especially bearing in mind that a good queen should live on average three to five years. There are beekeepers who have had queens live up to six or even seven years, though this is exceptional. Since about 2000, however, many beekeepers at home and abroad have reported that queens today don't seem to live very long. Many hardly survive their second season, with a large number of new queens each year failing and being superseded as soon as they are mated. It's not unusual to find supersedure cells in March or April where a stock is rearing a new queen, or supersedure cells over the winter so that the beekeeper begins the new season with an unmarked queen in the place of last year's marked queen. Beekeepers are also finding new queens laying well who suddenly disappear, or a few drone cells appearing in capped worker brood – a signal that something is wrong with her laying. Within weeks these queens are superseded or simply disappear. These things are happening as much with foreign queens as with British ones, and it's difficult to explain why. Some people blame bad beekeeping, or the weather, or birds taking queens, but these are often the experiences of people who've kept bees for decades, and the same problems occur in good seasons. Moreover, birds have always been a threat to queens on mating flights, but there's no earthly reason why more of them should now fall prey to birds than in the past. Who knows? Except that 2000 was just after varroa had arrived in Britain; are we now experiencing the long term effects of varroa viruses on our bees? We know, for example, that queens infected with viruses lay fewer eggs because their ovaries are smaller. A queen laying fewer eggs will probably be perceived as failing by the colony and will be superseded.

Perhaps this high turnover of queens in recent years, together with the standard advice now given to beekeepers to change a queen after her second season, explains the rising number of imports. It's another way in which we're not working with nature, but against it. Are our own actions responsible for the problems we're having with queens? For all our efforts to fix the traits we desire and to mitigate the possibility of swarms that reduce the honey harvest and put us in danger of litigation from our neighbours, are we inadvertently disrupting something about our queens? They are part of the delicate precision mechanism of a colony, producing queen substance that we now know is literally handed from bee to bee, fanned around the hive like a series of texts. Who knows, but our selective breeding for certain traits might be tampering with the most important trait of all, which is the production of queen substance? Or could it be that fixed comb hives with their large number of frames is also part of the problem? In nature honeybees have fewer combs, which are also full of holes, and this might allow better passage of air and pheromones around the colony. Bees that are further away from their queen in frame hives might not, for some reason, be detecting a strong enough presence of queen substance and are making the decision that a queen is therefore failing.

Of course what we don't know about imported bees is at what age they have been harvested and how long they might have been banked before being dispatched to a buyer. There's evidence now that the age at which a queen is harvested and is introduced into a colony and how long she has been banked in a queen cage corresponded to the acceptance rate of queen -introduction and the rate at which queens are superseded. The optimum age for harvesting is between three and four weeks and mated queens who have settled into laying are far more likely to be accepted by a colony. Queens banked for any length of time are also more likely to have injuries caused by workers, such as bitten legs and antennae, such injuries making it more likely that a queen will be superseded within a few months. There was much wisdom, it seems, in Brother Adam's advice about queen introduction – that the most important factor influencing acceptance is the condition (or maturity) of the queen. He did it often enough over the years that I think we can trust him when he assured us that: '..the acceptance of a queen is not determined, as hitherto generally assumed, by ''colony odour'', but by her behaviour. A fully mature queen, one that has been laying for a considerable time, will have lost her original nervousness and will behave sedately and calmly.' (8)

Then there's the added factor of all the chemicals poured into many hives by beekeepers and the chemicals brought back on foragers and returning drones that include pesticides and pollution. Are these perhaps blocking or interfering with queen substance and other pheromones and preventing them being read by the bees throughout the colony? We have no idea at the moment how much these factors might be working together to produce the ubiquitous problems we're experiencing.

Honeybees are sensitive - we know this. Evidence has been accumulating for some years that workers can detect different patrilines in eggs and larvae and that they tend to select eggs and larvae for raising queens from the rarest patrilines. Why is such sensitivity important to them? They can also detect infected brood so that it can be removed. They can detect things about queens that we can't see or begin to understand at the moment, such as the amount of pheromone she is producing, how well she is mated and when she is failing. It seems to me that honeybees are telling us something important about what we are doing wrong in modern beekeeping. If queens are living shorter lives and are superseded or disappear even when we think they are laying well, something is going on that isn't quite right. If a queen is laying well, her brood will also produce pheromone. Is this pheromone interrupted by factors such as the size of the hive, or the build up of chemicals and pesticides in old wax after two or three years? None of these might be the main cause of the problem. As with *Isle of Wight d*isease, there might be many other contributory factors, and any one of them might be the tipping-point. Who knows? But it seems likely to me that the time the queen problems kicked off is a smoking gun; it corresponds with the arrival of *varroa*. With *varroa* have come all kinds of viruses, such as deformed wing virus (DWV), and they are bound to affect queens just as much as they affect drone sperm counts, drone vigour and flight. It's interesting then that we're seeing a proportion of queen cells with immature queens dying in them, and a proportion of virgin queens with deformed wings; typically with one wing at right angles, or stunted, or absent altogether. These queens can't fly and so can't mate either.

A century ago even Tickner Edwardes' progressive b*ee master* was critical of imported queens. When asked for, '...one royal maxim of success above any other,' (9) he replied: '... beware the foreign feminine element. Let British beekeepers cease to import queen bees from Italy and elsewhere, and stick to the good old English Black...above all, though she is not so handsome as some of her continental rivals, she comes of a hardy northern race, and stands the ups and downs of the British winter better than any of the fantastic yellow-girdled crew from overseas.' (10)

Perhaps the issue of appearance and aesthetics was a factor in some of the earlier examples of queen importing, before the demise of the Black bee made imports a necessity. In March 1915 a letter appeared in the *British Bee Journal* from L.A.V. Rye, dated 27 February, asking for beekeepers' experiences of various crosses, as he was considering introducing *Ligurians* and *Carniolans* into his apiary. A response in the same year referred to crosses made by T.B. Blow of Welwyn as early as 1887 who had been importing these races. His letter reporting his results focused as much on their beauty as on their more positive functional traits.

I'm sure it is too simplistic to blame foul brood and Isle of Wight disease exclusively on imported foreign queens. It's likely there were other complex factors involved. What about the introduction of the synthetic fertiliser – ammonia, in 1910? Apart from the environmental consequences of ammonia at that time, mechanisation also meant that farmers had stopped growing clover for animal feed, which was also a natural way of fixing nitrogen in the ground. Consequently, this would have reduced honeybee forage. Mr. Woodley observed and wrote about this in his column.

The story of the invention of the first synthetic fertiliser is a cautionary one. Fritz Haber and Carl Bosch were the German chemists who first developed the Haber process of making ammonia from nitrogen and hydrogen in 1909. Air was forced into an iron tank under heat and pressure. Hydrogen was added which broke the nitrogen's bonds, rebonding them with hydrogen atoms to create ammonia. The liquid that came out of the process was liquid fertiliser.

Haber's motto was we only want one limit, the limit of our own ability, a chilling example of the dangers of science divorced from morality. During the First World War, when Germany was losing, Haber suggested to the German high command that he had a brilliant idea for producing a chemical weapon, in violation of the Hague Convention of warfare. By adding ammonia to chlorine he developed an asphyxiating gas that on 22 April 1915 killed 6,000 troops on the battlefield. By the end of the Great War 100,000 had been killed by the gas and many wounded. In 1918 Haber was even awarded the Nobel Prize for chemistry. His wife, Clara, (also an eminent scientist) and convert to Christianity, was a pacifist and later shot herself in protest against her husband's involvement in the development and use of gas warfare. Their son emigrated, but later in life also shot himself.

Take heed, supporters of genetic engineering. What started as a noble endeavour to increase food production, in the service of life, unleashed death on an industrial scale. When much of our modern food security is in the hands of just a few modern agrochemical multinational companies, the story of Haber and the first synthetic fertiliser does not inspire me with confidence for the future. Clearly, *the limit of our ability* coupled with moral relativism are a toxic mix.

I reach out and touch the stone bee trapped like a fossil on the Woodley grave. I run my fingers across the cold carving, as though reading Braille. Although the bee is disfigured by lichens, I can still see enough of the image to recognise that it's a queen.

Towards the end of his life William Woodley lamented the mismanaged methods of modern beekeeping, but not beekeeping itself. What exactly were those misapplied methods, I'm wondering again? I don't believe he was fundamentally critical of the frame hive at the end, despite his call to return to skeps as disease ravaged British stocks. Perhaps it was the obsession with seeking non-swarming bees, although his writing at times seemed to approve, in principle, of that objective. Or maybe it wasn't the obsession with non-swarming bees in itself, but the consequences of it, which was a growing enchantment with foreign races of honeybee and the need to import queens to satisfy their promise of fecundity and increased profit. Whatever specific methods he alluded to, it seems to me that the common denominator was the increasing use of foreign queens at that time.

The increasing use of foreign races since the 1850s seems to me to have been an obvious blind spot in apiculture in William Woodley's life-time. Beekeeping had already gone down a cul de sac for half a century by the time Isle of Wight disease took hold and, ironically, the main consequence of the disease was that beekeeping afterwards was forced to continue down this cul de sac because imported bees were by then the only option left for the survival of the craft and for the nation's food security. The misapplied methods to which Woodley referred were becoming ossified because beekeepers could by then see no alternative.

In the same way today our modern methods and habits of mind in beekeeping have become set in stone, with many beekeepers seeing no alternative to mite treatments or the use of imported queens. We have gone so far down that cul de sac that there is nowhere else left to go that does not further compromise our native honeybees.

I turn back to the car and notice that the side road leading to the little church of St. Nicholas, Beedon, is a cul de sac. I wonder if William Woodley might have smiled at the irony.

# MINDING THE BEES

## Chapter 9
### The Times Remain The Same

'IMPORTED ITALIAN QUEENS. 6s each.
E.WOODHAM, Clavering, Essex.' (1)

*British Bee Journal 1904*

Low Sunday. Heavy hail whitens the ground after lunch. Then we wake up on Monday morning to light snow that has turned the world into a late Christmas card, white blossom sitting like scoops of ice cream in the plum and greengage trees on the south side of the Abbey Church.

A male chaffinch lands on top of the chicken coop as I'm collecting eggs after lunch, calling *pink, pink* repeatedly- the sound he makes at this time of year when he's separated from his mate. Another way to recognise the chaffinch's usual song is to think of the descending note as a waterfall, ended by a short, rude rasp, the bird equivalent of blowing a raspberry. On the subject of bird song, there's still no sound of the cuckoo.

The Call ducks are laying, on and off, and a couple even started sitting, but then abandon the eggs in their shed as the Arctic conditions bite. Two chickens have been broody for the last few days, so I put them in brooder cages and slip several duck eggs under them; an insurance policy, as the ducks missed the chance to nest altogether last year and might miss again this year if these conditions continue. Chickens will hatch duck eggs quite happily and will look after ducklings as diligently as if they were their own. A chicken has even been known to foster kittens.

Although the weather is cold, I decide when the snow thaws to have a quick look at a couple of frames in the centre of each hive and to give all the stocks some pollen candy in case their food stores run out during the cold spell. There are two hives with marked queens from last year, and brood, but the queens are not yet in full lay, probably due to the weather. In fact, they've still hardly started laying at all.

The other hives are interesting. One shows a few emergency queen cells. Unlike swarm cells that tend to be on the bottom edges and sides of the frames, supersedure cells are often in the centre of the frames, are the same age and are grouped close together, but I have seen them on the edges of frames. I see two of them, but between one and three is usual. Emergency cells can be more numerous and are spread across more than one frame, while they are often at different stages of development. It's obvious to me that they must have started replacing the old queen back in March, as these cells are now sealed. Hopefully by the time she's ready to mate the weather will have warmed up and there will be drones about. After sixteen days, from egg to fully formed queen, she will emerge, but has only a narrow window of about three weeks to be mated. If the weather remains too cold or wet, or it's too early in the year for drones, she won't be mated and the colony will become hopelessly queenless because they will not have a fertile queen or any other eggs or larvae from which to raise a substitute queen.

Even in 1911 there were articles in the *British Bee Journal* that illustrate how young our current knowledge of bee biology is. Back then it was still believed by many that the queen was mated within the hive, while the idea of multiple matings was still controversial and scientifically unproven, as evidenced in the *British Bee Journal* by A.W. Salmon's letter of 1911:

'The possibility of queens being fertilised in the hive has so often been discussed that many bee-keepers are, no doubt, still hopeful of proving that such can take place, and a few still think they have changed the hypotheses of early philosophers into actual facts. Such experienced men as Mr. Cowan and his colleagues of various nationalities, who have devoted a lifetime to research in a scientific and practical way, have undoubtedly proved conclusively that fecundation cannot take place except on the wing.' (2)

In the fourth hive there's capped drone brood and eggs, but no sign of a queen. In this situation I might have laying workers, where the ovaries of workers (usually suppressed due to the presence of a queen) become stimulated to produce eggs. Workers, however, are infertile, so any eggs they lay will be haploid eggs and will develop into drones. In this scenario a colony will also be hopelessly queenless, having missed the opportunity to raise queens from eggs or young larvae.

I've seen the phenomenon known as laying workers before, and usually you see multiple eggs in a cell, but here I'm seeing eggs laid singly in each cell, and that's a good sign that a queen is present. Something else is going on then. I look again at the drone brood with its domed caps. There's no worker brood. An infertile queen will only lay haploid eggs and these will all become drones, so unless worker brood appears by the time I next look, it seems likely that this queen was replaced over the winter and isn't mated, in which case this stock is hopelessly queenless because of what beekeepers call a drone-laying queen. Having missed the small window of opportunity to mate, such a queen is now unable to mate successfully at all.

At the moment I can only wait. I will wait for the hive with queen cells to hatch a new queen.

The weather should be in her favour by then and the colony will then be queen-right. The suspected drone-layer in the other hive will be confirmed in a week or so when more capped brood appears. If the capped brood is smooth and in worker cells then we have a normal queen who is mated and is laying fertile eggs. If, however, the brood is all domed, it will confirm a  drone-layer who will have to be removed. If laying workers are the problem, however, this can be sometimes be reversed by repeatedly donating brood from other stocks; the brood pheromone will eventually suppress the worker ovaries, but it takes time and you need enough brood from other stocks that you can take some every week. I don't have many stocks and it's early in the season, so there might not be enough brood to donate, in which case it might be impossible to save the stock. With laying workers there's no point donating a frame of eggs and larvae in the hope that they might raise a queen, because the colony already thinks it has a queen.

I've read that these early season problems with queens are become increasingly common. The queens in my own stocks were all raised last season, so they are young and shouldn't already have failed over the winter or be failing at the start of the new season. It suggests perhaps that last season's queens didn't get properly mated. As the old bee master reminds us: 'Bees have exceptions to almost every rule. While other creatures seem to keep blindly to one pre-ordained way in everything they do, you can never be certain at any time that bees will not reverse their ordinary course to meet circumstances you may know nothing of.' (3)

I have been alerted to another early season problem by an email from the local BBKA about the threat of foul brood, a bacterial disease that can enter hives from bees drifting or robbing neighbouring hives (Ligustica is especially prone to drifting, so would be a problem in this situation). The advice is to keep entrances small, reducing the risk of robbing. Foul brood attacks the larvae before they are sealed over, turning them from pearly-white to a yellowish colour. The dead brood produce a foul smell - hence the name foul brood.

Since before the beginnings of the BBKA foul brood has been a problem. There were a number of attempts at the end of the nineteenth century and the beginning of the twentieth century to introduce legislation to control the disease, without success. The 13 April publication of the *British Bee Journal* for 1905 shows the results of County beekeepers' votes for and against the proposed *Foul Brood Bill* that year. The majority voted for the bill, but the weight of voting was against, if you consider the number of hives represented by each vote. Those who owned more hives tended to be against the bill. The explanation for this vote is clear from a letter by J. N. Langwathby in the *Journal* on 13 April 1905: 'It is well known that in the majority of cases foul brood flourishes most in small apiaries...but to the owner of a large number of stocks, the advent of an inspector, with legislation to back him up, commanding an inspection at some inconvenient time would be indeed serious...' (4)

William Woodley also opposed legislation, as an owner of a large number of stocks. He was forthright in his view, published in the *Journal* a number of times, exemplified by his *Notes By The Way* column from 7 December 1911:

'I have to thank Mr Crawshaw for his report on what the Bee Disease Act committee are doing. Since the "Tickner Edwardes" draft bill appeared in the 'Smallholder' I have had several letters on the new 'terror' to bee-keepers, which will meet with our most strenuous opposition, and I trust bee-keepers will carefully read and study the effect that such a measure will have on our industry before they append their signature to the so-called petition – to condemn the straw-skep to extinction. As regards the BBKA draft of the bill (FB), there will have to be included a clause exempting large apiaries from compulsory inspection.' (5)

In March of the same year his criticism was founded more specifically on the way experts in Ireland had conducted inspections of large apiaries after the Irish Bee Disease Legislation had been introduced:

'Page 89 of our journal shows how the Irish Act is being administered; the prominent bee-keepers are made the scape-goats by the inspectors, who, when once appointed, are masters in everything pertaining to bees. They can choose their time to inspect them, even in July just in the midst of the honey harvest. Consider what it means to a man with one or two hundred hives, to have his whole apiary upset by a hive-to-hive overhaul when every hive is supered. Why, it would cause a loss of several pounds (£ s. d.), as the upset would mean a day's work lost to every colony examined, or, in short, 100 days' work lost to the owner of 100 hives ; and then, if the inspector should find one of the hives infected, he would have the power to prevent the owner from disposing of the produce of the ninety-nine healthy stocks — perhaps a ton or two of honey.' (6)

He drew further attention in his *Notes By The Way* of May 1911 to the dangers posed by such experts: '... and it seems to me that it will require very serious consideration before we shall willingly allow travelling experts to handle our bees after they have been in contact with all and sundry bees, and old appliances.' (7)

Woodley received much criticism and negative publicity for his views on *Bee Disease legislation*, but these notes from 1911 make it clear that his opposition was not merely self-interested. The terror to beekeepers which he feared included the threat of extinction to skep beekeeping from the legislation. Skeps could not be inspected as easily as frame hives, or managed if foul brood were found in them. In this event skeps would have to be destroyed by the expert carrying out the inspection. The skep's destruction would also be the destruction of a significant part of the poor cottage labourer's income and economy. The skep's extinction would contribute to the eventual extinction of the already declining cottage labourer's rural way of life that had been unchanged for hundreds of years, intrinsic to the traditional cottager as, 'The Beedon children's garland...perambulated the village when "going a-Maying" ' (8)

We know that he was also friendly with many local farmers and rural labourers who kept skeps and who would have been the first to be affected financially by the intervention of experts from bee legislation, such as, 'My neighbour, the young farmer mentioned,whose bees were catalogued for sale, has decided to take them with him into Kent, as he is moving to a fruit farm at Kingswood.' (9)

There were other practical reasons to oppose the legislation; that perhaps the most frequent objection to proposed legislation against foul brood was the danger of spreading the disease by those who, in the event of such a law, would be responsible for carrying out the law's provisions. For some beekeepers, however, the extinction of the skep could not come soon enough, and it was an issue of hygiene. Just as today beekeepers are divided over the subject of chemical treatment against varroa, at the beginning of the twentieth century they were divided over the skep and the modern frame hive. As with the modern problem of varroa, the cause of the division back then was also disease. When Isle of Wight disease arrived after foul brood, it reinforced the need for good hygiene in beekeeping, as neglect and poor hygiene seemed to go hand in hand with both diseases.

Mr.Woodley would have been aware of the hostility sometimes expressed in the *British Bee Journal* about the skep, in which letters and articles appeared regularly on the issue, such as L. Illingworth's of 13 March 1919, 'I am not ashamed to admit I dislike the skep, and that I look forward to the time when the frame-hive will be universally adopted.' (10) Others thought that the modern frame hive was the root of the problem and that the skep was a possible solution, as the following opinion expressed in a letter of 6 March 1919: '...possibly the skep may solve one of the problems connected with the "Isle of Wight" disease. Bees in a skep seem to live charmed lives. The skep versus the bar-framed hive, which will win the day?' (11)

Others were equally scathing about the frame hive, blaming Isle of Wight disease on modern methods. They observed that the disease seemed mostly to happen in frame hives. Some had only known of one infected skep and witnessed that it had been infected by bees robbing frame hives that had died out from the disease. The argument over the issue raged on in the *British Bee Journal* for years, with many vehement in their claims that in the time of the old skep Isle of Wight disease had been unknown and that its arrival had coincided with the arrival of the frame hive.

Others managed, like Mr. Woodley, to appreciate the virtues of both traditional and modern beekeeping methods, such as the French beekeeper Madeleine Maravel, described in a 1915 edition of the *British Bee Journal* using her skeps to increase the honey crop in her frame hives. Her system began in April when, bit by bit, she moved her skeps until they were close to strong frame hive colonies. When a strong nectar flow started Madeleine took the skeps away during the day, re-siting them elsewhere. Foragers returning to the skeps would be diverted into the nearest frame hive where they would double the foraging force and the honey crop. It was a variation of the same technique that beekeepers employ today with nucleus boxes after making a split.

The times remain the same! Foul brood is back, along with the ubiquitous varroa, and beekeepers are still arguing whether the modern methods or the more 'natural' ways are better. Moreover, we are more aware today of the problems of chemicals and sprays, such as insecticides, in the environment, pollution and the use of miticides in our hives. Mr. Woodley even suggested in March 1911 that Isle of Wight disease might been connected to the use of crop sprays, 'It is suggested that the " Isle of Wight disease " was started with charlock-spraying in the Isle of Wight.'(12) Perhaps he was also thinking about the introduction of the first synthetic fertiliser in 1910.

I find myself grateful for small mercies, as Mr. Woodley was throughout 1911, expressed in January that year, 'We have escaped so far, I believe, that insidious foe called the "Isle of Wight disease".' (13) Again, in May he was grateful, '...we have another cause for thankfulness in that we have in this district so far escaped the "I.O.W. "disease...' (14)

I don't know about climate change, but a century later, even the weather is replicating the conditions that caused the 1915-16 outbreak of Isle of Wight disease. Back then the weather in March varied from mild to very cold – as it has done this year throughout March and April. In 1916 the bees emerged in a weakened condition, just as they have this year. Easter was warm in 1916, as it was this year, after which it turned cold again, just as has happened this year, confining the bees to their hives. The times remain the same, it would seem!

Not surprisingly, the majority of correspondence in the *British Bee Journal* in the years up to Isle of Wight disease (and especially afterwards) was against skeps. Most *British Bee Journal* readers were BBKA members and had embraced the new methods of frame hive beekeeping, regarding skeps as old-fashioned and their owners as uneducated in modern scientific apiculture. After Isle of Wight disease the skep was perceived by such beekeepers as positively dangerous, harbouring diseases and impossible to manage (unlike frame hives) in the efforts to eradicate foul brood and the worsening plague of Isle of Wight disease.

Occasionally there was a letter in the journal recognising the financial difficulty of getting the working class labourer to switch from skeps to frame hives. Harry Clarke, of Allesley, near Coventry, suggested in his letter of 30 October 1903 a method of selling frame hives to the labouring class of beekeeper in instalments, but the agenda was clearly to eradicate skep beekeeping for good, '...so that in a few short years the old fashioned skep would be superseded by the frame hive in our cottage gardens, to the advancement of the industry.' (15)

Throughout March and April we have had unusually cold weather, despite dry, sunny and mostly settled conditions from a high pressure system. Ice-cold air has predominated, bringing the kind of cold that settles deep in your body long after you come indoors. Whereas we've had a winter with barely a frozen night or day, nearly every morning in April so far has been covered with a heavy frosting. The white of the blossoming blackthorn across the meadow is almost indistinguishable from the frost that sits over the world at first light.

Scrolling back through Mr. Woodley's *Notes By The Way*, there's something reassuring about the number of times every month and every season that he reports the vagaries of the British weather,  'An old Wessex saying is that if the sun shines before 12 o'clock on Candlemas day winter is not half over.' (16)

With our discussions about global warming and its effect on our weather, we forget that British weather is always inconsistent and unreliable, and that seasons don't behave invariably in the same way from year to year. Mr. Woodley also wrote about cold, wet and late springs, and summers that prevented the bees producing a good honey crop. A number of times he reported his good wife having reminded him that the weather he was complaining about had a precedent back in 1888. Mrs. Woodley not only had a good memory for such things, but also the reassuring effect her husband needed. The impression given when reading between the lines of such reports is that she regarded the inconsistencies and extremes of the English climate as entirely normal. Nevertheless, many beekeepers can recognise themselves in Mr. Woodley, imagining that he, like us, came in from time to time from his apiary and sat, head in hands, in his cottage kitchen at Stanmore, complaining about the effect of the weather on the bees; as on 26 May 1898 when his *Notes By The Way* lamented three good bee days in what had so far been a cold, wet and sunless month.

Indeed, 1911, although an exceptional year for honey, was not a good year for his swarms which we know he packaged and sent as far afield as Ireland and Scotland and from which he made important income. By June 1911 he complained: '...but the continued honey flow does not conduce to swarming. Every empty cell is used by the workers for storing surplus, thus restricting the brood-nest. I hear of very few swarms.' (17) In the same month he wrote: '...a wet, cold, and sunless period ; and as to the bees, alas ! I  have no good report to make of progress in the apiary. As to swarms, I am not getting any.' (18)

By July the swarm situation was a lost cause, 'The continued heat has not started swarming, so that I expect the season for this is now over for1911.' (19) This was serious for Mr. Woodley, expressed that month in his *Notes By The Way*: ' ...hopes day after day for swarms to issue, so that he may fill his orders or replenish his empty hives; and the contrariety of things  mundane will often give the small apiarist swarms galore, coupled with empty supers, and  the large apiarist only a small percentage of swarms.' (20)

At the moment it's neither wet nor sunless, and looking at the long range forecast more settled conditions look like continuing to the end of April, with Atlantic air soon to replace the bitter Icelandic air of recent weeks. There's no doubt about it, but this has delayed the spring expansion of the brood nest in the hives. That's a good thing, up to a point, because you don't want brood that the bees can't cover and keep warm. On the other hand, from the end of April until the middle of June and the start of the main nectar flow is only six weeks, and it takes six weeks to produce a forager – three weeks developing in the nest, and three weeks as a house bee. Bees produced in the next couple of weeks will be the ones who will begin to make the honey harvest, and if the queens don't start laying in earnest fairly soon there won't be enough foraging bees to produce a crop this year.

My memory is not as acute as Mrs. Woodley's, but I reassure myself that I've been in this position before. A few years ago the *Beast from the East* brought several inches of snow in late March, and by mid April the bees had hardly started building up. In May it warmed up and the expansion was explosive. Once a queen really gets into her stride with egg-laying she's like a machine-gun, laying about two to three thousand eggs a day. At that rate she can increase the colony by at least fifteen thousand bees in a week. In a fortnight that enough bees to trigger swarming.

Mind you, it's times like this when you don't want Italian bees. If you have Italians you'll have been feeding them syrup since February, for a start – just to keep the large population alive. In these cold conditions they'll have been confined, and any stored syrup can begin fermenting, causing dysentery. Unable to make cleansing flights, the bees defecate in the hive, spreading dysentery and other diseases across the combs. This was probably one of the factors that caused Isle of Wight disease when bees were confined again after a warm spell of weather during Easter of 1906 and again in the spring of 1916; the larger the confined colony, the worse the dysentery is likely to become.

The other concern I have is that this weather might have resulted in the death of many wild colonies that will have run out of stores in March. If that's the case, there will be fewer swarms of the locally adapted bees that I want to collect. It might also turn out to be a less swarmy year for managed colonies too because of the cold, late spring and the delayed spring expansion of the stocks, especially if the next several weeks also turn out to be cold and wet. Swarming is the result of hive prosperity, but in years when colonies can't become prosperous because of poor weather, there won't be as much swarming. That could be bad news not only for the survivor stocks in the Douai apiary, but for my efforts at catching wild bees too.

Looking back through the *British Bee Journal*, there is compelling evidence of similar anomalies in the weather and the seasons over a hundred years ago. I read that in 1893 a drought began at the end of February. By mid June a large apiary reported having had no swarms issue, while the *Editorial* also mentions a scarcity of drones. In other places swarms were seen unusually early – 23 April at Lower Breach Farm, Ewhurst, Nr. Guildford, and at Wonersh, Hants, on 2 May. Farmers were unable to sow spring corn, as the ground was too dry and hard, while white clover flowered a full month early, starting 7 May. The main nectar flow happened weeks earlier than usual too, with beekeepers taking supers of honey in April and reporting April swarms. In 1911 Mr. Woodley ended the beekeeping year in October by observing that after the summer drought, '... our water supply is running very short.' (21)

Our own weather begins to warm up at last, with flowering currant festooned with full pink blooms, constellations of golden lesser celandine by the garden pond, and dandelions. In the potting shed a robin has hatched some grey, downy chicks in a green plastic flowerpot on its side on a low shelf. Tree buds begin to burst into stippled yellow-green against duck-egg-blue sky. sixteen degrees Celsius – good weather at last to make an inspection of the hives and to monitor progress.

One hive looks very strong and has built up brood since the last inspection. The queen is laying well and the colony is calm and easy to handle. The defensive hive needs a little smoke to settle them. The queen is black, with dark tan bands alternating with the black along her abdomen. She looks like a Carnie. Maybe these bees have some Carniolan genes? Many of the workers have that distinctive greyish look of the Carnie.

The hive with the queen cells last time I looked now has a small virgin queen who has a caramel-coloured abdomen. She looks Italian and many of her workers have more yellow on them than those in the stock I've just inspected. With any luck she'll get mated in this good weather. Because of her small size, she's probably newly emerged, and that should give her another couple of weeks to mate. We'll be well into May by then so she should manage a mating flight.

Finally, I turn to the hive that looked suspiciously as though it might have laying workers or a drone-laying queen. Laying workers tend to put multiple eggs in a single cell, but because workers can't reach the end of the cell these eggs are often stuck to the cell wall, nearer the top of the cell. Eggs in this hive, on the other hand, are on the floor of the cells, and often in twos or threes – which looks like a drone-laying queen, though I haven't seen her yet. The multiple eggs in cells isn't typical of a drone-laying queen, however; it is more typical, in my experience, of a young queen who hasn't yet settled into her laying, and if that's the case it tends to sort itself out quite quickly.

Beekeeping is more than manual work or a therapeutic craft. You have to use your brain too. I'm always thinking and questioning. And a lot of beekeeping is problem-solving, which is also half the fun, so I go over my options in this situation:

*Option 1:*

*Assuming I'm dealing with laying workers, shake out all the bees on the ground, some distance from the apiary. ( I have no spare nucs from which to donate many frames this year or a queen.) Laying workers can't fly, so the foragers will return to the apiary, leaving the laying workers and younger bees stranded. Bees returning to the apiary will find their way into the nearest hive. The problem with this option is that I'm not sure I have laying workers, and this strategy results in the loss of a hive that might be fixed in another way.*

*Option 2:*

*Unite the colony with another stock using the newspaper method. The problem hive's brood box is placed on top of a queen-right stock's brood box, with a sheet of newspaper in between. Over twenty-four hours the two colonies will nibble through the paper and unite. The queen-right stock's queen pheromone and brood pheromone will flood the affected stock and suppress the workers' ovaries. Once again, this assumes I have laying workers. If the problem is a drone-laying queen, the uniting method will result in the conflict and battle to the death of two queens. What if the good queen is killed by the drone-layer? Then I still have my problem but I'd now have lost a good queen and endangered a strong stock.*

*Option 3:*

*Start from the conclusion that, whatever the cause, I'm dealing with drone-laying. I first need to rule out laying workers before I can assume that I have a drone-laying queen. This also buys me time to see if the queen is just newly-mated and hasn't yet settled to her laying pattern. The first stage is to donate brood into the problem hive, with nurse bees attached, and overwhelm the bees in the problem stock. Brood pheromone should suppress worker ovaries, if enough brood is donated over a couple of weeks, after which they should eventually raise a queen cell and the problem will be solved.*

I decide on option 3 and donate two frames of brood at various stages of development with adhering bees. I'm careful to make sure I don't transfer the queen from the nuc.

*(Note to myself, to check for queen cells in several days.)*

I have wondered over the winter and recent weeks if there's anything else I could be doing. One consideration was that I might change to small cell foundation, but having researched it I'm unconvinced that it is worth all the effort. I thought about it after reading claims by some beekeepers that small cells have helped their bees develop resistance to varroa, but I have not found anything more than anecdotal evidence for this.

The standard cell size, if you buy wax foundation for the bees to draw out, is 5.4 mm. Small cell foundation is 4.9 mm. In the US it is controversial because of the issue of Africanised bees that have spread in recent years; Africanised bees are smaller than European bees and, as small cells result in regression to a smaller bee, the argument against it is that it makes it harder to distinguish between the two.

In the UK Africanised bees are not the issue, so I gave small cells some thought. Advocates of the system say that smaller bees can land on smaller flowers and can take advantage of a wider variety of forage, which sounds plausible. They say that in humid areas bees on small cells will remove excess moisture from honey, but that's not an issue for me. A more interesting observation is that brood on small cell frames hatches 24 – 48 hours earlier than from standard cells. That has implications, it seems to me, for varroa control, as we know that varroa favour a longer period of development, which is why they prefer to reproduce in drone brood. The claim made by those who use small cells is that varroa struggle to reproduce in these cramped cells.

But does that mean fewer mites in small cell hives? Not according to a number of studies. According to *Thornes'* informative website, scientific studies have been carried out by laboratories in the University of Georgia, the Florida Department of Agriculture and Ruakura Research Center in Hamilton, New Zealand. From these studies there is no clear evidence that small cells work when it comes to varroa. In fact, mite levels were even found to be higher in some cases because small cells give you a higher density of brood per frame.

Small cell beekeepers argue with the studies that have been done, that they have not been done over a long enough period, for a start. They don't suggest small cells are a complete answer to varroa control, but that it helps. They also make the claim that the system helps bees' resistance to other diseases transmitted by the mites because the mites find it harder to feed on the bees' fat tissue in a small cell (it was previously thought that mites fed on the host's hemolymph / blood), and fat tissue is where bees' immune system is located.

There might be something in the claims these beekeepers make, but changing to small cell is a lot of effort and extra expense, and it can take two or three seasons before the bees have regressed uniformly to the small size. When you begin introducing small cell foundation older, larger bees will apparently ignore the smaller cells or will attempt to enlarge them, so you have to wait for the bees to transition to the smaller size and to build up the number of regressed bees.

On balance, it might be that hives with small cells have manageable mite levels despite using small cells and not because of it. Beekeeping skill and management, and the genetics of the hive might all be important additional factors accounting for the survival of bees on small cells. What seems to be important, however, is not how many mites there are, but survival; it's possible to have a high mite count and yet the bees are surviving.

The last point is also why I am not counting mites any more, as I used to in January or February especially. Firstly, there's no point, as I don't intend medicating the bees anyway. Secondly, the mindset this approach can foster is that we have to do things as a matter of routine when we might not actually need to. I wonder how many beekeepers count the mite drop and default to medicating without asking themselves why their bees are surviving and if they might continue surviving? I wonder how many beekeepers calculate a low mite load and think their hive is safe but it crashes anyway? For those beekeepers locked into medicating or who for some reason can't see their way to going treatment-free, a compromise, it seems to me, is to medicate if mite levels seem to be very high, just to be on the safe side, but not to treat out of habit, regardless of the mite load. If there is a heavy mite load, one area to look at is the number of bees you can see that have Deformed Wing Virus (DWV), a virus transmitted by varroa. If there are few, if any, bees with DWV that might suggest they are coping.

Like Mr. Woodley and the old skeppists, the perfect bee for me is the one that best meets my needs and the kind of beekeeping I want to do, and in my case that means I want bees that survive. About the only thing I can do at the moment is to continue watching for swarms to rise this year.

I've realised also why William Woodley didn't restock with Italians after Isle of Wight disease wiped out his apiary in 1915 and 1916, and why he opted instead for the Dutch and French bees of the same *Amm* race as the native Black bee. Firstly, *Ligustica* would not have served him, primarily because she had a reputation for being unreliable at swarming, while swarming is an impulse the old skeppists and Mr. Woodley's business absolutely relied upon. For the skeppist it afforded the new stocks, to over-winter and from which to make next year's production stocks, while for William Woodley the native bee issued the swarms he needed to meet his orders. It was an economic choice, every bit as focused as Brother Adam's commercial aims. Brother Adam dismissed the native bee as of little value commercially because his idea of a perfect bee was based on the production of extracted honey. For William Woodley the Black bee was the perfect bee, at least partly because he sold swarms as well as honey. The second reason he preferred the Black bee is (as also observed by Brother Adam ) that she finished section honey with incomparably fine white cappings compared to the finish made to sections by any other race. We know this was very important to Mr. Woodley because his honey sections won prizes on the show bench.

# MINDING THE BEES

## Chapter 10
### The Blind Old Bee-Man

'GIVING UP BEEKEEPING: few stocks of Goldens,
hives and sundries for sale – I. H. ANDERSON,
128 Castlenau, Barnes. S.W.' (1)

*British Bee Journal 1915*

Perhaps among the wisest words about beekeeping I've ever read are from Tickner Edwardes' book *The Bee Master of Warrilow*, in which the Bee Master advises beekeepers: '...yet he must never lose sight of the main principle, of carrying out the ideas of the bees, not his own. In good beemanship there is only one road to success: you must study to find out what the bees intend to do, and then help them to do it. They call us bee-masters, but bee servants would be much the better name.' (2)

Today we talk much more about working with, rather than against nature, and this essentially is the wisdom of the Bee Master. I completely agree, moreover, with the need to study, which was one of the aims and the great achievement of the British Beekeepers' Association. In defence of their desire to eradicate traditional skep beekeeping, maybe part of their motivation arose from a not unreasonable view that a degree of ignorance might have been synonymous with old fashioned methods of beekeeping. In the last years of the nineteenth century when *foul brood* became a problem, and in the early twentieth century when Isle of Wight disease blighted British beekeepers almost everywhere, an explanation for these problems was necessary before anyone could hope to find a remedy. It was easy to blame skeps for the spread of diseases because they are much harder to inspect, and when modern frame hive beekeepers were generally members of the BBKA who read the *British Bee Journal* and were abreast of the latest knowledge and methods. But if you read through the *British Bee Journal* in the years when *Isle* of Wight disease was at its worst, readers proposed all kinds of ideas, observations and experience for tackling or even curing the condition, such as one reader who experimented successfully with administering salt water to affected colonies. Even those who considered themselves modern beekeepers were clutching at straws and stabbing in the dark. Occasionally someone pointed out the obvious observation that skeps were much more hygienic because old comb is removed with the honey crop and new comb is built every year when a swarm is hived.

Over the centuries it's been close observation and study that has certainly advanced beekeeping because it has advanced our knowledge and understanding of how honeybees work. One giant in the area of scientific research was the eighteenth century Swiss entomologist Francois Huber (1750 - 1851), who has been greatly overlooked (he was praised in the first pages of the very first edition of the *British Bekeeping Journal*). He was not only remarkable for his powers of observation and discoveries about honeybees, but also because he was blind. With his assistant, Francois Burnens, who acted as his eyes, Huber made the most ground-breaking original discoveries, such as that queens mate in the air (not actually proven until the twentieth century); that drones are killed at the end of the summer; that eggs are converted into queens through diet and feeding; that bees use antennae to communicate, and that wax is formed and secreted through an abdominal ring. His observation hives opened up like the pages of a book - an appropriate symbol for the work of study and research that they assisted and the necessity that all beekeepers *read* their bees and study.

Huber is important today not only because we acknowledge him as a founding father of modern bee science, but because there is an equally important lesson he has to teach us about the use of our eyes. Ironically Huber saw more about honeybees, despite his blindness, than many of us are able to see today. Not that modern beekeepers are ignorant; many are very informed and attend lectures and read books, often exchanging information with other beekeepers. But I question whether modern beekeepers are looking in the right direction, and I wonder if we all have blind spots in our beekeeping? Just as beekeepers in the past, such as Huber and Mr. Woodley saw the bee colony as a complex system, not unlike a watch mechanism, beekeepers today do well to see their bees as part of a much bigger, more complex environmental mechanism. Our eyes are opening, through increased awareness of the importance of climate, pollination and modern farming methods, to the intricate ways the ecology of the world fits together like the different parts of a clock.

We know that a small amount of dirt or dust in a watch mechanism can cause a breakdown, and we are aware of such an undeniable breakdown in the environment, with endangered species, extinctions of species and even entire genus groups, habitat loss, and the effects of pollution and intensive monoculture in agriculture, to name but a few. In recent years we have had more awareness and appreciation (helped by Covid) of nature and biodiversity, with serious discussions about re-wilding and about how to bring farming and the environment into closer harmony. Our beekeeping operates in this complex, finely-tuned ecology, of which we are a vital component. The dirt in the mechanism ranges from neonicotinoides used in oilseed rape production, to the importation of foreign queens, and the way we use chemicals in our lives. We are, in many ways (and not always through plain ignorance) still as blind as Huber to the ways our honeybees are adversely affected by these complex and interconnected variables. A century after it happened, we still don't really understand fully what happened with Isle of Wight disease. Nor do we know how drones find DCAs for mating. We don't know the full implications in our hives of outbreeding depression caused by high levels of imported queens every year. We don't fully understand epigenetics either, or the epigenetic factors that accompany transgenics. In many ways we remain as blind as Huber.

One of the reasons I find William Woodley so interesting and important today is because he represents the beginning of modern beekeeping that was made possible by the eyes of Huber's investigations, and yet he also represents a blindness that we share a hundred years after him, which is a humbling reminder that we don't know everything there is to know and that we're not really as fully in control as we want to believe we are. We are not masters of everything, as *Covid* has taught us, and the way to progress is through a new ecology that includes a human ecology based on the humility of service to each other and responsible stewardship of the earth.

This leads me to perhaps my most controversial views in this book, having done a lot of thinking and study. Firstly, no one reading what I have written could doubt that I love the natural world and am endlessly fascinated by the wonders of nature. It will be obvious to anyone that I care about the environment too and that I don't deny the existence of certain environmental issues, but that I don't espouse the apocalyptic narrative propagated by groups such as The World Economic Forum or by our politicians. I hope it will also be evident that I am a thinking person who questions and examines everything. One of the most important aims I have is to get more people discussing how we arrived at where we are in beekeeping and where we might be going. To do that, many assumptions and ideas need to be carefully and critically examined.

For example, we need to define what we actually mean by *progress*. We have to define what we mean by terms such as *master* and *servant*. We need to be aware of the origin of sciences such as *ecology* and the subject of environmentalism and what we actually mean by environmentalism. We need to be aware also that just as there is bad music and good music, bad art as well as good art, there is also bad science that can masquerade as good science. We need to be aware that sometimes the science and the politics are hard to tell apart.

I shall illustrate my points with a very powerful example from history, which I think most people today will agree serves as a reliable reference point from which we must learn the lessons of history, and from which we *re-member* the changes in beekeeping a century ago. My illustrations begins with the Third Reich, which took root in one of the world's most scientifically advanced nations in the 1930s. At the time Nazism took hold, Germany had won about half the Nobel Prizes that had been awarded. Nazism latched onto the late nineteenth century romanticism of the *volkisch* movement, a vision of social change that combined racist thinking with nature mysticism, calling for a return to the land and to the simple natural life. In 1867 the German zoologist Ernst Haeckel coined the term *ecology* and developed the science of the relationship of organisms with their environment. In particular he used Darwin's evolutionary theories to legitimise his unacceptable ideas about nordic racial superiority as scientific fact rather than the personal opinions of a distorted mindset that they were. These ideas directly influenced the development of Nazi ideology, while ecology helped give scientific credibility to the volkisch movement's embedded racism.

Hitler was also influenced by Fritz Lenz, a geneticist, who in 1917 had written an essay *Race as the Principle of Value: On a Renewal of Ethics*. It was in Hitler's library after its publication in 1933. The essay argued in particular that technology had alienated humanity from nature. Lenz admitted that his nasty little essay contained the key features of National Socialism. In 1913 there had also been an outline of many of the themes of the modern environmental movement by the philosopher Ludwig Klages, viewing progress and technology as essentially destructive forces that had destroyed Germany. Environmentalism at its roots then was anti-Capitalist and anti-individual freedom.

This combined hatred of technological progress and Capitalism, viewed as forces that had destroyed Germany, combined in Nazi ideology which then began to identify with nature politics and green policies (but of a very dark-green!). The Nazis, for example, were the first nation to create nature reserves. They explored renewable energy, and had plans to build huge windmills to generate electricity and to create green employment. The Nazis experimented with organic farming (ironically and perversely even in one of the concentration camps). Their policies included land use planning, with a view to addressing food security, as Hitler worried that Germany would run out of food. Hitler also directed the planting of oaks around the Reich. The autobahns were built thoughtfully to avoid disturbing rivers and forests. In 1933 Hitler signed an animal protection law and placed wolves under his protection, although there were no wolves in Germany. Hitler hated hunting and was a vegetarian, as were many Nazis. A transport policy envisioned people using trains instead of cars, and plans for synthetic fuels that would reduce pollution.

I happen to be vegetarian, and I happen to love nature, and I am aware that none of this example should remotely suggest that today's environmentalists are necessarily Nazis; that would be crass, simplistic and offensive to many. The point I am making is that the most technologically advanced, rational and scientific nation adopted (largely uncritically) a particularly pernicious and evil ideology that led to World War and the slaughter of millions who ironically were treated worse than animals in a state in which animals had more rights than a Jewish person. How could this have happened?

It is important that we examine the motivations of politicians and ideologies – even those preaching the reasonable idea of ecological sustainability today. We are told we are led by science and by the facts, but so were the German people by the Nazis' indoctrination of racial superiority. But there is also bad science, especially if it is attached to politics and ideologies. If funding is only given to those whose outcomes suit a political or ideological agenda, for example, the result will be bad science and misinformation.

It is important too that we examine, for example, the influential vision of politicians and ideologues such as those unelected Technocrats who attend the World Economic Forum in Davos; who view themselves as a messianic elite with all the answers to the world's many crises, under the slogan of *saving the planet*; we must ask *who* is included in their vision and, perhaps more importantly, who is *not* included. If we accept everything uncritically because on the surface most of what is presented seems to be for the common good and matches shared values (like saving the planet) the pernicious underbelly of unhealthy ideologies can creep in under the radar. Before we know it the Fourth Industrial Revolution could become the Fourth Reich.

There are ideological themes of the modern environmentalist movement that I don't espouse because they express an unhealthy antipathy towards humanity, such as its overall apocalyptic narrative and dogma of over-population. There are those, like Greta Thunberg, who rant irrationally about our use of the earth's resources and shout at us that modern technological progress and lifestyles are destroying the earth, demanding zero carbon emissions overnight and crazy targets that can only be achieved at the expense of people's lives. I don't subscribe to this simplistic and dangerous view, and I see it as a new green tyranny not far away from the green tyranny of the Nazis. In his book 'Green Tyranny' Rupert Darwall quotes one of Hitler's aides who reported that Hitler had once said he wasn't interested in politics but in changing people's lifestyles. There are some environmentalists today who express the same view.

We are in many ways better off today than decades ago or a century ago because of technological progress and Capitalism. Our lives are much more comfortable and we enjoy better health and education. But we need to travel by car and plane, at least up to a point, so what are we going to do while we still need to use petrol and diesel; we need a healthy birth rate to maintain enough people in the economy to buy the goods we produce and to provide the wealth and the services to look after the elderly of the future; we need fossil fuels, at least at the moment; what are we going to do instead – burn trees?

Progress, technology and Capitalism are not the real problem. It's also a mistake to separate the laudable quest for renewable energy from the world's present global geopolitical situation and the need many nations now perceive to shift the global balance of power as soon as possible after the mistake of finding themselves too reliant on Russian gas or other foreign supplies of fossil fuels. Nations are in a hurry to produce their own renewable energies largely because of the shift in global geopolitics, but they can push forwards the necessary agenda (which is going to hurt in the short term at least) by convincing enough people that the main motivation is to stall an imminent environmental apocalypse.

It is, in my view, a dangerous environmental dogma for other reasons too, to repeat the mantra that we are destroying the world, a dogma that the World Economic Forum and its globalist idealogues use to create a permanent sense of alarm and crisis. Some suggest that this will make it easier for them to impose a global management system that already threatens our fundamental freedoms. I don't know if that's going too far, but the worst aspects of human nature remain the real cause of global problems such as lack of food. We do well to look within ourselves for the answer (not the Technocrats at Davos, many of whom have caused problems such as the economic crisis, to which they now boast the solutions). Neither is it enough to seek structural reform only, as though all the evils in the world are all external to ourselves.

Neither do I subscribe to the green ideology's doctrine that we must bow down in humble submission to nature as its servants. There are those who would have us kneel to Mother Nature in one form or another. From that we also begin to kneel to a globalist elite. My view of this is that Creation exists to serve humanity, not the other way round. That is not a free pass to exploit and over consume, especially if that involves the exploitation of the poor. Lordship doesn't mean *lording it over*, and dominion doesn't mean *domination*, provided there is a good human ecology of stewardship, of care for the world and care for each other. But my human ecology does not start from the Marxist premise that we are servants of nature rather than masters or that our purpose is purely to serve the state.

The Fourth Industrial Revolution has begun, in which some are re-imagining everything; our relationships with each other and the environment; how technology can improve our lives; how to synthesis the biological, digital and technological sciences; how we might biologically augment the human; new political and economic models; how we might all live better, happier lives. It all sounds like a melting pot of shared values towards an agreed common good. But we have to examine the ideologies and the ideologues underpinning this momentum. There are, for example, some who are already talking in terms of present humans being the last generation who are not augmented by robotics, gene-editing and digital technology, and that in the future those like us will be regarded as sub-humans beside the new human species of the brave new world order. Does this ring any alarm bells...?

What has this to do with minding the bees, you might well ask? Who knows at the moment, except that a century and a half's obsession with perfecting the bee for the maximum economic gain might become an extension of the obsession that is already gaining momentum - of perfecting the human being, and the definition of the perfect global management system; and it is conceivable that the same ambition to augment humans might be applied to the struggling honeybee because we want to save the bee. But it could be attached to a new ideology that will propose a solution to the world's problems of food security, pollution and saving the planet. It could even be done in the name of environmentalism, on the back of bad science, preached by ideologues who convince us of facts that might be nothing more than pernicious, post-human, nihilistic opinions. In short, we need to look over our shoulder, read the road ahead and proceed with caution. We do well to open our eyes and examine the evidence of over a century of modern beekeeping, not with the view that technological progress is inevitably damaging, but so that we progress in the right way; not without regard for the environment but wary of an ideology to which it has been attached since the nineteenth century and its underlying political and philosophical assumptions.

Some might argue that the alarm bells I ring about the possibility of a transgenic honeybee are groundless, on the basis that it would be such a stupid idea that surely no one would attempt it. I hope I'm wrong, and that the idea doesn't gain any traction with environmentalists, governments or Big Business. It's important, however, that we sometimes have the courage to set ourselves up to be proved wrong. After all, if I am proved right I will hardly celebrate. The reason why I think it morally right enough to risk looking stupid is that history is full of bad ideas that went badly wrong, and they are often predicated on the best of intentions.

Take the global situation with crop fertilizers at the time of writing this. To satisfy the more extreme environmentalists, government targets and the need to adapt to changing global geopolitics, we and other countries have reduced our use of fossil fuels such as natural gas (or banned mining methods considered 'not green'). We have also set unrealistic targets to lower our emissions by certain dates that are unachievable. Countries also switched to reliance on Russia for natural gas to meet their green targets and to save the planet, which has left them vulnerable and in a hurry to create alternative renewable energy.

Natural gas, however, happens to be a key ingredient in making ammonia, required for global crop fertilisers. Consequently, there is now a serious global shortage of fertiliser needed to produce the crops that can feed the world, with no suitable and sustainable alternative. Food prices are already being driven up while crop sizes will likely become smaller. A bigger population than Marxism killed in the last century could now face the real possibility of serious food shortages caused by diminishing crop production due to a shortage of fertiliser. The problem is much bigger than a grain shortage caused by war in Ukraine, the usual narrative we hear. That might please certain extreme environmentalists or those who predicate their ideology on the dogma of the world's overpopulation as the cause for every problem, as though humanity is a disease or parasite on the earth; but that premise can take us to all kinds of dark places; like who gets to decide that? Who gets to decide who has to go or be sacrificed? The cost of this starting point of saving the planet at any cost is usually people, starting with the poorest who are the first hit by rising prices and the first to die of cold and famine.

We don't therefore necessarily begin a conversation about transgenic bees from the starting point of anything that seems particularly controversial or malevolent, as long as it's couched in terms of environmentalism and saving the planet. The fertiliser problem illustrates, however, that we don't have to propose our good idea for saving the bees and saving the planet within the context of an obviously evil political ideology or experiments on human beings for the idea to be deeply flawed and dangerous. History is littered with similar good ideas that started out with the best intentions of serving the common good. Nazism was one of the cleverest and most dangerous examples of this. Like Hitler, we can begin innocently enough by claiming that we're not interested in politics but in changing people's lifestyles.

Someone will probably at some stage propose that a transgenic bee is a good idea and that it will save the bee, be good for the environment and will safeguard food production. I suspect it will be a biotech company. It will use the environmentalist arguments that its proposal will help save the planet and feed a hungry world. If that still sounds too far-fetched, there is now a serious 1.5 billion dollar startup biotech company with a de-extinction programme to bring back the dodo and the woolly mammoth as early as 2027. That lifts the idea of a GM bee out of the realms of fairytale and into a disturbing vision of reality. The list of arguments about why de-extinction is a great idea include that it will help protect and restore the mammoth's ancient environment and biodiversity, and that it will protect the Asian elephant with which mammoth DNA will be hybridised, none of which are actually provable anyway. But is this the most sensible use of such resources that might better be used to save other species from impending extinction? All for the sake of hybrids that won't be real mammoths or real dodo's anyway?

Quite apart from whether or not the science can at the moment pull it off, there are numerous reasons why de-extinction is a really stupid idea and full of unknown risks, and not that far away from the reasons why releasing GM bees would be just as stupid. What about the moral hazard that if we start thinking we can resurrect extinct species we might not be so careful about their preservation and protection? Would the resultant 'mammoth' actually be the same animal that roamed a few thousand years ago? Would the mammoths created become sick because the earth is so different from the environment they once inhabited, with new diseases to which they would not be immune? Would they also carry viruses in their DNA that might unleash new plagues? What about the environment they require (before you get a pet you need a cage for it, and we would have to create a whole new habitat in which to cage our experiment)? What about the unknown consequences of epigenetics? What about the many unknowns that we might not even be able to imagine?

The goal of saving the planet is arguably not an absolute value if it disregards people either, and if it ends up at the expense of humanity. Good ecology should always begin with a human ecology that every person matters. How much of a biotech's motives are really primarily about saving the planet and how much are they likely to become focused on financial profit?

There would be enormous commercial possibilities from creating a dodo hybrid or a GM honeybee, just as there are from creating fluorescent zebra danios. I suspect that where there is money to be made, that will always drive any other proposed motivation or perceived good that can come from the idea.

Apart from the obvious political and economic dangers associated with privatising a GM honeybee  by Technocrats, governments or Big Business, what other risks might there be? The following list is by no means exhaustive, but it serves to illustrate what a minefield we wander into if the proposal is ever made:

1. GM bees would randomly cross-breed with existing bees (managed and wild) and we have no way of knowing what characteristics (such as aggression or allergic reactions to stings) might result  in subsequent generations from such crosses.

2. Could hybrids between GM and other populations of honeybee become a problem, as Africanised bees have become in the US?

3. If GM bees could be edited to disable their ability to interbreed with other populations of honeybee (eg self-destructing in the second generation), could that strategy misfire and transfer the gene for self-destruction to other populations in the second generation?

4. Would GM bees create new allergic reactions to bee stings?

5. What consequences might there be if GM bees go feral and descend into random out-crossing among themselves and local populations?

6. If GM bees are engineered for resistance to certain pesticides or pollutants, could those toxic stresses trigger unforeseen and undesirable epigenetic changes?

7. How might GM bees affect hobbyist beekeepers? Would they affect our stocks? Would the GM bee be imposed upon the hobbyist, for whatever reason?

8. What unknown epigenetic modifications could occur once GM bees are released into an environment? Could the range of modifications be as varied as the range of conditions / environments into which a single GM type is introduced?

9. Will a GM bee be a 'one-size fits all' type, adaptable to many environments? If so, could we still expect epigenetic changes triggered by different environments?

10. Would a privatised GM bee be patented and become a monopoly, creating a mono-strain to which everyone must sign up?

11. Could GM bees out-compete and pose a threat to other populations (like Africanised bees)?

12. Before out-breeding local populations of bees, will hybrids between them and GM bees lose constitutional vigour, becoming susceptible to new diseases? Could this susceptibility be inherited by locally adapted bees, affecting hobbyists?

13. If GM bees are permitted to cross with other bee populations (including wild bees) could we lose the unique genetics of the different races?

14. In engineering a GM bee, who gets to decide the criteria for building the perfect bee? What characteristics might be selected and discarded?

15. Could hobbyists' bees become more difficult to handle because of GM genes mixing with our bees?

16. Could Big Business weaponise the GM bee to eliminate the competition of what might be regarded as inferior and diseased populations or races of honeybee by vectoring diseases or pathogens to them to which GM bees are immune?

17. Could profit-driven Big Business or a Technocrat use the GM bee to drive hobbyists to extinction unless they accept a mono-strain of GM bees?

18. Would a patented GM bee eventually be presented as the only option for hobbyists as well as commercial beekeepers, in the name of ridding apiculture of diseases and pathogens?

19. What would happen to beekeepers' enjoyment of raising their own queens if they were prohibited from breeding patented GM queens?

20. Would a privatised GM bee eventually become a monopoly which could seduce the hobbyist with the promise of a 'perfect' world in which their bees are bomb-proof against diseases and pathogens while producing bumper crops of honey in all seasons and conditions?

21. Would the rights of hobbyists who prefer non-GM bees be protected?

22. Initially closed populations of GM bees would have limited gene pools. Eventually the gene pool would require new blood, to maintain genetic diversity, but from where? Could we find that a GM bee has out-bred all other non-GM bees or caused their extinction and that we have no genetic bank from which to introduce new blood? Would we find after some decades that we have an increasingly inbred GM bee, with no new material from which to draw new blood, and that the GM bee ends up imploding because of its most well-known achilles heel – inbreeding and lack of genetic diversity?

23. What blind spots do we have with regard to transgenics?

It's beyond the scope and the intention of this book to fully assess the degree to which William Woodley should perhaps have seen the deficiencies of the traditional skep. It's the work of another study to appraise the full extent to which he came to a view towards the end of his career that frame hives and modern methods might have been partly to blame for Isle of Wight disease. In my view, from the evidence of his *Notes By The Way*, Mr. Woodley certainly had his own blind spots. His simple background and loyalty to working class labourers around him were probably factors originally accounting for his opposition to legislation. Perhaps he saw that legislation would have made beekeeping inaccessible to all but those who could afford it when the origins of cottage beekeeping had been an economic necessity for those who couldn't easily make ends meet. As well as growing up in a hamlet with his great aunt, alongside rural Berkshire cottagers, his early career as a Chieveley grocer would no doubt have acquainted him with ordinary people who struggled to put food on the table.

Mr. Woodley, however, represents for me much of what is best in human nature. In a modern society in which our management system selects for conformity to its own vision of a post-human (or transhuman) future, Mr. Woodley represents the vestiges of the pluralistic society we once were. Today you're only really free to agree. Independence of mind such as Mr. Woodley's is increasingly cancelled and dismissed while the views of ordinary people today are generally ignored in political debate. In contrast, it's refreshing to read the debate and differing views (even heated at times) that passed through the *British Bee Journal* for decades, even allowing someone like William Woodley to become isolated rather than merely cancelled outright for his views. The debate that occurred in the old pages of that journal over Bee Disease legislation, the skep and modern methods of beekeeping, have come to symbolise for me all that was best about this nation and all that remains best about human nature. There was genuine freedom of speech and freedom of ideas back then, grounded in real and meaningful human relations. Today these are the mere ghosts of an age that was built upon genuine merit, principle and goodness.

William Woodley, despite charges that he might have had his own blind spots, or mixed motives, looked around him and saw a moral issue about real people, about whom he cared deeply, in real communities like Beedon parish, who were being ignored, dismissed and left out by progress. He saw it because he lived in an age before today's moral inversions, of real and meaningful relationships and the strong human bonds of kinship and friendship that created the kind of human interactions that are ceasing to exist today, in social media and the alienating environment of modern life. That's why I believe that the antidote to the nihilistic void that all the proposed solutions of the World Economic Forum can't ever fill exists in the kind of close communities in which William Woodley was deeply rooted: the family, neighbours, local farms, beekeeping association and town allotments, faith communities, to name a few. It's in these helpful communities that we find and build the strong and meaningful relationships and human interactions that are the antidote to the globalist vision of transhuman utopian equality in which the individual is actually emptied of everything that is good and meaningful in human life.

It's also on the local level that we can take control ourselves and begin to solve the issues in our beekeeping, by working together, before a technocratic, globalist elite or an agrochemical giant (or both) propose their own dark-green solutions to our problems, just as they are doing for every other crisis. Those solutions will not end well because their view of human beings is empty of everything that carries meaning. This is why I am also passionate that more beekeeping associations work together, helping their neighbours and fellow beekeepers to develop locally adapted bees, to raise their own high quality queens and to work with nature to solve the issues that bees and beekeeping are facing. Arguably, the worst crisis we all face is the eventual interference, in some form or another, of those who think they can save the planet at the expense of our basic freedoms. You can be sure that somewhere down the line someone will have a bright idea for saving the bee, if we leave it to them, and it won't end well. They promise stability and security against the permanent alarm of imminent environmental catastrophe, but there will be winners and losers with their solutions, and there will be many left out, just like the old skeppists in the First Industrial Revolution.

I have to admit, there's one quotation, from an article 'Prominant Bee-keepers', in the *British Bee Journal* in November 1909 about William Woodley that has caused me to pause and reflect on this issue of what he saw or chose not to see: '...assisted the chief bee-man of the place – who like the great Huber, was blind – in recovering swarms from tall trees, the boy mounting the trees and being ''shown'' how to manage by the directions called out from below by the blind old bee-keeper.' (3)

How could the irony of such an introduction to beekeeping be lost on anyone who knows how differently he saw the issue of skeps and Bee Disease Prevention legislation compared to a great many other beekeepers? The implication of this detail of his biography is that the blind bee-man, like the great Huber, had something important to teach the young Woodley. As beekeepers who never know everything about the craft, perhaps it's good for us to remember that we always have something more to learn. The lesson of Huber and his assistant, or young Woodley and the old blind bee-man is, surely, that beekeepers can and should learn from science, from each other and that we should listen to the wisdom of those who might see things differently or who have noticed something others, or we, have not noticed.

Is it fair to suggest outright that Mr. Woodley might have been wrong to resist *Bee Disease Bills* that were proposed throughout the time he was writing for the *British Bee Journal?* Certainly, from our position today, from which we accept the need for at least some legislation to protect bees and beekeeping, it's easy for us to criticise Mr. Woodley for apparently not seeing the obvious and the inevitable. It's easy for us, as it was in his time, to accuse him of protecting his own interests as a businessman who didn't want government interference regulating or interfering with his livelihood. In fairness, however, his resistance from 1905 to the *Bee Disease Bill*, and to other Bills that followed, that led later to criticism of him for causing the unregulated worsening problem of Isle of Wight disease, would not have helped much with that specific disease. Legislation was aimed solely at foul brood. It was clear after the *Irish Bill* was passed that legislation only dealt with foul brood (which was curable) and that it would probably not have been effective against the mysterious Isle of Wight disease. In fact, the government was probably more to blame for not intervening and for failing to address the problem until 1919 when bee stocks were so seriously diminished that food security had reached a crisis point.

What Woodley could see perfectly well was that legislation would have put an end to the '... now despised straw skep'. (4) That mattered to him, for all sorts of reasons, I dare say, apart from the obvious and repeated defence of the cottager, '...swarms have issued. This is the chief source of income for the cottager who still continues on in the old style of beekeeping.' (5)

It's interesting to note that by 1913 his fewer published arguments against legislation were no longer defending country neighbours, because there weren't many country folk near him keeping bees any more; probably largely because Isle of Wight disease had destroyed their bees while technological progress had emptied the bee gardens of their tenants! 'Now the countryside is bereft of bees, the modern farmer...does not trouble to keep them, the cottager, ''like master, like man'', follows suit, and has practically given up beekeeping...we have one agricultural labourer only who keeps bees in Beedon, and it is much the same in the surrounding parishes.' (6)

Human beings are complex though. Our motives are often mixed, and I suspect Woodley's concern for his neighbours and their poor wages was coloured by memories of *mindin' the bees* for his great aunt, and catching swarms under the guidance of the old blind bee-man. Perhaps this is why, in some of his *Notes By The Way* before 1900 he described how he still made use of a number of straw skeps in his apiary, despite having adopted the modern frame hive. Why was this, I wonder? It seems obvious to me that he retained a deep affection for the skep as well as for the country folk among whom he was raised with their traditional methods of beekeeping and their rustic ways.

No one could accuse Mr. Woodley, of all people, of resisting progress. Not only had he adopted the frame hive on an industrial scale, but there is evidence that he was perfectly well-disposed towards the Beekeeping Association's experts, and welcomed their visits and advice: 'When one of our experts calls, depend on it, we talk bees!...Mr. Flood told me of a case of foul brood breaking out in a village some miles from here and of the application of naphthaline in the hives...' (7) He had even been invited to become an expert, which he had refused: '...for your humble servant (the writer) has received more than one invitation t swell the ranks, but modesty for one reason and excess of work already act as restraining forces... Then there is another reason...I have no parchment, no degree!' (8)

In 1912 the draft *Bee Diseases Bill* was published, and from that point we can detect a marked change in William Woodley's correspondence in the *British Bee Journal*. His *Notes By The Way* departed increasingly through the year from his usual friendly advice and assessment of the season, becoming more political and outspoken in his opposition to the Bill. In one edition he admitted that even his friend, Mr. Heap of Reading, was appalled by his opposition to legislation. In his responses to many letters and the draft Bill, Mr. Woodley pointed out again and again the futility of *Foul Brood Acts* in other countries and how similarly useless legislation would be against the Isle of Wight disease. He cited a report in May from the County Associations - that foul brood appeared to be burning itself out, while he wasted no time suggesting that it had done so without the help of any legislation. He criticised the Bill as unfair because it would not allow notification from beekeepers or compensation for action taken by inspectors. Again he cited the futility of foul brood legislation in other countries. He drew comparisons between the careful way inspectors of other livestock diseases operated and the way inspectors would be allowed to move from dead and diseased bees to another apiary, spreading the disease as they went in a way that would not happen in any other area of agriculture. He argued that lessons should have been learnt from the Irish legislation which seemed fairer in its treatment of beekeepers.

In May 1912 came what seems to have been the beginning of the gradual end of his years as a correspondent for the *British Bee Journal* as his views became increasingly outlawed and out of step with the views of the editors and many other readers of the journal. His title to *Notes By The Way ( A Closing Note on the Bee Disease Bill)* suggests an increasing friction with the editors of the publication with whom he did not see eye to eye over legislation. I wonder if he had been instructed that this should be his closing note on the subject?

He insisted repeatedly that the Draft Bill should include notification by beekeepers rather than responses to infection being left to local authority bylaws. Tellingly, the piece was followed by the editors' footnote, in which they contradicted William Woodley's opinions and expressed their belief that the correspondent's concerns were groundless. As the editors' footnotes were always deemed a final word on the matter, this seems to have been a clear attempt to silence Mr. Woodley once and for all and to shut down further support for his stance. This seems to have finished Mr. Woodley's regular correspondence, as his *Notes By The Way* continued through that year's editions, becoming less frequent in 1913. In May 1913 he had a final rant about the 'despised straw skep', (9) the 'spread of bee diseases by misapplied modern methods...' , (10) the cost of frame hives and equipment which labourers could not afford, and the demise of beekeeping in Beedon and the local parishes. Mr. Woodley, 'scourged' (11) by the beekeeping world, seems by then to have been beyond redemption, although, in the spirit of free speech, he did return briefly to voice his opinions and to defend himself some years later. You were free to disagree back then, up to a point. The editors today would surely have simply cancelled him outright at a much earlier stage, and he would have been destroyed – not by a bee disease, but by those with whom he had the independence of mind to disagree.

No one ever began beekeeping more ignorant or more blind than I was when I started! I could have done with the boy Woodley's blind old bee-man to guide me, and the wisdom of Huber as I was first inspecting my hives. I had no idea what I was looking at; the difference between cells of pollen and larvae, of capped worker and drone brood. And I think I had tunnel vision for at least the first few years from an obsession with preventing swarming and being seen to be successful solely by the criterion of taking a decent honey crop.

My admiration for William Woodley is that he examined and criticised the beekeeping dogmas of his time and challenged its ideologues whose project was the ruthless pursuit of their view of technological progress at the expense of the cottager who was ultimately left out of their vision. The new aims of modern beekeeping had intended the extinction of the skep from the start, and foul brood and Isle of Wight disease probably helped justify their arguments. William Woodley was courageous enough to speak out, to generate a debate and to be isolated by his principles, but he always had one eye over his shoulder for those who were being left out, while being unafraid of technological progress. We need people like that in every age.

It was the first swarm I took that began to open my eyes. It was not so much a 'falling from my horse' moment than a slow conversion; but, like the apostle, I remained blind for some time afterwards, though there was a slow dropping of the scales from my eyes. That was the turning point in my beekeeping and it sent me in a direction that has turned out to be the complete opposite of the course taken by Brother Adam, a brother monk in a sister monastery of the same English Benedictine Congregation. While I still have enormous interest and admiration for his astonishing commitment, knowledge and achievement in producing the world-famous Buckfast bee, I have become increasingly critical of that project, to the point of completely rejecting the Buckfast strain as a bee that I would want to keep, given my understanding of the situation of bees and beekeeping in the twenty-first century.

The swarm came from an air brick leading to the roof cavity of Kelly's Folly, the building where I once had an office and where the sacristy is located, with some guest rooms on the top floor. Built in 1923 by Abbot Edmund Kelly, it was intended as our community Chapter House, though it never fulfilled its purpose and became something of a folly. A few years ago the roof needed replacing and the workmen found in the compartments of the roof cavity the remains of at least a dozen honeybee nests. At different stages, for some decades, colonies had been living in at least one or two nests in the roof cavity at any one time.  There was another colony above the parish priest's office in another building nearby, and when the roof was repaired every colony was destroyed and the air bricks all sealed over. I was unhappy with the decision at the time, but was powerless to stop it happening, and so the wild bees in our buildings all came to an abrupt end.

*Kelly's Folly*

My own swarm that issued a year before this from Kelly Folly were very dark bees, not like the Italian-looking Buckfasts with which I started. Those had come from another monk at another monastery. The swarm from our buildings was quite defensive, swarmed the following year and was eventually requeened with eggs from a gentler hive. It cured the temperament problem, but ended the genetic pool of those bees. I realised too late that I had lost what were probably locally adapted bees that might also have been varroa tolerant. What might have contributed to their survival was a gradual reselection of the genetic traits of the native Black bee latent in wild stocks, in the way described by Tickner Edwardes.

Why would this swarm capture prove to be such an important experience? Firstly, I began to miss the excitement of seeing swarms issue from Kelly's Folly every summer. I also started to think that the Abbey had lost something special and that Kelly's Folly even looked rather sad and abandoned without its bees. It became clear, after speaking with older members of the community, that bees had been living, almost uninterrupted, in the building at least as far back as the 1980s, but probably before that. I was informed that our beekeeper at the time, Fr. Robert Biddulph (Biddy) used to collect swarms every year from the building. What interested me was that the bees from Kelly's Folly were surviving, and had survived through the 1990s and the 2000s, without beekeeper intervention after varroa had become a problem in Britain. Why, I wondered, had the bees in Kelly's Folly survived throughout those decades, before there were commercial treatments against varroa? Why had the bees in Kelly's Folly not died out?

The other point of interest was how dark those bees were. By then I had discovered the story of the native Black bee and the controversy over whether or not it had survived Isle of Wight disease after World War I. I knew it was unlikely that the wild bees in our buildings would be pure native bees, but I thought it was possible that they could have a high percentage of native bee genetics simply through natural selection of the traits needed for survival in Britain – and those traits are the *Apis Mellifera mellifera (Amm)* which is the native Black bee.

The bees in Kelly's Folly had captured my imagination, almost to the point of obsession because they seemed to be bees that might hold the answer to the problem of varroa. If that was true, these were bees that didn't need chemical treatments to keep them alive – treatments that make varroa stronger and that don't solve the problem. On my own doorstep I had discovered the possibility of locally adapted bees that were surviving the mess made by modern beekeeping. When I read that a similar genetic adaptation had happened to native bees in the Arnot forest in New York State in the years after the arrival of varroa, it strengthened my intuition that the bees in Kelly's Folly might well have adapted to varroa in the same way.

From that gradual awakening I stopped being interested in beekeeping purely for honey production. My eyes had opened to the fascinating world of bees not merely as a factory for honey, but as a complex, finely-balanced mechanism, beautiful and delicate as the workings of a watch. Once I understood this analogy and that honeybees can be upset by the smallest of disruptions, just as a watch mechanism can be upset by dust and dirt, the foundations of my early beekeeping collapsed.

Did Mr. Woodley's ideas change about the frame hive too? Did he begin to wonder if, for all the progress of modern beekeeping, perhaps it was more to blame for foul brood and Isle of Wight disease than most others could see? I wonder if his concern for the cottager's livelihood that depended on skep hives was only one reason why Woodley opposed *Bee Disease Bills*? Did he perhaps come to believe, ironically, having been at one time Britain's biggest bee farmer, that the answer to the problem of bee diseases was to return to the old ways of the skep? It's easy to see his views on proposed bee legislation as short-sighted, but there were not many writing in the *British Bee Journal*, or who read it, who saw any merit in skeps. Occasionally, after World War I someone, such as W.T.D. 20 February 1919 argued in favour of their use, having inherited five skep stocks, 'I took them over five years ago, and changed over to the frame hive...but since changing over I have had to fight with disease.' (12)

After Isle of Wight disease and the end of the Great War, however, Woodley's views were clearly out of step not only with most beekeepers, but with the editors of the *British Bee Journal*, as the following *Editorial* makes clear: 'One thing yet is lacking, ie legislation to eliminate, as far as possible, the menace of disease being spread by the ignorance, supineness, or cupidity of a few people...the Government are evidently convinced that it is necessary to protect those who are prepared to carry out beekeeping on orthodox, and up-to-date lines.' (13)

Mr. Woodley was now regarded as a heretic in the world of apiculture whose new orthodoxy preached evangelically for the conversion of all beekeepers to the frame hive. Skeps and their advocates, together with the rustic, simple cottagers who had used them for centuries, were now classed as ignorant and even immoral. If Mr. Woodley was neither if these, he was at least regarded as greedy and protective of his own economic interests.

From the end of the Great War there were many articles and letters in the *British Bee Journal* about Isle of Wight disease and the need for legislation. In fact, editions particularly from 1919-21 were dominated by these issues. Having been hit twice by Isle of Wight disease and having restocked with Dutch bees, William Woodley started up again. There are adverts for his Dutch bees for sale in the *British Bee Journal* from this time. He was by then only a few years from the end of his life and already in his 70s, so it's reasonable to assume that he could have simply decided to settle into retirement. Maybe it was time for him to stop commercial beekeeping anyway, though it's hard to see how he could have continued writing for the Journal. It's ironic, however, how Mr. Woodley, having embraced the industrialisation of beekeeping with modern methods and equipment, was suddenly out of date and out in the cold. I get a sense of a sinking enterprise at his Stanmore and Worlds End apiaries, perhaps no more than a hobbyist-sized operation towards the end of his life.

Perhaps he sat at the kitchen table at that time, head in hands, lamenting the course of his misfortune in becoming what he called the scourge of his craft. I like to imagine his good wife, practical and matter-of-fact, just as she was about the weather, telling him there's nothing new under the sun and that the times remain the same. Or did she console him that history would be kinder to him than his own times had been?

# MINDING THE BEES

## Chapter 11
### The Perfect Bee

'ITALIAN QUEENS DIRECT FROM ITALY-
Address, E. PENNA,
Bologna, Italy.' (1)

*British Bee Journal 1910*

As May begins there's a gradual sense of the progress of spring. There's an inner radiance in the month's new greens that is the fresh vigour of new life. I stand under the apple trees near the library, their red blossom, opening to pink and white, unreal as cake decorations. Sitting outside the Abbey church before Matins, a distant, but distinct *cuck-oo...cuck-oo....* The cuckoo call is now one of the most exquisite luxuries of May – to be savoured. Like all luxuries, it leaves you desperately hungry for more. I savour the calls, like tasting freshly-cut asparagus, new potatoes or a sweet strawberry warmed by a hot June day. High in the trees wood pigeons, salmon-pink with dawn light, clatter through green stipple.

May is the volatile teenager of the year; full of the bright vigour and freshness of early summer, but moody, with days that begin under cobalt sky and often decline into lambent green and the solemn anticipation of rain. Down in the apiary there are signs of hope, however, with young bracken shoots suddenly appearing through the old, rusted leaves of last year's growth – one metre tall croziers that by the end of the month will have formed a high hedge, hiding the apiary from sight.

In the monastery's guest garden, Allium *(Purple Sensation)* flowers are open, heads green and red like giant versions of those traffic-light lollipops I sucked in my childhood. As their pompoms expand, the flowers turn purple. The first crimson pea flowers of Common vetch *(Vicia sativa)* appear in the lengthening meadow grass, and there are swathes of blue Germander speedwell *(Veronica chamaedrys)* also known as Cats eyes and Bird's eyes speedwell. In Ireland it was once sewn into clothes as a charm to protect against accidents. Along the country lanes Cow Parsley *(Anthriscus sylvestris)* is starting to flower, with laced, creamy-white umbels. Queen Anne's lace, wild chervil or keck, this iconic flower of May's verges is a relative of the carrot.

For all the progress in modern beekeeping techniques, I wonder what progress we have actually made. It seems paradoxical to me that the more progress we seem to make in scientific knowledge, the more this makes me question the apparent progress of the craft. Is it progress that we have produced bee strains and procedures to maximise our honey crop at the expense of so much else? Varroa has weakened our bees, imported stock has compromised our local stocks at a genetic level at the very least, while we are also in real danger of losing the distinct genotypes that for tens of thousands of years have been genetically adapted in the most finely tuned ways to their particular regions. For centuries uneducated cottagers kept their bees in primitive skeps and they never had the problems that seemed to begin in the late nineteenth century.

Is it really progress that we have industrialised the honeybee to the point that we have compromised its ability to function at its best in the wild, so that we now have stocks that are neither wild nor really domesticated? And if the honeybee can never really be domesticated (which is my view) then it can't be truly feral either – feral meaning a domesticated species that has returned to the wild. Returning many of our existing stocks to the wild is like letting loose an aviary of budgerigars – they'd survive for a while, but they'd never completely adapt to the wild in Britain. How many swarms of Italian bees, for example, escape into the wild in Britain and are never caught by a BBKA member and consequently die off in their first winter, not least of all because they have become accustomed to chemical treatment against *varroa*; not to mention the real danger that the larger size of their winter cluster puts them in danger of starvation from lack of adequate winter stores? I suspect there are not a few such cases, especially if one in fourteen British hives is currently queened by a foreign queen, of which *Italians* are probably among the most popular choices?

That tension between a more natural form of beekeeping and the progress of the craft that filled the editions of the *British Bee Journal* between the end of the nineteenth and the beginning of the early twentieth centuries is still with us. So is the question of what really constitutes progress in beekeeping. Tickner Edwardes' introduction to *The Bee Master of Warrilow* tends to view progress as an inevitable force for good, and its opposite in terms of something merely picturesque and poetic. Although he admitted that modern hives are no more beautiful than a, '...Brighton bathing-machine' (2) and that the old methods are 'quaint', (3) the 'reverence' (4) he advocated towards traditional old fashioned beekeeping does come across as rather condescending. That view is as simplistic as the ideas of the old skeppists he criticised. But the, '...good-humoured contempt' (5) he seemed to express towards old-fashioned beekeeping was also unashamedly brutal in its dismissive assessment of, '...the old methods, which – to say the truth – were often as senseless as they were futile.' (6)

Like so many who openly preached for the end of the skep, Tickner Edwardes regarded the old-time skeppists as ignorant and superstitious hangovers of, '...the astounding delusions of Medieval times.' (7) Tickner Edwardes, who seemed to lament the, '...passing away' (8) of the old bee-gardens and the, '...rapidly diminishing class' (9) of the skeppists, actually regarded them as those who,'...obstinately shut his eyes to all that is good and true in modern bee Science.' (10) It's hard to read his so-called reverence for the old methods as more than a sentimental weakness to which he reluctantly admitted when in the next sentence he wrote, 'The advantages of modern methods are too overwhelmingly apparent' (11) and 'The old school must choose between the adoption of latter-day systems or suffer the only alternative – that of total extinction at no very distant date.' (12)

This was in 1907 when *The Bee Master of Warrilow* was first published, and the threat of foul brood had just been overtaken by the first wave of Isle of Wight disease, which had yet to peak towards the middle of World War I. If this was the voice of the modern beekeeper before Isle of Wight had done its worst, it's understandable why Mr. Woodley and his supporters became so unpopular by 1912 when British bee stocks were diminishing to the point that by 1919 there was a national crisis.

As late as 1914 even the *British Bee Journal* seemed to express this lingering love-affair with the old ways, despite the editors' clearly stated support for legislation and modern methods and Mr. Woodley's departure as a regular journal correspondent. We find an article in the journal on Mr. Martin, a shoemaker of the old kind, and his old-fashioned quaint Cotswold apiary, with a photograph of him standing behind several hives of various designs, ranging from skeps and a modern frame hive, with hybrid hives in between them: 'Starting in the spring with four skeps and the frame hive, he has had five swarms.' (13) He was a cottager of the old kind too, by then a dying breed, 'Being a shoemaker of the sort that "make to last," most of his time is spent in a workshop adjoining the cottage which is to the right of the picture.' (14)

MR. A. MARTIN'S APIARY, PRESTBURY, NR. CHELTENHAM.

*Mr. Martin and hives BBJ 1914*

I wonder if the editors simply saw this article and photograph as a convenient piece of propaganda for the inevitable triumph of modern methods, as Mr. Martin represented exactly the kind of old cottager that many accused of holding back the progress of the craft. Here was an example of someone in the same rustic tradition as Owen Thomas a century earlier, but Mr. Martin was clearly making the gradual transition to the frame hive. Even the photograph seems to reinforce this view, the various old-fashioned, Heath-Robinson designs moving from left to right in the picture towards a modern white frame hive. The desired evolution of beekeeping is as clear from this photo as its allusion to the development of crouched apes into upright *Homo sapiens*.

Today we probably have a more nuanced view of scientific progress, understanding that science and progress are not always synonymous, and that scientific progress can, in some cases, represent anything but progress. Our view of nature and what is natural have also changed so that we regard them more today as values we have a moral duty to preserve and protect for their own sake as well as for our benefit. This, together with an awareness that for too long humans have taken advantage of the planet more than we have been good stewards of it, has made us more sceptical about the idea of technological progress previous generations took for granted as something always intrinsically good. There are countless examples of this, from the accumulation of plastics in the oceans, to the effects of non-organic farming on the environment. Less discussed is our reliance on metals,minerals and opals for our smart phones and tablets (to name two technologies) which are mined in hazardous conditions by some of the poorest and most exploited people in the world.

Recently I read of a study published in the US that has found radioactivity in American honey dating back to Cold War Weapons tests in the 1950s. Radiocesium, a fission product of nuclear blasts, has remained in soil where it has been taken up by plants in the same way they take up potassium. It is now present in honey in levels one hundred times higher than in other foods, helped by the way bees make honey, which uses concentrated nectar, thereby also concentrating the radiocesium from foraged plants. They also found (no less alarming) antibiotics in honey. This simply must have some implications for bee health as well as our own, and for issues such as Colony Collapse Disorder. And would Tickner Edwardes argue today, I wonder, that nuclear weapons and fission have always represented progress because they are based on scientific advances?

Similarly, what would he make of the, '...latest improvements in apiculture' (15); chemical controls for varroa, instrumental insemination of queens, and the importation of queens that are nowhere near adapted to our climate, to mention just a few issues? What would Tickner Edwardes' assessment be of beekeepers like me who would welcome an opportunity to begin again by developing a naturally selected bee that is adapted to our conditions? No doubt many beekeepers today would regard my views as retrograde and perhaps fanciful. But I ask the question: in what direction now do we progress in beekeeping in a way that

Mr. Edwardes might recognise in our own time as both good and true? Unfortunately, whereas he regarded the old school as doomed to extinction, I regard the school of modern beekeeping as already living on borrowed time, with ways that are just as senseless and futile in the long run as the old methods sometimes were. Skeppists did not experience a build-up of chemicals and pests and diseases in old comb because their system relied on the building of new comb every year, and the native Black bee was especially good at building that new comb, even in poor seasons. The old beekeepers relied on swarming too, which creates brood breaks and smaller colonies, both of which are advantages today when managing varroa mites. Old fashioned beekeepers did not artificially keep weak or diseased stocks alive either; their system naturally selected strong stocks because it relied on swarming, and only strong stocks swarm. To learn from these old ways might be some progress today, and it doesn't mean we all have to return to skeps to do it. It does mean, however, that we might do better by managing frame hives with this old bee knowledge in mind. There is a place for the best of the old wisdom alongside modern technology and science, rather than resorting to the darker shade of green that some environmentalist espouse.

Sometimes perhaps we have to go backwards in order to move forwards. Our Douai beekeeper for years – Fr. Robert Biddulph (Biddy) – went backwards in his beekeeping before his death in 1995, as far as I can surmise from his records. Those records have enabled me to piece together some significant stages in his beekeeping which are interesting because they show a development in him as a beekeeper. When he arrived at Douai Abbey in the 1950s he was already a beekeeper and was keeping Italians, of the Ligurian type favoured by Brother Adam. I know from his records that he was experimenting in the late 1950s with instrumental insemination, so he was trying to keep his Ligurians genetically pure, as far as I can see. When he went up to the Lake District for parish work in the 1960s those bees went with him. He'd have found Italians quite unsuitable, however, in the cold, wet north of the Lake District winters, and I doubt they lasted very long. By the time he returned to the Abbey in the 1970s he was probably keeping local bees collected from swarms, as those who remember him say that from the 1980s he was catching swarms issuing from Kelly's Folly. These bees are recalled by the monks who were here at the time as swarmy and defensive – certainly not Italians.

Biddy also wrote some rules from his vast experience of fifty years as a beekeeper: his Number One Rule was never to take swarms if they are not your own bees, as they carry the risk of disease. But at the end of his life he certainly was taking swarms – every year, from our buildings. I assume he used them to boost the honey crop in the apiary, by uniting them with his production hives. Why, however, would an experienced beekeeper break such an important rule?

*Fr. Robert Biddulph*

I think it was because the bees in Kelly's Folly must have been free of disease to have survived year after year without beekeeper intervention. I know they were there continuously, at least throughout the 80s and 90s, until a few years ago when the roof of Kelly's Folly was replaced and the bees at that time were destroyed; but it's not a stretch of the imagination to suggest that they might have occupied the building during the whole of the 1970s as well, or even as far back as the 1950s – or earlier, with occasional vacancies!

Biddy is an example of a beekeeper who went backwards, as far as I can tell, from the forefront of scientific progress and new methods in apiculture, to working with locally adapted stock. It's my own view (though I haven't found firm evidence for it) that Biddy probably had advice originally from Brother Adam of Buckfast, as the Ligurian was one of the foundational races of the Buckfast strain, and Brother Adam might well have taught Biddy about instrumental insemination. If Biddy was keen to produce a Douai version of the Buckfast strain, it seems likely that he would have consulted Brother Adam. Yet, by the end of his life, Biddy had clearly adapted to a different way of beekeeping.

Biddy had to adapt and change his methods for more than one reason; moving up to Cumbria as a parish priest would have put an end to his project with Ligurians with the different climate and the smaller space available compared with life at the Abbey. No doubt as a parish priest he also had less time than he'd had as a monk at the Abbey. But I wonder if the change in his circumstances was also because of the advent of varroa in the last years of his life back at Douai Abbey. Perhaps he recognised that the bees in Kelly's Folly might have been less at risk from the emerging blight of varroa for which there was at that time no treatment? From docile, high-yielding, imported Italian bees to lower production, defensive local stocks, he had, to many beekeepers, gone decidedly backwards after fifty years of beekeeping. In fact, I believe he was experienced enough and meticulous enough to know exactly what he was doing. In my opinion, he might even have been ahead of the times, already doing in the 1990s what beekeepers like me ought to be doing now, a quarter of a century after his death, by advocating locally adapted bees.

In 1913, as Isle of Wight disease ravaged Britain, there were beekeepers writing in the *British Bee Journal* who were also going backwards in order to make progress. One example was H.H. Brook of Altrincham who wrote, 'I have experimented with foreign races, only after trials extending over some years, to decide to clear them out entirely and keep to "blacks".' (16) His objection to *Ligustica* was that they were unsuitable for section work, not finishing their cappings to the high standard of the Black bee, while *Carnica*, when crossed, produced the worst-tempered bees he had ever experienced. In contrast, he had a very different experience of Black bees: 'I have blacks which winter well, build up into strong stocks and in good time, gather nectar if it is to be had, cap and finish their honey beautifully, and are certainly much more pleasant to handle than most crosses...' (17) L.B.W. of Somerset was similarly experienced with Italians and various hybrids and also decided in 1913 that,'...a good strain of the native bee is far superior to any foreigner.' (18)

What is even more ridiculous about today's situation is that even European Dark bees and the Buckfast strain can be bought as imported stock. Buckfasts come in from Romania or Germany, Dark bees from Greece, with Carniolans bred in Slovenia; all imported queens, some of which could easily be bred in Britain, if these bees are what British beekeepers want. But even the idea today of a racially pure genotype is controversial, and there are beekeepers who question the sense of concentrating on producing a race such as the native Black bee. The criticism sometimes made (and I am coming to this view myself) is based on an understanding of the complicated genetics of honeybees and how they reproduce. In essence, bees have developed polyandry and haploid males as a strategy for maintaining the genetic diversity needed in a colony. Without the genetic diversity gained by queens mating with at least a dozen or more drones from different colonies a stock won't have the full complement of traits needed for survival. The progression of worker jobs is not fixed along an inflexible time-line, but is triggered by a certain plasticity in the superorganism of the colony, enabling some workers to jump to new tasks or even to return to a former task if the colony requires it. It is genetic diversity that gives a colony this plasticity to adapt to the need for different tasks. In turn, genetic diversity requires polyandry, and polyandry needs large, well-populated Drone Congregation Areas, with drones from many different hives. Polyandry ensures genetic diversity because the resultant family in a hive contains many subfamilies who all have different skills and resiliences which make the stock more adaptable for survival.

Going backwards in beekeeping, however, isn't an absolute value either. A case in point here is the native Black bee. There are beekeepers dedicated to breeding the native bee with a view to bringing them back, but even this is controversial and might not be progress in the right direction. To breed native bees (as with any pure strain or race of bees) you have to rely on isolated apiaries where the bees can mate naturally without risk of interbreeding with different races, or you can artificially inseminate queens. These two options, however, are not without their own set of problems: isolated apiaries might not contain sufficient drones to ensure genetic diversity, even if they are quite large apiaries, especially as native bees can easily fall back on mating in the vicinity of their own apiary.

Instrumental insemination is just as much a problem because it relies on the breeder's choice of drones, which will always be random and no substitute for drones selected by natural competition in Drone Congregation Areas where only the fittest and strongest drones are selected to pass on their genes – something a bee breeder can't judge in the same way bees can. Even if a breeder knows the genetic pool of a selected drone, there can be no way of knowing how such a drone might perform in the highly competitive situation of a DCA. Moreover, bee breeding for the selection of certain characteristics isn't as easy as it sounds, and there are other variables involved in the bees' performance, such as the skills of the beekeeper and the environment in which the bees are kept. Heavy selection for a particular trait can also result in the loss of other desirable traits along the way. For example, it has been found that heavy selection for hygienic behaviour can result in bees that are unable to process toxins.

I'm not a scientist, and these opinions can be debated. Bee breeders who are passionate about native bees argue that we would do well to return to a bee that is genetically adapted to Britain, and that does sound like a laudable objective. But is it realistic? The native bee was adapted to a Britain that for centuries was not much changed until a hundred or so years ago. With the loss of habitat, such as most of our wild flower meadows, the increased size of towns and cities, and the rise of intensive agriculture, modern Britain is very different from Britain a century ago when William Woodley was a beekeeper. Add pollution and the increased use of insecticides, not to mention bee diseases and pests, and it's questionable whether or not the traits selected by native bees up to a hundred years ago would still make them the best-adapted bee for the UK today. Make no mistake, I love the idea of restoring our native bee to its proper environment, but how much dogma is spoken about this, based on something that sounds like a good idea – but it could be based on little more than a poetic and picturesque fantasy. The attempt to return to a racially pure native Black bee could prove to be just as damaging to native bees in the long run and as fanciful as the de-extinction of the dodo.

It's a sobering thought that right up until the middle of the nineteenth century the different genotypes of honeybee in Europe and other parts of the world were racially pure, and as such were intricately tuned to local conditions. It was only in the second half of the nineteenth century that beekeepers in Britain began to experiment with Italian and Carniolan bees – in the same way that Victorians developed a taste for exotic flora. Indeed, Brother Columban, Brother Adam's mentor, was not averse in the last decades of the nineteenth century to importing Italian, Carniolan and hybrid bees from Germany. When Brother Adam took over the apiary and Isle of Wight disease wiped out most of the hives at Buckfast it was the Italian and native bee crosses and Carniolan crosses that he observed had survived, and these were originally imported stocks.

Moreover, adverts in the *British Bee Journal* show that there were others who had attempted to hybridise bees twenty years or more before Brother Adam started. In 1905 an advert appeared in the *British Bee Journal* for F. W. L. Sladen's *Golden Prolific* and *Hardy & Golden Prolific Italian* and *Black bee crosses*, the results of years of breeding in Ripple Court Apiaries near Dover. Another such breeder was Samuel Simmons who advertised the *British Golden* and *White Star bee*. Sladen's advert offered a range of fertile queens from Britain, Italy and America: Italian queens at 5/6; Golden Italian 1905 queens at 6/6; Carniolan queens at 7/6 and British queens at 3/6.

By 1929 Brother Adam was also advertising fertile Buckfast queens at 10/6, of the leather-coloured type, though these were not yet the Buckfast hybrid for which he became famous. The description of the bee in this advert fits the Ligurian which became a base bee in the Buckfast genetics. According to Brother Adam, however, the first Buckfast cross was made before 1920 between *Apis Mellifera ligustica* and *Apis Mellifera mellifera*, from those survivor stocks at Buckfast Abbey after the ravages of Isle of Wight disease. Clearly, however, importing and hybridising the different subspecies had been happening decades before Brother Adam.

Until that time Western European honeybees had largely existed in their many subspecies in isolation from each other, as they had done since the last Ice Age. The Italian bee was isolated by the peninsula bordered to the north by the Alps. In Sicily too a separate subspecies formed, closely related to the Italian. Northern Greece developed the Macedonian bee, while in the south *Apis Mellifera cecopria* evolved. Crete had its own *Apis Mellifera adami*, while in Cyprus *Apis Mellifera cypria* evolved. Another subspecies, ruthneri, established in Malta. The European Dark bee, (of which the British Black bee was a type) *Apis Mellifera mellifera*, spread its territory to Germany, France and Britain after the Ice Age, while the Iberian bee was isolated in Spain from other subspecies. Further east, occupying the Balkans and Slovenia, the Carniolan bee *Apis Mellifera carnica* established itself in the same way, adapting in finely-tuned ways to different environments and climates. In central Caucasus the Caucasian bee *Apis Mellifera causasia* was closely related to the Carniolan. With the Crimean and Ukranian honeybee, each of these genotypes were in very slight ways adapted to different conditions over thousands of years through natural selection of the traits best suited to each location.

Now, as throughout the last hundred years, we can order queens from all kinds of exotic and far-flung places, having bees in our hives that are completely unsuited to our conditions, interbreeding with each other and with native wild bees to create mongrel stocks in which we see anomalies such as queen failure and supersedure of queens in their first season. Add varroa, a multitude of viruses, and a lack of drones with the required health and vigour for successful polyandry, and queens that we want to lay flat-out in a double brood box, and we wonder why we're having problems with our bees. In an attempt to provide racially pure queens, many such imported stocks are bred in isolated apiaries, as some breeders are now doing with native bees in Britain. Alternatively, they are artificially inseminated, so that beekeepers receive beautiful-looking foreign queens such as the caramel-coloured Italian, but are actually buying bees that lack the genetic diversity or hardiness needed for survival.

Recognition that the native *Apis Mellifera mellifera* had survived Isle of Wight disease was slow after World War I. Occasionally there is a letter in the *British Bee Journal* by a beekeeper claiming that their bees survived, but the view promoted by Brother Adam was widely held well into the twentieth century that no native Black bees survived here and that those claimed to have survived were imported Dutch bees of the European Dark variety. Some lone voices flew in the face of the prevailing dogma that *Amm* was extinct in Britain, such as the naturalist and entomologist, Beowulf Cooper (1917 – 1982) who was employed as a researcher and advisor by the Ministry of Agriculture, Fisheries and Food. Beowulf recognised the dangers of hybridisation caused by large numbers of imported bees.

He wasn't the only one to voice this view, though it was not frequently suggested. Occasionally there were others, such as Terry Theaker, a Lincolnshire beekeeper, who claimed to have pure strains of native bees that had survived Isle of Wight disease. On 27 July 1963 there was a meeting (including Beowulf Cooper), at Mr. Theaker's apiary at Leadenham, Lincs, and a coming together of these lone voices who recognised the survival of indigenous *Amm* in Britain. The *Village Bee Breeders' Association* (VBBA) was formed in 1964 to conserve, study and restore native bees to Britain and Ireland. The name *Village Bee* referred to the tendency of *Amm* to form small colonies. Later VBBA became BIBBA (Bee Improvement and Bee Breeders' Association).

Despite my reservations about the feasibility of restoring *Amm* as a pure race in Britain, it would seem that in Ireland especially their work has been extremely important. It is now understood from modern research, based on mitochondrial DNA analysis, that Mr. Cooper, Mr. Theaker and BIBBA were correct and that Brother Adam was wrong – the native British Black bee did not become extinct with Isle of Wight disease. Not only this, but Ireland's *Amm* bees are distinct from *Amm* strains across Europe, such as the Dutch bee. This again flies in the face of Brother Adam's claim that any native Black bees in Britain after 1915 were of Dutch or French origin following the Government's restocking programme.

*Amm* is actually of a distinct lineage from *Carnica* and *Ligustica*, its close neighbours, and is more closely related to Iberiensis, the Spanish bee. Experiments in Poland have shown that *Amm* will actually breed true and will deselect *Carnica* sperm, which is fascinating. In Poland the two genotypes exist in different regions, with a hybridisation zone, on which the research focused. *A Carnica* queen was inseminated instrumentally with 50% *Carnica* sperm and 50% *Amm* sperm, the process repeated with an *Amm* queen. It was found that the *Amm* queen used only the *Amm* sperm and bred true, whereas *Carnica* used sperm from both genotypes. That is interesting because it suggests that in natural mating where there are many genotypes present with their hybrids *Amm* might be selecting a higher number of *Amm* drones for mating, while also selecting only the sperm from *Amm* drones even if mated with non-native drones. Other studies in Tasmania showed that queens will not mate with weak Ligustica drones. Perhaps these studies go some way to proving that, despite extensive hybridisation, honeybees might be more able than we have previously thought, to select the genes needed for their best chances of survival. At the very least, it seems that *Amm* can.

It might also explain why *Amm* has survived and is present to a high degree in the DNA of many wild colonies in Britain. It's why I believe that progress lies in the direction of locally adapted bees. Such bees will not only be better adapted to climate, length of seasons, winters, diseases and pests, but they are also likely to have selected traits such as the *Amm* ability to select for *Amm* genetics – simply through natural selection, which will bring to the fore the traits most likely to be successful. Most importantly, these bees will have the genetic diversity to ensure survival.

One Polish Professor from Kracow had some interesting statistics from research into breeder queens distributed in Poland: 7% had nosema, 14.8% had injuries (some both infected and injured) while 20.4% were of substandard quality. In another study of breeder queens in the US all queens investigated were infected with Deformed Wing Virus (DWV). These statistics are additional problems with imported queens that I hadn't considered, together with the statistic that only 1% of beekeepers in Britain are professionals. That means most of our imported queens are bought by hobbyists (and are possibly below standard, as breeder queens are in Poland) with problems not only of diseases that can be seen, but with unseen viruses too, and are weakening the gene pool of native adapted bees. It's likely also that many of these queens don't survive their first season or winter, which means more repeat orders for imported queens the following spring. That's good – if you're a bee breeder who profits from the sale of queens!

Another point I picked up from a BIBBA lecture is that many commercially produced queens are created in colonies in batches of forty or more queen cells – far more than would be produced even in a swarming colony. If these are also emergency queens they might have been fed less, as larvae up to three days old can be used by bees to raise queens; that could shorten their intensive feeding period by as many as a few days, which means these queens are rushed by colonies before the cell is sealed. How many queens like these are imported from disreputable breeders only interested in mass production and maximising profits? Being substandard, such queens would probably be superseded in their first season, despite (to the beekeeper's eye) looking beautiful, laying well and being young queens. We can't tell what is wrong with these queens, but bees can.

It's very important that BIBBA are making this information known, but other information from them does not persuade me that restoring the native Black bee as a pure race would ever work. One factor to consider is that even with *Amm* as a distinct genotype there are many ecotypes, with traits specific to particular regions and environments; remember the *Amm* bees of Landes, France, which were adapted for a heather honeycrop? The problem with BIBBA's breeding lines is that breeders will usually select the desirable traits for their own purposes, just as Brother Adam did with the Buckfast strain, but these are not always the same traits desired by every beekeeper. Brother Adam bred a bee that does not produce much propolis, which makes it easier to carry out inspections, as the bees don't glue everything together. But some beekeepers, such as myself, believe that propolis plays an important role in the honeybee's health, when you consider how they propolise the entrance and the inner walls of the nest and the antibacterial properties of the substance. Moreover, nature does not select for docility, low propolis production or low propensity for swarming, while for wild honeybees improved productivity isn't a genetic driver. Nature selects traits for survival. This means that managed native bee strains from breeder lines that are bred in isolated apiaries, or are artificially inseminated, are not necessarily selected for survival. In this important regard they are no more desirable to me than the Buckfast strain. Nor are they likely to be adapted for survival in different regions of the British Isles. Native bees for Scotland will not necessarily fare better in the Midlands, and Cornish stocks will not be adapted for life in Berkshire any more than a good breeding line from Germany will automatically perform as well in Britain.

In my view the best way forward is towards the development of local adaptation, even if that includes near-native bees rather than pure *Amm*. If *Amm* queens are found to have mated more often with *Amm* drones, this suggests that native and non-native bees can co-exist and segregate to some degree. It should be possible then to find locally adapted bees with a high percentage of *Amm* DNA because queens are to some extent avoiding non-native drones.

In time the most desirable traits for survival will be repeatedly selected and we could end up with near-native bees that are locally adapted. In short, I don't think the drive for a pure race is desirable or really possible without compromising genetic diversity which is itself a key survival mechanism. More important, I think, is that bees have the genetics for survival in a particular environment or region.

The emphasis on near-native and locally adapted bees does not rule out the importance of the work done by bee breeders to maintain and improve strains of the native Black bee or any other subspecies. We're in real danger of losing many of the European genotypes and we don't know when we might need this biodiversity. The problem is simply that bee breeding alone, through instrumental insemination or isolated apiary mating doesn't provide sustainability when the progeny are taken to other places where there is constant hybridisation with imported bees. Bee breeding is certainly helpful when integrated with a bee improvement programme which involves enough beekeepers to sustain the results of the breeding and improvements.

In 2020 BIBBA shifted its emphasis  away from purely native bees and launched an initiative which aims to involve every beekeeper, regardless of their starting point. The one wise condition or rule they insist upon for their *National Bee Improvement Programme (NatBIP)* is: '...that beekeepers should aim to avoid the use of imported, or the offspring of recently imported bees.' (19)

The idea is that individuals, working alone or in groups, begin to select from their local stocks in order to develop and refine bees that will be ecotypes with traits adapted to their area. In time, with enough participants, sufficient drones will be produced of the same ecotype that all queens in an area will be able to mate with drones of their own kind. As NatBIP explains in its introduction, 'The Programme is designed to promote the improvement of local bees and the development of local ecotypes and to avoid further input from imported bees.' (20)

NatBIP recognises that a, '...stable and sustainable future for beekeeping'  (21) is only possible if there is a realistic alternative presented in place of the reasons why so many hobbyists import queens of other subspecies. This initiative is an excellent response to an experiment by the COLOSS (Prevention of Honeybee COlony LOSSes) honeybee research association examining survival qualities of imported bees compared to local bees around Europe. COLOSS highlighted the benefits of using local bees and advocated: '...the conservation of bee diversity and the support of local breeding activities must be prioritised in order to prevent colony losses, to optimise a sustainable productivity and to enable a continual adaptation to environmental changes.' (22)

Essentially the initiative is about working with nature rather than against it. Left alone, what would happen to our native stocks? Without imported foreign queens the effects of hybridisation would begin to dilute in local populations simply through natural selection.

The best genes for survival (which are likely to be mostly *Amm* genes already latent in hybrids) would become reselected by survivor colonies, producing local strains based largely on the native bee. This suggests the direction to be taken by beekeepers, so if we work along these lines it's perfectly possible for us (and breeders) to refine and develop the characteristics we want in order to benefit from beekeeping – such as docility, productivity and varroa-resistance. The difference from many current breeder programmes of other subspecies is that those characteristics will be fixed together with survival traits rather than at the expense of them.

Of course, such a programme will not necessarily be quick to develop or to reap results. It's a work that will probably take many years just to convert the critical mass of hobbyists needed to begin to make a difference. There will probably always be some beekeepers who will remain faithful to their allegiance to particular methods and to a favoured subspecies or strain such as the Buckfast or F1 Buckfasts. They will still want imported Carnies, Buckfasts or Italians for the short term benefits of high yield, despite the fact that research does not confirm the idea that foreign bees are better, unless viewed purely from the perspective and the goal of maximising honey production in a particular good season. Then there's the issue of bee farmers; though a tiny percentage of beekeepers, the sheer numbers of their bees is enough to affect a whole area. If only a few hobbyists are involved with developing locally adapted bees while a bee farmer nearby is flooding the area with Carniolan or Italian or Buckfast drones, it will be very difficult for the hobbyist to maintain progress with the bee improvement programme.

For me, and I'm sure many other beekeepers, the National Bee Improvement Programme's aims seem to present what Tickner Edwardes described as a third class of beekeeper: '...upon which the hopes of all who love the ancient ways and days, and yet recognise the absorbing interest and value of modern research in apiarian science, may legitimately rely.' (23)

What, I wonder, would William Woodley have made of this, '...spirit of progress' (24) in modern beekeeping? We know that he preferred working with *Amm*, even when beekeepers in Britain were beginning to keep Carnies and Italians, and when breeders such as Sladen and Simmons were creating their hybrids. We know that Woodley adapted to foreign Dutch bees after Isle of Wight disease destroyed his stocks, because there was no alternative. We know from his adverts at that time that he was selling Dutch bees. He was, after all, a businessman and he had been threatened with economic ruin by Isle of Wight disease. Despite this, I think he would have been appalled to see the effects of hybridisation on the scale that we now see among our stocks and the increased importation of foreign queens by hobbyists who are nearest in spirit perhaps to the old cottagers with their skeps and bee gardens, and yet who are not driven like the cottagers of old by the economic necessity of honey and wax production or of making a business from beekeeping. I can't help but think that Mr. Woodley would probably have welcomed the idea of some kind of return to a simpler world that works both for honeybees and for beekeepers. I'm sure he'd probably also approve of the idea of locally adapted bees that are better suited for survival. And if there were a chance that he might catch a wild swarm of near-native bees as he had done as a boy in his great aunt's bee garden in Stanmore under the direction of the old, blind bee man – this, I don't doubt, he would probably regard as a welcome bonus.

In fact, there is evidence from his August 1912 *Notes By The Way* that he blamed foreign bees and modern beekeeping methods for the increase in bee diseases such as foul brood and Isle of Wight disease: 'Then we shall have hopes of a healthy race of bees, such as was  generally found in cottage gardens before the advent of foreign bees and modern bee-keeping.' (25)

He wasn't alone in beginning to form such views. Others, such as H. Campbell of Norfolk were also looking in the direction of foreign imports and asking if Isle of Wight disease had been introduced by bees that were immune to it in their native environment:

'It is obvious that there are two possible answers to the question as to how the disease came to the island — and only two. Either it was brought from some other country, or else it was generated spontaneously in the Isle of Wight... ...Now, taking the first alternative, if the offending micro-organism reached us from abroad, we are justified in asking why we have had no news of its ravages among foreign bee-keepers; and here again, as it seems to me, there are only two answers. Either it came from some country that is not, so to speak, kept permanently under the microscope... or else it was imported from some district whose native bees have become immune by selection.' (26)

What would Brother Adam have made of all this? In charity to Brother Adam and his extraordinary dedication and love of the honeybee, not to mention the admirable and clever development of what he considered to be the perfect bee, his times were not the same as ours. It's easy to be critical from the advantage of a retrospective view of him, but he stopped keeping the bees at Buckfast before varroa really became a problem. And yet we must not avoid criticism of him simply because of his unquestioned stature in the beekeeping world. He was wrong about certain assertions he made – that Isle of Wight disease was caused by acarine; that the native Black bee was wiped out by Isle of Wight disease in Britain; that we could improve on nature and produce the perfect bee (whatever that means).

Certainly Brother Adam made mistakes and was aware of his reputation and protected it at all cost. It is little known that he had a problem with American foul brood which he kept quiet and didn't declare. According to the BIBBA video, *Beekeeping at Buckfast – Past and Present* (Feb 2021), he attempted to treat it himself. He was found out and the Ministry of Agriculture, Fisheries and Food (MAFF) descended on Buckfast and ended up having to burn hundreds of hives in 1998. But Brother Adam liked a challenge, and perhaps he'd have changed direction and looked for an opportunity to develop a bee that was resistant to varroa. With the benefit of later research and scientific knowledge of the effects of hybridisation I wonder if Brother Adam might have changed some of his views about what constituted the perfect bee and if he'd have changed some of his practices such as isolated apiaries and artificial insemination? What, in particular, would have been his opinion of imported queens, given that this is almost certainly how varroa was introduced almost all over the world?

Like Biddy, our beekeeper some decades ago at Douai Abbey, Buckfast has also now changed direction in its beekeeping.Buckfast isn't any longer producing Buckfast bees; nor are they exclusively about maximising honey production, but their apiary is now playing an important role in educating the public about the issues facing modern beekeeping and the western honeybee. I understand that people (probably some beekeepers among them) frequently contact the apiary at Buckfast and are astonished (even horrified) that Buckfast Abbey no longer breeds or sells Buckfast bees. This in itself conveys a strong message about how beekeeping itself now inhabits a very different world from the one known to William Woodley, Brother Adam, or even the more recent Fr. Robert Biddulph of Douai Abbey

# MINDING THE BEES

## Chapter 12
### The Best Propagandist

'WANTED; a healthy stock or early swarm of Dutch or Italians -
PATTINSON, Hollythwaite, Windermere.' (1)

**British Bee Journal 1919**

Swarming time for the heathland bees in a skep bee farm in Lower Saxony in the 1970s. I've watched the film on Youtube so many time, but it's always just as exciting. The beekeepers sit on boxes in front of the skep stand in the home apiary on a hot June day, watching the stocks. In a season on the farm as many as several hundred swarms might issue, and they all have to be caught as they leave the skeps so that none escape. These are the stocks that the bee farmer will package with a new queen and put on a train to their buyers.

The men know from experience when the stocks are about to swarm by the movement of the bees at the skep entrances. They walk along the stand, covering holes that might be used as alternative escape routes for the stocks, usually with a thin cord that they tie loosely around the gap at the base of the skep. Occasionally they tip a skep on its side to look inside. A cluster of fat, sealed queen cells at the base of the combs is another sign that the bees are ready to swarm.

As each stock rises the bee men fix a long catching bag, about four or five feet long and a foot wide, to the flight entrance at the top of the skep. It's made of material like net curtains or stockings, and is transparent. The foot-end tapers to a point that is fixed to a pole, secured vertically in front of the stock so that the bag is stretched out horizontally in front of the skep. As the bees emerge they fly into the bag and collect at the far end, near the pole. The beekeeper can see their progress through the transparent material, and when they appear to have completed their departure he detaches the bag and hangs it under the eaves of the stand, in a shady spot where the cluster can settle down at the top of the bag.

I am inclined to agree with J.N. Kidd of Stocksfield, who wrote in August 1913 *On Swarming*, 'It is the charm of beekeeping...It is the best propagandist, the best advertiser of beekeeping, the thing that spreads the ''bee enthusiasm'' more than anything.' (2)

May. I am able at last to begin sitting outside the Abbey Church at 6.00 am, before Matins, in the cool, overcast light. In the field opposite a hare appears one morning and sits upright, long ears tipped with black; walks with a casual see-saw motion across the field...pausing... posing on haunches like an old-fashioned rocking-chair...walks again, tilting back and forth until it's out of sight. A kite rises on wings, terrible and prehistoric, passing behind the trees; out of scale by one hundred and fifty million years. Three jackdaws *jack, jack* under the beech trees near the Abbey gates, swaggering in grey hoodies.

I record them in my notes. You never know when notes might be useful – perhaps someone in the future might find them in our archive. I record on 19 May a bumblebee nest in a bird-box on the side of an outbuilding. There are several bees inside, near the entrance – white-tailed. There are three families of bumblebee in the UK: White-tailed, Yellow-tailed and Ginger-tailed; these ones are *Bombus hypnorum*, the Tree bumblebee, which often nest in a bird-box.

21 May: after breakfast a Common Cockchafer lumbers along my sunlit windowsill. It flew in two nights ago and thumped around like a clumsy plastic toy powered by an elastic band, then hid itself. Here it is – about an inch long, with a black body, brown wing-cases, fanned, light-brown antennae and a pointed abdomen. I watch it move like some slow, rusty wind-up toy from another era.

*Cockchafers* are members of the scarab family which in the UK refers to any beetle of the subfamily *Melolonthinae*. The name cockchafer derives from the late seventeenth century use of the word *cock* as expressing vigour and *chafer*, meaning a type of insect with a propensity for gnawing and damaging plants. Their larvae are called *rookworms*, because rooks are said to be partial to them. The Common cockchafer or May Bug, May Beetle, Doodle Bug, Spang Beetle or Billy Witch *(Melolontha melolontha)* was key to one of Nikola Tesla's early inventions; as a child he once made an engine by harnessing cockchafers.

You never know indeed when such records might be useful. Thomas Owen, the eighteenth century skep-maker and beekeeper kept a record of his swarms, year after year. Most modern beekeepers would not want to record any of their swarms because swarming has come to be regarded as the result of poor beekeeping and inexperience. It's why I spent my first few years as a beekeeper regarding swarm prevention as something of a holy grail. Gradually I improved and began to lose fewer swarms, until one year I eventually had no swarming, so that now I know how to stop swarming in its tracks, or (even better) can prevent it altogether. That first year I had no swarms gave me a greater sense of triumph than the successful honey crop that resulted. Yet beekeepers from the past, like Thomas Owen, until the advent of the modern frame hive, actually relied on swarming to increase their stocks and to take a honey crop. To them swarming was as expected as the months and the seasons. Their only concern was to catch a swarm before losing it altogether across a field or the local village.

When you stop and think about it, it's paradoxical that we're now trying to do the very opposite of what beekeepers used to do, to achieve exactly the same result: for them a swarm was how you made your honey crop, while for us a swarm represents the end of our hopes for a crop. In 1905 there was even a quaint charm printed in the *British Bee Journal* for catching bees, no doubt just for a bit of fun and a smile at the old superstitions of simple rustic folk, but it highlights the lengths people once went to in order to catch a valuable swarm:

"Charm for Catching a Swarm of Bees. "

Take some earth, throw it with thy right hand under thy right foot, and say:

'I take under foot, I am trying what earth avails for everything in the world, and against spite and against malice and against the wicked tongue of man and against displeasure.' "Throw over them some gravel when they swarm, and say:

'Sit ye, my ladies, sink,

Sink ye to earth down;

Never be so wild

As to the wood to fly.

Be ye as mindful of my good, as every man is of meat and estate.' (3)

What it comes down to is essentially the commercialisation of bees in order to maximise their productivity. Whereas the old skeppists were content to take a dozen jars of honey from a skep, modern beekeepers often want to take a minimum of one hundred pounds of honey from a single hive. This means having highly productive queens and lots of workers; the modern obsession with yield is also a major reason for the problem of imported queens. It's no wonder the Buckfast strain is popular when you consider that the record yield from a Buckfast hive (according to Brother Adam) was four hundred pounds of honey.

Recently I was reading an online discussion between beekeepers, about Buckfast bees, with the beekeepers comparing their ideas and experiences of the strain. One beekeeper wrote about his beekeeping and justified his reasons for importing Buckfast queens, using the analogy of sport fishing: he posed the question: who wants to catch a small salmon? In other words, his sole reason for importing Buckfast queens was competitive. I've heard many reasons why people keep bees and most people are motivated by an interest in apiculture, even if their primary motive is honey production; but I've never heard of anyone keeping bees for the same reason that anglers go after giant fish. I'm not, in principle, against sport fishing for an elusive, giant carp, or a trout to bring to the table, as long as the hobby doesn't damage the environment or a species. Mostly, coarse fishing stocks are returned, to fight another day, and the fisherman returns home with a deeper appreciation of nature, having relaxed on a river bank or next to a glittery pond for several hours. The problem with keeping bees simply for the sport of beating last year's honey crop is not even justifiable, however, in the way it's justifiable for an amateur vegetable grower striving to produce a giant leek or onion. To maximise a honey crop a beekeeper needs a very large population of workers at the same time as the main nectar flow. There is one sure way to achieve this, and it's by using a prolific race or strain. Buckfasts were created to produce a very high yield – and they do. So do Italians and Carniolans. In the past the British Black bee, forming smaller colonies, was not regarded as commercially competitive with its foreign cousins. Yet it was often reported in the old journals that in a good year it did just as well, while in a poor year it was more reliable because it was adapted for our more unpredictable and changeable weather. If you were producing cut comb honey, however, no other bee could come near the Black bee's achievements.

The first objection I have with this kind of competitive or just plain greedy beekeeping is that such prolific queens must usually be imported. Even Buckfasts are mostly produced abroad now, and Buckfast Abbey's apiary has become more of an educational centre, with no queen production for commercial purposes. The next objection is that the large stocks produced by foreign queens produce very large brood nests, almost uninterrupted, from late winter to late autumn, with Italians perhaps having some brood even right through the winter. More brood means more varroa and more viruses vectored by the mites. Management of varroa means cutting out large quantities of drone comb and destroying drones, and the need for chemical treatment of the stock. As with most imported queens, the head of such a colony will probably not survive more than her first year, if not through natural means then through beekeeper intervention because they desire first year queens who are at their most productive. In this regard we treat queen bees in the same way commercial laying chickens are treated; for the first year they are commercially viable, but as egg-laying diminishes in the second year their returns diminish. It seems to me that in this kind of beekeeping the pursuit of high yields of honey is all about short term gains with long term losses.

Our relationship with the honeybee does include the reasonable desire for honey, and that's fine. But that's like mistreating a beautiful handmade watch with no regard for the delicate machinery of its mechanism. Take, for example, the complex component of the drone. He isn't of much interest to many beekeepers, except to trap *varroa* mites, which beekeepers do by cutting out the infested drone brood. The importance of drones is suggested, however, if you insert a frame of foundation into the centre of a brood nest in early summer. The bees will usually draw it out as drone comb, suggesting that they want to produce many more drones than we might think.

In fact, the importance of drone production has been demonstrated by Professor Thomas Seeley who has researched the production of drones in his work studying honeybees in the wild. Seeley found that, despite what one might think, honeybees invest equally in the production of swarms and drones. He uses the analogy that queens in swarms are produced like apples on a tree, and so he calculated the dry weight of the prime swarm and aftercasts compared with the dry weight of drones produced in a typical season. They were the same. That is a startling comparison, because it highlights the under-production of drones in many managed hives, mostly due to the beekeeper's interventions.

Perhaps it's time we lowered our expectations and approached our goal in a way that's closer to nature. Many of us are uneasy with photographs of intensively-reared chickens that are grown rapidly for the market at the expense of flavour and the animals' quality of life. We don't want to see intensively-reared pigs kept in small pens either, providing our food at the expense of animal welfare. Most of us (and though I'm vegetarian, I'm not evangelical about it) want to see our food produced in an ethical way that allows animals to live the best life they can, providing us with food of the best quality and flavour. We think this way about cows and chickens and pigs. We don't think this way about honeybees. Maybe we assume they already have privileged lives because they aren't kept in cages or pens and we don't take them to slaughter. We don't stop to consider the slow, invisible ways we might be slaughtering our honeybee stocks by reducing their genetic diversity, importing foreign stocks and bee breeding for every desirable trait that qualifies as the perfect bee except the one trait that really matters, which is survival!

The root of the problem with the modern frame hive in the nineteenth century was swarming, and it remains the root of the problem today, especially as it has become a dogma of beekeeping that to allow swarming is the key marker of bad management. William Woodley, the boy, was minding the bees perfectly well when he alerted his aunt that they were swarming. Today when we do everything we can to prevent swarming, our methods rather than our attitudes to swarming itself are often accompanied by anything but actually minding the bees.

I am reminded though that our distant forebears did not always encourage uncontrolled swarming either. In the nineteenth century John Mitchell Kemble discovered a charm towards the end of a larger Anglo Saxon text *For A Swarm of Bees* which was intended to prevent swarming:

'Settle down, victory-women, sink to earth,
never be wild and fly to the woods.
Be as mindful of my welfare,
as is each man of border and of home.' (4)

In later centuries Christians (and, one assumes, monks) would have turned to prayer rather than magic, as evidenced in the Old High German Lorsch Bee Blessing from the ninth century. It was discovered in a manuscript of a copy of the Apocalypse of Paul in the Vatican Library, a text from the monastery of Lorsch, in Germany, famous for the Lorsch Codex. It has striking similarities with the Anglo Saxon charm, suggesting that the prayer had its roots in a pre Christian Germanic culture:

'Christ, the bee swarm is out here!

Now fly, you my animals, come.

In the Lord's peace, in God's protection,

come home in good health.

Sit, sit bees.

The command to you from the Holy Mary.

You have no vacation;

Don't fly into the woods;

Neither should you slip away from me.

Nor escape from me.

Sit completely still.

Do God's will.' (5)

Key to attitudes towards swarming was the importance of managing the situation, whether swarming was permitted or not, so as not to lose your bees over the fields where other people could take them. This management issue was important for different reasons while Isle of Wight disease was raging. William Woodley wrote in July 1912 that he had sent swarms all over Britain that season, but that some buyers were becoming nervous and had cancelled their orders, for fear of importing Isle of Wight disease into their apiaries. He had even admitted earlier in that year's editions of the *British Bee Journal* that if there was evidence of a trade in diseased stock, and that this was spreading the problem, then the trade should stop. No doubt Mr. Woodley was referring to the unscrupulous sale of obviously diseased stock, but what he probably didn't consider was the possibility that stocks might have been diseased but asymptomatic, and that even his bees might have come into this category.

At the risk of sounding fundamentalist about the topic (which I'm not) let me clarify that I'm not actually advocating that we allow all our stocks to swarm every year, as the skep beekeepers used to. Quite apart from anything else, most beekeepers don't have someone to mind the bees for them during the swarm season (although since Covid many are working from home again, as skeppists like Mr. Martin, the shoemaker, did and can mind their own bees). There are all sorts of good reasons too, apart from securing a reasonable honey crop, why it's no bad thing to control swarming. If you live in an urban area or near someone who is easily upset or is allergic to bee stings, it's in everyone's interests to avoid the aggravation of a swarm. During the pandemic lockdowns there were restrictions about meeting others, and people were guarded about strangers coming onto their properties; you don't want to be chasing down your swarms in that kind of situation. There is also the danger of litigation in a litigious culture if your swarm results in people being stung, though most people have no idea how harmless a swarm is once it settles. Enter N, to tell a couple of stories...

I was standing one day in the grounds after Matins and Lauds (morning prayers), listening for the cuckoo, in the last week of April. (*The Douai Magazine* records in the 1940s that the average date for hearing the first cuckoo was 12 April.) By now the month was moving effortlessly into May as though through a gradation of filters. There was a grey-green haze through distant trees of new leaves appearing, like the rapidly stippled undercoat of a Bob Ross painting (I could almost hear Bob Ross' breathy *There!*) *Clematis montana* nodded fat pink buds like whimsical dashboard accessories. Bashful buds, clustered in tight red blushes, were opening cautiously into the full white confetti of apple blossom, ready to celebrate a spring whose arrival this year had all the prerogative of a late bride.

I heard a cough in the Arctic morning air. The hat appeared first, dancing over the top of the hedge between the lane and me, a hat to get anyone noticed. An elderly gentleman appeared with two dogs – mongrels suggesting some whippet in one of them and a Jack Russell in the other. The man was slim and sprightly, with grey collar-length hair, a grey beard and eyes blue as forget-me-nots. As he walked, the hat almost had a life of its own, adorned with several long, jaunty pheasant feathers and a front rim lined with several three-inch yellow canines. I'd seen him several times before, walking his dogs along the lane, but we'd never spoken. Robin Hood came to mind – grown old disgracefully, and determined to stand out. He was wearing predominantly dark-green, with a grey, sleeveless jacket full of pockets – the kind you see fly-fishermen wearing.

'Morning,' we exchanged across the hedge.

'Lovely, isn't it?' he smiled, with a nod of the merry hat.' Lovely to be out, enjoying nature.'

I agreed, and asked if he'd heard the cuckoo yet, which he hadn't. I told him I'd only heard a distant one. A conversation followed in which he wondered where all the skylarks had gone. I informed him that we have yellowhammers on the meadow and he told me he was a retired ornithologist. He advised me to go down to the canal to listen for them because they like to use the nests of the reed warblers.

We digressed and I learnt a little more about him before we got onto the subject bees. At this point N (he'd also introduced himself ) told me two little stories about local swarms. He asked me if I knew the Major in the village? I didn't. Anyway, N told me, the Major's bees had swarmed one year into N's garden and he could stand right under them with his nose up against them and they didn't bother him. The second anecdote was less amusing, but illustrates the general ignorance of people about honeybee swarms and why beekeepers really don't want their swarms ending up on other people's property; apparently a swarm landed in an inconvenient part of a local gastro pub's garden, preventing some people from sitting down – so they burnt the swarm. I don't know if *they* were the publicans or the customers, and N didn't elaborate, but it was enough information.

It's true that swarms can cause all kinds of upset, and that isn't just a symptom of the modern world. There's an amusing little press cutting from the *Essex County Chronicle* of 28 June 1902, published in the *British Bee Journal* in July that year. The incident happened at Castle Hedingham, Essex, when a swarm settled and built a comb of honey up in a large elm tree near the home of Mr. V. W. Taylor J.P. A resident had decided to smoke the swarm out and to claim the honeycomb, so he lit a fire under the elm. The tree caught fire, but was dealt with by the resident – or so he thought. It was seen again on fire at midnight:.'...by Inspector Wapling and P.C. Frost, who, after two hours' strenuous labour, thought they had managed to extinguish the flames. The tree, however, was found to be burning on the Sunday about midday, when several members of the village fire brigade attended and established a bucket service. Happily their efforts were successful. The bees were still about the tree on Tuesday, when some of them showed signs of their recent experiences.' (6)

Swarms can also bring a sense of consolation and comfort. In 1911 Betsy Powner wrote in *The Daily Mail* about the vigil a swarm of bees kept after her grandmother's death, settling on her bed curtains until she passed away, after which they all went back outside through the window. They didn't return to the hive though, but disappeared.. Other tales tell of bees following hearses to the grave, while in France hives that had lost their bee master were consoled with one of his garments draped over them. Less convincing is the amusing anecdote told to me several times by one of the Douai community, that when Biddy, our beekeeper at the time, moved up to the Lake District his bees followed him up the motorway!They did, in a manner of speaking, but by the more prosaic method of a van.

In the nineteenth century this rhyme was said, accompanied by jangling keys:

*Honey bees, honey bees, hear what I say!*

*Your Master J.A. has passed away.*

*But his wife now begs you will freely stay,*

*And gather honey for many a day.*

*Bonny bees, bonny bees, hear what I say.*

My favourite rhyme is from John Greenleaf Whittier's Telling the Bees 1858:

'Before them, under the garden wall,

Forward and backfired

Went drearily singing the chore-girl small,

Draping each hive with a shred of black.

Trembling I listened, the summer sun

Had the chill of snow;

For I knew she was telling the bees of one

Gone on the journey we all must go.' (7)

Honeybee swarms have been both good and bad omens in folklore for centuries in many cultures. In 64 AD swarms settled on the Capitol in Rome, unsettling citizens, while Emperor Claudius' death was supposedly foretold by a swarm in his camp. Pompey's defeat at Pharsalus was allegedly augured when a swarm settled in his standard. Throughout Christian Europe swarms  have traditionally been regarded as portents of death. For example, if a swarm lands on a dead tree branch, it is said that the tree or the swarm will die. If a swarm settles in the house, folklore warns that someone will die. A queen bee landing on you is certain death, apparently, though this seems to contradict the traditional belief that if a swarm settles on you (as is supposed to have happened to the infant St. Ambrose of Milan) it bestows eloquence. It seems to have done in his case. A swarm contains a queen, so presumably when a swarm lands on someone a queen lands as well. Contrastingly, John Evelyn's *Diaries* for June 1662 relate that the king forbade the disturbance of a swarm in his ship. Similarly, in French folklore tradition a swarm alighting is the harbinger of a prosperous year. In 1520 Ludovicus Vives was appointed Professor of Oratory at Corpus Christi college (The College of Bees) and was welcomed by a swarm settling in his study. He became known as the Mellifluous Doctor.

All this aside, enjoyable as it is, swarming is simply a natural sign of a colony's health and prosperity, suggesting to us that a swarm is a beautifully elegant mechanism, whose intricate workings have only recently been discovered, while much of its workings remain unknown. The fission that occurs when a colony swarms is coordinated by what biologists call swarm intelligence, or the collective behaviour of individuals with each other and their environment within a self-organised system. It is also a thing of beauty, as described by *The*

*Bee-Master* of *Warrilow*: 'From early morning they hung in a great cluster all over the face of the hive, and down almost to the earth beneath; a churning mass of insect-life that grew bigger and bigger with every moment, glistening like wet seaweed in the morning sun.' (8)

This describes the beginning of swarming, when, '...in the cluster itself there was an uncanny silence.' (9) This silence and bearding behaviour might start a couple of hours before swarming, and it is the calm before the swarm. On the beard interesting behaviours start to happen, mostly between and from older bees in the colony which make up the 75% of the colony that leaves with the old queen. Close observation reveals activity which is crucial to swarming, though the precise sequence of events culminating in the swarm decision are not yet known. According to experiments done by Professor Thomas Seeley of Cornell University NY, bees in this beard preparing to swarm begin running, waggle-dancing, piping and buzz-running. These activities begin gradually increasing, until about sixty bees in the beard are piping and a few hundred bees are waggle-dancing. What's it all about? This consistent pattern of behaviours across swarms has been identified as related to the search for a new nest site, and it happens invariably just before the swarm departs.

Tickner Edwardes believed the, '...faint, shrill piping sound' (10) was the old queen working herself into a '...swarming frenzy' (11), but experiments have shown that workers are making these noises and that the signal will continue for an hour before the swarm starts, so it's a key signal. All these bees buzz-running, piping and waggle-dancing are nest site scouts returning and recruiting other scouts to inspect their favoured nest site. Occasionally a scout will waggle in favour of a particular site with a falling number of recruits, and these bees are bumped, to stop them dancing. Meanwhile, scouts leave to inspect different potential sites, as they are recruited by others, so that every scout individually inspects every nest site choice and returns, in turn recruiting other scouts to inspect their choice. In this way the scouts vote on their choice of nest site – a remarkable example of swarm intelligence. The super-organism of the colony makes a key choice that will determine its survival, and individual members are involved, working together towards this decision in a way that is startlingly close to the way a human brain makes a decision.

As the chosen nest site wins the democratic vote, some scouts remain, guarding it, and fight off competition from other colonies. The precise sequence of events that happens next is unknown, but the trigger seems to be the number of bees piping, together with a dramatic increase in mobility, velocity and buzz-running. Waggle-dancing, however, is not one of the triggers. What happens next is best described as explosive. Owen Thomas, the eighteenth century Welsh skeppist and skepmaker, recorded the verb *rising* for swarming stocks. *The Bee Master of Warrilow* captures the beauty and excitement of the drama: '... a peculiar stir would come in the throng of bees cumbering the entrance to the hive. Thousands rose on the wing until the sunshine overhead was charged with them as with countless fluttering atoms of silver-foil...within the hive the rich bass note had ceased; and a hissing noise, like a great cauldron boiling over, took its place, as bees inside came pouring out to join the carolling multitude above.' (12)

Others record the drama of swarming in more prosaic anecdotes: in *The Times* letters in August 1967 Irene Lloyd advised what to do if you come across a swarm in a woodland. She advised shaking the bees over yourself and while walking home, brushing them off with a long feather. Why, I wonder, go to all the bother when you can just as well walk around them with a courteous observation of social distancing? Knowing my luck, if there were ever the remote possibility of such a scenario happening in my life, I'd get all the way home without ever finding the feather. Lady Plowden's letter to *The Times* the same year has her own helpful tip. She relates the compellingly plausible tale of a gardener who was wearing a cap when a swarm landed on his head. He went slowly to the front of an empty hive and lay down in front of the entrance; the swarm duly entered the hive.

It's this drama and beauty of the natural world (including its folklore and country tales) that was inextricable from country life until fairly recent history that we have subordinated to our preoccupation with high yields. It all comes down to the extra space we've given bees in frame hives; in skeps the stocks were naturally smaller because *Amm* is not as prolific as Italians or Buckfasts, for example, so swarming was a sign of congestion and that the stock was ready to fission. Modern hives, on the other hand, are designed to hold a much larger stock, which is needed if you keep a prolific strain (and you want to avoid swarming), to maximise the honey crop. Moreover, a second brood chamber (or *deep*) may be added to make the stock even bigger and to give them more space. If stocks are on the point of swarming, frame hives allow the beekeeper to change the circumstances of the bees so that they think their prosperity has changed. Brood and stored honey can be taken away, along with queen cells, while empty frames can be inserted into the brood nest, into which the queen can lay. None of these manipulations can be done with a skep – but that was the whole point! The skeppist wanted stocks to swarm in order to take a crop of honey, whereas if the stock of a frame hive swarms it won't make a surplus that satisfies the expectations of the beekeeper. It might yield the same crop as a swarm from a skep, but beekeepers can't take it because the larger colony of bees will need these stores. The large population of bees the beekeeper has created is precisely to produce what the stock requires in addition to a surplus that may be taken off. It all comes down to the size of the stock, as the Bee Master explains: 'But if swarming be prevented, and the stock re-queened, artificially every two years, we keep an immense population always ready for the great honey flow, whenever it begins.' (13)

We do well also to remember that swarming is a sign of the health and vitality of a colony. When a stock fissions, it's because it has reached its most prosperous; it has the maximum amount of brood in the nest, honey and pollen stored, queen cells and drones – none of which are intrinsically related simply to the size of the colony, but to its vigour and health. When a stock swarms it does so in an optimum state for securing its survival. The problem beekeepers have created by preventing stocks from swarming is that we're losing touch with the most reliable indicator of our stocks' health and maximum fitness for survival. Beekeepers will select their breeder stocks for a variety of desired traits, such as brood production, honey yield and docility, whereas the old skeppists were continually selecting for survival by virtue of the skep system. It's another of the many ways that modern beekeepers keep poor stocks going artificially that are fundamentally unfit for survival, regardless of the other apparently desirable characteristics they might possess.

How can modern beekeepers learn from this lesson of the old ways, without returning to the skep? How might we expect a reasonable honey crop without a vast population of bees that in turn encourages a vast population of varroa mites that then require chemical treatments and a reliance on the fecundity of foreign queens?

To start with, we need to lower our expectations and change our priorities. When we buy a chicken from the supermarket we make similar choices: do we want the biggest bird at the lowest cost that has been reared intensively in less than satisfactory conditions; or do we want flavour and value for money from a smaller bird reared more naturally, that has had good quality of life? Do we want the cheapest eggs from chickens kept in small cages, fed just enough to produce the egg, or do we want dark-yolked eggs that taste as if the chicken that produced it has supplemented its daily ration with free range forage such as worms, weeds and insects? Once we can see this parallel with beekeeping, we might be satisfied with a smaller honey crop from stocks that are better suited to survival in our locality. By changing our mindset we would find that we no longer need foreign queens.

The question might be asked: do we need to allow our stocks to swarm (and is that always a good thing anyway?) or might we take advantage of the frame hive by swarming our stocks artificially through splitting them. This way we take advantage of some of the benefits of swarming while remaining in control and good management of our stocks? Moreover, splitting, or making an artificial swarm, is a very good way of raising our own locally adapted queens.

❦

A news story appears that on 26 April a swarm of bees settled in a Greggs Bakery in Lincoln, a reminder to me that swarms can happen early in the season and that I need to get a move on with placing my bait hives. It occurs to me also that I've been thinking about the duration of the swarm season as typically spanning May to early July, but I need to think instead of catching an intense period when swarming tends to happen, and that varies slightly every year. As Mr. Woodley reminds me from the *British Bee Journal* in July 1902 when he recalled that season's swarming, 'Swarms came in a rush, and stopped almost as suddenly.' (14)

This intensity was reported in the same issue of the journal by A.A.A .Abbottsan, who had fifteen swarms issue from four straw skeps and five frame hives, and a total of twenty-one swarms, including others he took. These happened in fairly rapid succession: 24 May, 10 June, 18 June (they issued daily from this date), 24 June, 26 June, 1 July. In the same season, however, there were reports of early swarms in April. One was reported  from 24 April in Worcester, with another somewhere else on 30 April .

Incidentally, while looking through the journal on 26 April for other examples of early swarming I came unexpectedly upon a sad little notice reporting the death on this date in 1904 of Mrs. Anne Woodley (Annie). I recalled that her name appears at the top of their shared gravestone, indicating her earlier death, and I read somewhere that he remarried – a woman he had known since his childhood. It was a remarkable coincidence to stumble upon the notice of her passing on the anniversary of her death, and all the more poignant to read of William's apology to readers for not having been able that month to write his *Notes By The Way*. In a statement Mr. Woodley wrote, 'Her removal will make a great void in my life, how great time alone can tell.' (15) The brief notice explains that Mrs. Woodley had been unwell for three years. Annie had died in her early fifties.

I have an image of A.A.A. Abbottsan's frantic time chasing his rising stocks, and it reminds me that this is exactly the kind of exhausting activity that modern beekeepers seek to avoid. Not least of all, our lives are probably much busier than in the past and most beekeepers today simply wouldn't be able to keep pace with the demands made by his stocks on A.A.A. Abbottsan in 1902. Even the traditional Mr. Woodley remarked on the subject of swarming, 'No wonder that we, as honey producers, so earnestly hope to secure a non-swarming strain of bee some time.' (16)

No less dramatic than the explosion of a swarm is its settling. Within minutes, '...the glittering, gauzy atmosphere of flashing wings' (17) descends upon a suitable tree, branch, hedge or post where the cluster starts to form: 'The rapid growth of the swarm-cluster was always one of the most bewildering things to watch. From a little dark knot no bigger than a clenched hand, it swelled in a moment to the size of a half- gallon measure, growing in girth and length with inconceivable swiftness, until the branch began to droop under its weight.' (18)

No such desire for non-swarming bees from me this season though; I'm depending on swarming bees to build up my stocks of locally adapted bees, and I don't want to miss the swarms. And I must admit, there's something rather exciting about the prospect of swarming this year. I'm relishing the prospect.

Weather plays an important part in triggering swarms. They often happen on warm, sultry days after rain. At the moment the weather is turning showery, having been settled and very dry for weeks. On 10 May a subtropical plume is set to bring warm weather (if the models are correct!) I'm guessing that in normal years after such a change swarms could start arriving rapidly. I'd be surprised ordinarily not to take my first one by the middle of May when the weather looks like being warm, but that will also depend on how the bees are building up. And I'm beginning to realise that in beekeeping there isn't an ordinary or normal year anyway.

Victor writes in the *British Bee Journal* of a late vicar who asked his two reluctant sons to watch his beloved hives for swarming. When swarming looked likely the sons took it in turns watching, with a chair, a jug of beer, tobacco and a bucket of water and syringe. As swarms began emerging they would syringe them with water, persuading the bees of an imminent downpour and to abandon their departure until another day.

The weather this spring has been similar to conditions in 1891 when as late as 28 May William Woodley reported, 'We have had a considerable rainfall during the last few days, which was badly wanted...and the crops required rain.' (19) For us too April was unusually dry, but also the frostiest for sixty years. And it's been cold, despite the high pressure and settled conditions. The bees have been able to forage during sunny afternoons, covering the creamy candles of the laurel blossom, but often in temperatures below the seasonal norm. I suspect we might even be saying as late as the end of May, as Mr. Woodley did, that the weather has, '...dispelled our hopes of an early swarming.' (20) By 11 June he reported that he had not yet heard of any swarms.

Although my expectations of taking a decent swarm look unlikely for the next week and a half while the weather remains cold and wet, I'm thinking that I'd best get a move on setting my bait hives before the sub-tropical plume arrives, bringing warmer conditions. Who knows, but that might trigger one or two early swarms from wild bees, so I want to be ready on the front foot. Bees can swarm early, just as they can swarm very late, like the swarm that Mr. Plumb took from a post box, reported in the *Essex County Chronicles* on 3 July 1891.

Though beekeepers tend to regard swarms as a bit of a headache when they come from their own apiaries, we still appreciate a healthy prime swarm coming our way for free. A.J. Gorwyn informed the *British Bee Journal* of a swarm taken that issued from a skep, that weighed just over three pounds. Augustus from Sandy Ford, near Paisley, reported, '...a good natural swarm', (21) while in the same year an article on French hives testified to them, '...giving good swarms' (22). Despite Mr. Woodley's claim that he had heard no reports of swarming by early June of 1891, the journal reports differently on 4 May: '...a very fine swarm of bees issued from a straw skep in the apiary of Rev. E. Davenport, of Stourport, Worcestershire. This, so far as we have been able to ascertain, is the earliest swarm of which we have any record for the present year.' (23)

*Successful bait hive designs*

Supposing swarms appear and settle on my boxes, can I claim them as my own? Modern legislation about honeybees tends to focus on the management of diseases, but ancient law dealt more with ownership rights of swarms and wild honey. Legislation around beekeeping dates back to 1500 BC in parts of the world such as northern Turkey, but it was the Romans who laid the foundations of laws about animals in many countries. They ruled that honeybees are wild animals, but that hived bees belong to their beekeepers. Owners had the right to collect their swarms from neighbouring land, as long as the swarm was kept in sight. Once out of sight, the bees were deemed to have reverted to the wild, and the owner could no longer lay claim to them. That's why Mr. Woodley's great aunt needed his help; it was all about keeping the swarm in sight, so as not to lose it to someone else.

The Old English document *Rectitudines Singularum Personarum* from 1000 AD records the rights and obligations of workers on an English estate. The tenants and workers are listed in order, the beekeeper or *beo-ceorl* ranked only second from the bottom, ahead only of the swineherd, if you please! The *Domesday Book* (1086) contains many references to the collection of honey from the wild. During William I's reign (1066-87) the law began to state who had rights to wild honey, and the punishments for stealing bees and honey. Two centuries later *Magna Carta* added to the legislation with the *Charter of the Forest*, sealed 1217 by Henry III, allowing people to forage natural produce of the forest, including honey and wax. It remained in law until 1971, covering Irene Lloyd and those with her in the woods in 1967 who might have wanted to shake a swarm over themselves and brush them off with a feather.

In conclusion, it's incredibly difficult (if not impossible) to second -guess what bees are going to do in any given season. Every year is different in ways that we tend to narrow down to the weather alone. In fact, the previous autumn and winter are also important factors, and some of them are not merely circumstantial but are down to the beekeeper. There's a saying in beekeeping that the beekeeper's year begins in August. That's when preparations should begin for overwintering the bees. These factors all affect how the bees arrive at the following spring, how well they build up, their general health and mite load and the prevalence of any other viruses such as foul brood and Deformed Wing Virus (both of which are on the rise).

A lesson beekeepers do well to learn, however, is always to be prepared, and especially to have equipment ready. This can even reap its own rewards, as it is quite common for stray swarms to enter hives left prepared for occupation by another colony, (which has happened to me.) It's essentially this wisdom in apiculture that I hope to draw upon by putting out swarm traps (bait hives). They work on the same principle as the attractive hive prepared for occupation in the above quotation. Everything will depend on where my six bait hives are sited, my timing, and the weather. Of course, it's wise to take advantage of this bad weather and be ready for anything afterwards, assuming that bees will always do something surprising.

I can't rely on those old fashioned charms, though one of the brethren informs me that Biddy had been a rather good magician and belonged to the magic circle. Not that it helped him stop his bees from swarming. I'm told, by those who remember, that Biddy's bees were very swarmy at the end of his life. He would tie little pieces of blue cloth to trees in the grounds – what Mr. Woodley called *bee-bobs*, to lure swarms. Joking and magic aside, all beekeepers have to learn skills that are equivalent to the magician's sleight of hand. We're all in the business of deception, tricking the bees into responses that helps us to manage them, especially so they don't end up doing the disappearing act on us.

# MINDING THE BEES

## Chapter 13

### Timing Is Everything

'STRONG, healthy Skeps of Bees for sale.
12s each. MULLEY.
Upton-on-Severn.' (1)

***British BeeJournal 1912***

Timing is everything in beekeeping. William Woodley became a bee farmer at the right time, taking advantage of the invention of modern frame hives and the opportunities opened up by the Industrial Revolution, such as the steam train and the motor car, both of which assisted his business. His other interest, in which he was self-taught, would have offered fewer possibilities, because watchmaking was declining in Britain. Having dominated the world in the craft since the 1600s, British watchmakers at the end of the nineteenth century were on the brink of a fifty per cent decline in markets as industrialisation de-skilled the craft with the introduction of mass production techniques. Britain's skilled horologists rejected the progress of technology with which Switzerland and America were mass producing clocks and watches.

It was a new era of cheap, fast production with which watchmaker craftsmen simply couldn't compete. Everything in the world of a skilled horologist is slow and meticulous; as though harnessing the means of telling time from the vast, ancient workings of the cosmos, the cog of their art engages with the universe at its own creeping pace. To be an horologist is to acquire characteristics that sit like the slow, deliberate and precise heartbeat of a watch at the centre of an accelerating world of activity.

Most people will say that the most important invention that has changed the world is the wheel. An horologist would argue that the wheel is but a servant, while Time is a Master; therefore, the clock is the invention that has most changed our lives. Regulated time sets the timetable and the pace at which we all live our lives. Ironically, and paradoxically, only the horologist can refuse to be mastered by this invisible tyranny, as they did when industrial watchmaking began leaving them behind at their work benches. The horologist, master craftsman, remains Master of Time (the original Time Lord), refusing to become its servant. Up to a point anyway; John Harrison, the eighteenth century inventor of the precision pendulum clock, was born on 24 March and died on 24 March. Now that's timing! It's as though he mastered time right up to his last moments, but not quite.

William Woodley had been attracted to watchmaking before he decided to become a professional beekeeper, and it's reasonable to assume that he possessed many of the characteristics that set horologists apart and that these were transferable to beekeeping. Watchmakers have to be problem solvers, with the ability to make close observations. They are determined, patient and persevering, prepared in extreme circumstances to discard whatever they've done and to start again. Mr. Woodley had these traits for sure, especially when he had to pick himself up after Isle of Wight disease threatened more than once to ruin his business.

He also came to see how every beekeeper is really a separate cog working in a much larger mechanism; fellow beekeepers who criticised his stance on skeps and Bee Disease Bills taught him that. Isle of Wight disease demonstrated in a brutal way that every beekeeper's craft is done in relation to everyone else's.

We know the consequences of a watch that doesn't keep time accurately, or an alarm clock that doesn't go off. A five minute delay is enough to miss a vital connection. In 1987, I went through Kings Cross Station minutes before the fatal fire that caused such tragedy that day. I was early because my plans that day hadn't worked out, but it turned out for me to be good timing after all. Another time, having just arrived at the monastery, I forgot that the clocks had gone back when I rose one morning and came down to the Abbey Church for Matins. I felt immediately disconnected, out of synch with the world of the monastery and the rest of the world. For some minutes, until I had worked out the problem, I felt lost and completely disorientated.

Covid, more than anything else in recent history, helped us rediscover our connections; ironically it did it by disconnecting us. We have rediscovered our connections with the rest of society (that there is such a thing as society after all), and our disconnections with the rest of the world as supply chains have been interrupted or travel has been impossible. We have also seen in new ways the connection we have with the environment and with nature, and have appreciated the ways in which we are, admittedly, at least damaging the mechanism of our own survival as well as that of many species. I pull back, however, from the apocalyptic dogma that we are actually destroying the world. I will admit that Covid certainly sent us to sit in our rooms like solitary watchmakers where we began to observe the minute and intricate cogs within which we find our movement and our life. Covid sat us down in the silence of our own solitude where we heard the tick of our existence as part of something larger than ourselves. That was not altogether a bad thing either.

May showers have produced a green gauze across the landscape. Nature has placed one of her nicest cliches in the potting shed in the form of a robin's nest. A green flower pot on its side on a shelf five feet off the ground is spilling its contents of moss and downy feathers, together with four speckled brown chicks. I take my camera out while watering plants in the greenhouse, but find I've missed the departure of the fledglings, probably by an hour or two at most. I hear them hissing for food in the nearby bushes and kick myself that I didn't take the photograph earlier. Timing is everything!

Eric, the tame pheasant, runs to me when he sees me carrying the bucket of poultry food. His mate is beginning to come closer too, often within a few metres of me. When I move away she joins Eric and pecks at the food I throw down for them. The nesting season for pheasants (like the swarm season) is not an exact science, with eggs appearing any time between April and June. This year it looks as if Eric's mate has not begun nesting yet.

In the old apiary room below the North block the broody hens are starting to hatch the first ducklings of the year. They're early – by a day. The larger Kahki Campbell eggs usually hatch on day 28, but it's day 27 for them, while the Call duck eggs are a day early at day 25. I often find that eggs hatch a day early under broody hens. I feel beneath a fluffy white hen and my hand makes contact with two soft, warm ducklings, with two more eggs beginning to unzip.

Down at the apiary the bracken is now five feet tall, obscuring the hives from view. Foxgloves (*Digitalis purpurea*) are beginning to flower, and there's pink Campion (*Silene dioica*) in the hedgerows or in the dappled shade of creamy Elder flowers (*Sambucus nigra*). Some afternoons I see a pair of Mallard on the monastery pond rippling the surface as they forage the shallows near the budding yellow flag irises (*Iris pseudacorus*).

Though it's still cold and showery I decide to put the bait hives in their positions. The cuckoo might have been late this year, the warm weather delayed and the ducklings early, but bees have a way of surprising. When one beekeeper claims that swarming is late this season, another will claim it's started early; the best thing is to be ready for whatever happens.

*Clematis Montana* is exploding into flower, first in small constellations of pink-white flowers, culminating in nebulae of tremulous stars by mid May. All the outward signs of the relentless progress of spring into early summer continue to manifest, though daytime temperatures remain well below fourteen degrees Celsius, often no more than eleven or twelve degrees. Then rain dumps down on us on Bank Holiday Monday and stormy winds thump in the darkness like heavy waves on rocks. This time last year I was already making splits from the stocks and I already had queen cells. At this rate I won't be able to think of making increase to the stocks until at least the middle of May, as the wet, cold conditions will remain (according to the long-range forecast) until 12th of the month. As for the hive with a virgin queen, I have no idea if she will be successfully mated if these conditions continue.

In cooler weather there is something known as apiary vicinity mating, in which queens will mate near or in their own apiary. *Apis Mellifera mellifera* is known to do it, which might be a way to keep *Amm* stocks pure in a breeding programme. If there are other hives in the area, a queen can mate with drones from nearby stocks and won't necessarily end up inbreeding with drones from her own hive. I'm hoping that the virgin queen I saw on 19 April in one of my survivor hives will have opted for vicinity mating and that there have been enough drones available. In these cold, wet conditions, it's unlikely that Drone Congregation Areas will be sufficiently populated yet.

With the swarm lures in place I'm hopeful that when the weather suddenly warms up from mid May I might attract my first swarm. Typically swarming follows a period of wet weather. Like the watchmaker, whose progress is slow and painstaking, requiring close observation, I must be patient. This is another shift in the mindset of my beekeeping; from an impatient obsession with mass production that puts the beekeeper's head in his exasperated hands, I have become more sanguine even in these difficult conditions. After all, it's nature, and I've decided to work with and not against her.

The wintry wind and rain continue, with sunny spells, but cold wind is stopping the bees foraging, so I have to keep an eye on their food stores. I'm keeping a good slab of pollen candy on the cover board of each hive, just to make sure they don't starve. It's the latest that I've ever given supplementary feeding to second season stocks that normally run out of stores by March or April but can usually begin foraging by then. For this reason, when it stops raining I decide to check the candy situation and to give them more food. While I'm at it, I pull a frame or two from each hive for a very quick inspection, being careful not to chill any eggs or brood. That would knock the stocks back quite seriously and could result in disaster, as winter bees have probably all died by now and new bees are constantly needed so that the colony can expand at a steady rate.

As individual beekeepers we would do well to see ourselves a little more like independent watchmakers; people involved in a slow craft that isn't reducible to an end product mass produced at the lowest cost. In fact, our beekeeping methods at the moment are at the very highest cost – to ourselves and to our bees. Our beekeeping doesn't happen in isolation from other beekeepers or from the environment. Every one of us, no matter how small we might feel, is a tooth in a cog in the most intricate, precise and beautiful mechanism of the cosmos. Just as the most intricate workings of a watch needs every tiny tooth and the space each side of it to function properly, so our beekeeping, be it only a single hive, is related to something much larger than ourselves. And we are called or permitted to master it, as we're permitted to master time as watchmakers, or to master beekeeping; that mastery is a lordship that is never *lording over*. Neither is it purely *at the service of*. Properly speaking, we are stewards who must be responsible.

For all my criticism of the Buckfast bee, there is something of the watchmaker's spirit that I admire in the work of Brother Adam. No one else could claim the title Bee Master before it is rightly applied to this quiet monk of St. Benedict, but with all his knowledge and technical achievement he was the first to say: 'When bees were still kept in skeps the term 'bee master' was in common use – with some justification for at that time the life and death of a colony was at the end of each season, determined by him. But we never really have had or ever will have a mastery over the honeybee...it is up to us to understand her ways and adjust ourselves to her truly marvellous nature, not attempting the impossible of 'mastering' her, but rather doing all we can to serve her needs.' (2) Amen to that, up to a point. I suggest that we can help provide and care for our bees, but we neither serve them, nor nature! I'd go further, however, because the life and death of a colony are just as much determined by the modern beekeeper as they were by the old skeppists. Every decision we make – to use or not use chemical treatments for varroa; to feed or not feed syrup; to import foreign queens or to use locally adapted bees; these, and every decision in the management of our stocks no less determines life and death than the decisions made by the bee masters of the past who ended the season selecting which hives to put over the sulphur pit.

What most inspires me about Brother Adam was his slow, determined craft over a lifetime, by which he produced something on its own terms that could be said was as beautiful and as elegant as a handcrafted watch. Brother Adam worked with meticulous care and precision over many years, selecting and discarding, refining and improving his bees in order to create a superbee that was analogous to the watches made today by independent, mechanical-watch makers. Such watches are accurate to within a second per month and, like a well made skep, will outlive their creator. Brother Adam spent hours in the solitude of his workshop, writing and recording his results, as a watchmaker makes his drawings, plans and calculations. As the watchmaker sits long hours, closely observing with his eyepiece, so Brother Adam sat at his bench at the precision work of instrumental insemination of selected queens, and the grafting of young larvae for queen rearing.

There were no quick results in his work, and though I criticise what the Buckfast bee has become in modern beekeeping, I do believe that Brother Adam set out on that life's work with the goal of serving beekeepers, in response to a crisis in beekeeping after World War I. Like him, and in common with many beekeepers, I want to *do things* with the bees, which is why I don't see myself embracing natural beekeeping with hives that are never touched. Treatment-free beekeeping isn't natural beekeeping because keeping bees in hives isn't natural and because it still involves managing an apiary by making increase through splits, catching swarms and deciding which breeder queens to use for raising new queens.

There have been all sorts of strange things going on with the bees this spring, and they're still going on, testing my problem solving skills to the limit. The strong, queen-right stock is looking healthy and continues building up, with eggs and brood at all stages of development. I pull out a frame, making sure it has eggs, and take it to the queen-right hive that has no eggs or brood. I'm wondering if the queen is failing and if the colony will supersede her, if they're given a chance with some new eggs and brood donated from another hive. With available eggs, they have a chance of raising queen cells if they want to replace the queen. Perhaps last summer she didn't mate properly, or she might be failing for another reason unknown to me but known to the bees.

Next I inspect the stock that has a newly emerged virgin queen. I calculate that it's two weeks since she emerged, which has given her enough time to get mated. There isn't time to have a good look because of the cold wind, but a glance at her and two frames in the centre of the brood box leaves me just as uncertain. She looks enlarged, which could be a good sign (newly emerged virgins are quite small and their abdomens are fairly short). Or perhaps she hasn't yet had time to mate because there aren't enough drones around. In another week, if she hasn't been mated she'll become a drone layer and will have missed the three week window for successful mating.

But there's a surprise for me in the fourth hive, the one where I saw multiple eggs but could not see a queen. I first suspected laying workers, but the eggs were on the floor of the cells, which laying workers don't do, as they can't reach that far into a cell; instead they tend to lay eggs on the sides of the cell. If eggs are on the bottom of the cell a queen is laying them, and new queens can sometimes lay multiple eggs in a cell before they settle to a laying pattern. When I open the hive again on 4 May I find eggs, with brood at various stages, and new, capped brood as well – and it's normal worker brood. I know now that I have a laying queen and that she's fertile and not a drone-layer. There are plenty of questions though; why has she only just started laying, and why have I not seen her yet?

It's most likely that last August she replaced one of the new season's queens I had already marked. There was enough time after that for a new queen to mate but I wouldn't have known about it because I stopped making close inspections or looking for queens after all the season's new queens were marked. That was probably a mistake, as I should have monitored them throughout the autumn, to see how they were laying. The new queen was transferred to this hive with her nucleus, without me seeing her because she was probably dark as well as unmarked. If you're looking for a marked queen who isn't present any more while her replacement is very dark, it's easy to overlook her. But why has she delayed laying until now? It's possible, I guess, that she has a percentage of *Carnica* DNA. That would cause her to delay brood rearing until enough nectar and pollen are coming in. Late spring build up and brood size in proportion to available forage is a Carniolan trait.

Perhaps the same diagnosis applies to the colony in which I've just donated a frame of eggs and brood. The queen looks perfectly healthy, but she's been laying very small patches of brood. It might be that she's behaving the same way as my overlooked queen because she's also manifesting a Carniolan trait. If the colony ignores the frame and doesn't make queen cells, the queen must be all right after all, which would suggest she will probably begin laying when the weather warms up. Since the end of March the daytime temperatures have been below the seasonal average, with only a couple of days when it has risen to fourteen or sixteen degrees Celsius. It's been the frostiest month for sixty years too, and even now, in the first week of May, it feels as cold as February. Perhaps it's Global Warming?

It's likely, I speculate, that the bees are waiting for warmer weather before the queens really hit their stride with egg laying. That tendency is going to be more pronounced if a queen has Carniolan genes, while a queen with a high percentage of Italian genetics will start brood production much earlier. Maybe what I'm seeing in this variation in queen behaviour is a reflection of the mixed races of our local bees, with some queens being mated to more Italian drones while others are mated with a higher proportion of Carniolans. If the bees are late building up, the late, cold spring is an additional problem though, regardless of the lineage of the bees. It's been the coldest April since records began, and the coldest May Bank Holiday ever. On 5 May the rain turns to sleet. My sister in Barnet messages me to tell me it's been snowing there. *Seriously?* I email her. It's all bad timing for beekeeping.

One of the things that has interested me is the suspicious timing of the so-called Isle of Wight disease. As I continue researching in the *British Bee Journal* more details emerge. All through the 1915 editions it's clear that the mysterious plague was spreading and reaching a crisis point. Along with foul brood, the Isle of Wight plague was frequently the subject of articles and letters in the journal well before World War I, but it does seem that it hit many beekeepers, including Brother Adam at Buckfast Abbey, and William Woodley, especially hard in 1916. Throughout the 1915 editions of the journal people wrote of their experiences of the disease, such as G. Ward in October. In the same month another article reported that the disease was more prevalent in Durham than in any other district. Others reported the loss of bees in whole villages or districts, such as John R. Truss, who wrote in the December edition that even the bees in the village hall roof had fallen victim. This latter point is interesting because it lends weight to Brother Adam's claim that all wild native bees were affected, as well as managed stocks. On the other hand, other stocks seem not to have been affected at all. Richard Ling, of Briston, Melton Constable, reported in December having 31 stocks packed up for the winter and no sign of disease, while, 'I have not heard of any "Isle of Wight"disease within 20 miles...' (3)

It seems to me, however, that one factor that helped spread the plague was the lack of knowledge and information available to those who did not belong to the *BBKA* and did not subscribe to any of the beekeeping publications. This was pointed out by J.S. Fry in September 1915: '...it must be well-known that there are -...two thirds of the so-called bee-keepers in this country who do not possess the Guide Book, to say nothing of taking the Journal or Record. In the face of these facts, how are bee-keepers to combat "Isle of Wight" disease or Foul Brood?' (4) It's certainly easy to understand from this view the accusations made at the time by modern frame hive beekeepers that skeps were to blame; most skep owners would not have bought the journals whose audience and content were prejudiced towards managed hives and modern beekeeping. An exception to this can be found occasionally, such as, '...a beginner' (5) in October 1915 who asked for advice on wintering a skep.

The problem was compounded not only by traditional beekeepers' lack of knowledge, but also by the inexperience of many modern apiarists who were new to frame hives and modern methods. Frame hives were also still relatively new and no one had the kind of experience of them that has built up over the last century. Many of the beekeepers who had switched to frame hives were also inexperienced, having only recently changed over from skeps. There was also the issue of the neglect of some frame hives, either through inexperience or by beekeepers who might have continued to apply the 'hands-off' approach of the skeppist, or because the owners of some hives were away fighting the Great War.

Some of these soldiers wrote to the *British Bee Journal*. In November 1915 a letter appeared from a soldier at the Dardanelles, written in September. In October Sergeant A.G. Atwell wrote in 'With Bees at the Front' about skeps and beekeeping in France. Similar contributions were made by soldier beekeepers such as Lance Corporal J.L. Tuckell in November. Given that there were clearly many beekeepers fighting the Great War, it's worth wondering what happened to their hives back home at that time? Were they managed in the soldiers' absence, or were some neglected, allowing disease and unhygienic conditions to proliferate?

The journal also speculated at that time about the contribution that a swarm season might have made in spreading the problem, 'Last spring it was a most exceptional year for stray swarms...' (6) Was this just an unusually swarmy season, or were more swarms escaping because of managed hives that were suddenly unmanaged during the war? When you add to this the particular weather conditions of the winter and spring of 1915-16, it's clear that no single factor can necessarily be attributed to what became known as a disease.

A growing view (which I share) is that something more than a disease was presenting throughout those years, (possibly viral rather than pathogenic) exacerbated by an unfortunate set of circumstances and timings, all happening at the wrong time and creating a perfect storm. If there had been no Great War, for example, I wonder if the problem would have simply trickled along in the same way as foul brood, instead of suddenly exploding during World War I ? If the spring of 1916 had been kinder and had not turned cold again after Easter (like this year!) keeping the bees inside, would they have fared better that year? What contribution would the limited availability of sugar during the war have made in the autumn of 1915 and 1916, particularly to the neglected hives of soldiers away fighting? Interestingly, in World War II sugar rations became available to beekeepers. But during the Great War what difference might it have made if everyone's bees had been given plenty of syrup before wintering? Would it have been another factor of many that might have prevented the bees entering the spring of 1916 in a weakened condition, making them more susceptible to the vagaries of the weather that year?

It's easy to see how people like William Woodley began to form the opinion that it wasn't skeps alone that were to blame, but modern beekeeping and frame hives. By 1915 Mr. Woodley's regular contributions to the *British Bee Journal* had stopped, but reference in December was made to his controversial views surfacing in the *Record* that October. By then he had shifted from merely defending the poor cottager and his skeps to a much more radical view of traditional beekeeping, as F. Rider wrote: 'Mr. Woodley in the *Record* for October, recommends that we return to skeps as the only means of saving the remnant of the British bees. Although I cannot follow him in this direction there is a clue in his arguments which is worth consideration. The skeppist overwinters his bees on honey, the frame hivist winters his principally on sugar syrup. We have for many years outraged nature in this respect, and I believe this plague is the result.' (7)

At last this provides solid proof that Mr. Woodley believed the modern frame hive was largely responsible for Isle of Wight disease. But why would he have believed that the answer was to return to skeps? I assume it was because skeps were a lot closer to the natural way bees live. The more artificial the method of keeping bees, the less natural and the more interventions are required by the beekeeper. These interventions could be regarded as disruptive in the same way that opening a watch mechanism every day could disrupt the intricate balance and movement of the parts.

That's not a strained analogy either; whenever we open a frame hive we alter the mechanism of the colony in minute ways that we don't register, but the bees do. The temperature changes in the brood nest and can take a colony a day or so to reset. The bees will begin gorging on stored honey in response to the smoke, a reflex that prepares them to escape a fire. Queen pheromone is diluted by allowing outside air over the exposed top bars while the roof and cover board are off the hive. The even distribution of this important substance will, like the temperature, have to be reset by the bees. When beekeepers move frames, either by removal for inspection, or to rearrange the brood configuration, or during some other manipulation, they disrupt the way the bees are organised, especially if empty frames are put into the centre of the brood nest, splitting the brood and the nurse bees. Any number of small manipulations can set in motion a whole chain of events that we hardly understand.

On the other hand, no one can deny that skeps and modern natural beekeeping are intrinsically less disruptive. Natural and skep beekeepers tend to disturb their stocks less frequently and only when necessary. The analogy here with opening a pocket watch to look quickly at the time, is apposite. Modern beekeepers, however, have perhaps become more like horologists who can't resist opening the back of their watches just to admire the movements inside.

The idea that we might be outraging nature is with us more than ever today, driving our prevalent environmental concerns, of which we are all aware, especially since Covid 19 hit us. Incidentally, there is a haunting letter in the *British Bee Journal* in those years of growing crisis making the point that the bee 'plague' prefigured another plague that would have similar consequences for the world's human population. It's a chilling letter, a precursor to what became known as Spanish Flu in 1918. History, they say, tends to repeat itself.

By 1916 outraged nature's new plague had visited her wrath more or less the length and breadth of Britain. In the report of their Annual Meeting the Northumberland BKA wrote: 'We regret to report that the "Isle of Wight" disease is in almost every district throughout the country...The stocks of a good many beekeepers have been wiped out.' (8)

Strengthening the case that some native Black bees survived (contrary to Brother Adam's conviction that none survived after 1915) there are reports in the *British Bee Journal* of the existence of survivor stocks. Arthur Throwse of 51, Eade Road, Norwich, reported no losses from his stocks. F.N. Colebrook, on Military duty, claimed to have cured his bees with a product called Izal. In June 1916 Frank Coe of Wisbech claimed to have also cured his bees, using pea flour. Others, however, continued to be at the mercy of the disease. Even William Woodley was hit during these years, writing again in 1917: 'The winter of 1915-16 reduced me to a few stocks and as the spring advanced these developed symptoms of "Isle of Wight" disease.' (9)

It was 1916 when some extremely important letters appeared in the Journal, suggesting not only that Isle of Wight disease was not new, but that under a different name it might still be around today. If this is so, then the times in which we now live bear a striking similarity to those a century ago when 'plague' assailed bees and humanity alike at the same time. A subscriber cited an extract from the *British Bee Journal* from 1 June 1879 describing symptoms identical to those of Isle of Wight disease: '...200 or 300 bees, not dead, but unable to fly, and they crawl about in front of the hive until they die...On examining them I find they are bloated.' (10)

F. Kenwood remarked beneath the extract that a similar case occurred some years previously. The subscriber's response was made to Herbert Patey's suggestion in the 23 April edition of 1916 of the Journal, that Isle of Wight disease had occurred decades earlier. He referred to a shoemaker from the village of Stokenham thirty years earlier: 'He told me he had kept bees close on forty years, but that at one period he lost the whole of his bees, six stocks. They had, he said, some disease, apparently in the wings. They came out and dropped off the flight board...and were unable to fly.' (11) He added, picking up on the moot point that connected the disease to sugar syrup: '...the beekeeper never fed his bees on sugar syrup and bar-framed hives were not in use in the district until some years after the period referred to.' (12) X.Y. Z. writing in the 22 June edition, 1916, added that British bees in the 1870s were afflicted by an ailment known as May Paralysis and that it was the same as Isle of Wight disease.

This season, as foul brood is increasing again, beekeepers have been informed that a virus known as Chronic Bee Paralysis Virus (CBPV) is also increasing. It seems likely that May Paralysis could have been CBPV, in which case Isle of Wight disease might not have gone away. No wonder treatments during the Great War didn't work. Unlike foul brood, which is bacterial, CBPV is viral – like Spanish flu and Covid. It wasn't until the 1930s, with advanced microscope technology, that scientists could even see viruses. Antibacterial treatments would not have cured a virus. In cases where it seemed to have worked I suspect it was effective on secondary symptoms alone, such as the weakening effects of bacterial infections and poor hygiene that made bees more susceptible to viral ailments.

Further, there would seem to be evidence that CBPV might have a longer history. Over two thousand years ago Aristotle wrote about black, hairless bees on the island of Lesbos. On the other hand, there were many things about bees he was wrong about; he thought the ruling bee was a king rather than a queen; he thought drones were a different species; and he conjectured that bees stored noises in earthenware pots and carried stones on windy days to stop themselves being blown away. But when he wrote about hairless black bees it was an observation rather than conjecture. Hairless and greasy or wet-looking bees is a symptom of Isle of Wight disease. Could Aristotle's description be proof that Isle of Wight disease, by any other name, is actually as old as the hills?

I wonder then if William Woodley was at least looking in the right direction when he suggested a return to skeps, but not because the skep itself was the answer or because frame hives were the problem; but could it be that by focusing the debate on how bees were kept and managed, everyone was actually missing the point, and that the real issue was the importation of foreign bees and the hybridisation of British stocks? A possible reason why

William Woodley had never known foul brood in skeps was perhaps because foul brood (and later Isle of Wight disease) were introduced by foreign races of bee that had immunity to them, whereas our native Black bees did not. When Mr. Woodley had started his beekeeping with skeps there were fewer foreign races in British apiaries, and the bees he worked with were all his favoured native Black. This is probably why he was so adamant that while he kept skeps he had never even heard of foul brood. I suspect he might have been right about that.

Were foreign races of bee the real problem all along? While it's perfectly clear that the rise of foul brood and Isle of Wight disease at the end of the nineteenth and the beginning of the twentieth century paralleled the change from skeps to frame hives, the problem of bee diseases also went hand in hand with the increased use of foreign races of queen and cross-breeds. Hardly anyone at the time made this connection. If our earliest record of CBPV is from Aristotle in the Greek world, the smoking gun for Isle of Wight disease might indeed be a foreign race of bee, such as the Cyprian, the Ligurian or the Carniolan, all of which had been imported from the mid nineteenth century.

What about the state of British stocks at that time – managed and wild? For half a century there had been imported queens coming into British apiaries, experiments with cross-breeding the races and random hybridising. We don't know the conditions in which foreign queens at that time were raised, but we do know their journey here was long and must have been stressful. We know that stress, the age of the queen and epigenetics can affect the performance and life span of a queen. The science of genetics was still young, while our understanding of bee biology was incomplete in basic areas. Selective cross-breeding and inbreeding efforts were only decades old, based on Mendel's laws of heredity, but as yet there was no understanding of areas of genetics such as epigenetics, the study of how certain genes can be switched on or off by environmental triggers. (Twins, for example, can share identical DNA and yet only one of them might develop a particular disease.) I wonder to what extent these epigenetic modifications were occurring in British stocks in the decades leading up to and during Isle of Wight disease, as bees from abroad had to adapt to a different climate and environment, and because of cross-breeding and inbreeding to select and fix desirable traits. Could epigenetic modifications have modified genes (which we now know can also be inherited), making our British stocks (including native Black bees) susceptible to diseases to which they had previously been immune, or to new diseases arriving perhaps with foreign bees?

In 1963 Leslie Bailey and colleagues isolated CBPV and debunked the view that had prevailed until then, that Isle of Wight disease was caused by tracheal mites, now known as acarine. There seem to be two distinct types of the disease. In Type I (more common in Britain) bees shake and tremble. In Type II they are hairless and darker, as though wet or greasy. In 2007 the disease was recorded in Lincolnshire, but by 2017 it was present in thirty-nine out of forty-seven English Counties and six out of eight Welsh Counties. Between

2011-12 the National Bee Unit surveyed nineteen thousand colonies and found only 0.7 % of hives were affected, which is minimal. By 2015 a new study identified that 16% of colonies tested positive for CBPV, with 46% of bee farmers being affected. This makes it a reemerging disease, which, according to studies, is increasing exponentially. From no cases in 2006 to 80% of English and Welsh Counties affected in just a decade from 2007-17. This is an alarming rate of increase, and it is reflected by studies from other countries globally. Alarmingly too, in 2017, 25% of tested colonies were positive but asymptomatic.

*Black robbers, mal nero, maladie noire, or schwarzsucht*; by any other name Chronic Bee Paralysis Virus manifests the same symptoms as Isle of Wight disease a century ago: bloated abdomens, flightless, hairless, sluggish and dying black bees falling to the ground under their hives. Spread by body to body contact, the virus seems to reach a critical threshold which amplifies it under certain conditions such as wet weather or wet seasons. It would seem that the virus can also be present for generations without presenting symptoms – hence those *British Bee Journal* reports of May Paralysis in the 1870s could refer to the same disease. It also correlates with that period of wet weather after the Easter of 1906 and, again in 1916, that confined the bees to their hives for long periods just before the last wave of the disease hit many beekeepers, including Brother Adam and William Woodley.

According to the *Bee Base* records, commercial beekeepers are 1.5 times at greater risk of seeing the disease in their apiaries, while those who import bees are 1.8 times more at risk of their bees becoming affected. This doesn't necessarily mean that imported bees are the source of the disease. Perhaps they are carriers, or they might simply be more susceptible to strains of the disease endemic in Britain. What is clear from a study by Newcastle University is that cases are concentrated in apiaries run by professional beekeepers rather than by amateurs. This and evidence that the disease is twice as likely to be present where honeybees are imported, does seem to suggest a smoking gun in the direction of commercial beekeeping. Moreover, commercial apiaries are larger and have bigger, more densely populated colonies. On the other hand, most UK bee farmers are relatively small (one hundred to two hundred hives) compared to the commercial apiaries of ten thousand or more colonies in the US.

This doesn't necessarily mean that commercial beekeepers are causing the disease to spread. More research is needed and is being carried out, to identify the stress factors that might trigger the disease, such as wet seasons. Field trials are needed and analysis of virus DNA before conclusions can be drawn. One positive area of progress, however, is that the Veterinary Medicines Directorate has identified two antivirals in lab trials which seem to reduce infection.

If wet seasons and overcrowding seem to be factors that cause the virus to erupt, then Mr. Woodley's opinion that beekeepers had to return to skeps in order to save the native Black bee could have been disastrous if it had been taken seriously. Skeps are smaller than frame hives, which means that in wet conditions bees are more congested in skeps if they are confined. On the other hand, perhaps crowding would have been less serious for native bees, with their generally smaller colonies, but larger populations of Italians or Carniolans that had built up in the spring would have been more seriously overcrowded in skeps during a wet May or June. And we know that some skep beekeepers even as far back as the 1870s were discussing in the *British Bee Journal* how to Ligurianise a skep of Black bees by introducing an Italian queen. I wonder if CBPV was already present, though asymptomatic, in such colonies during wet weather seasons well before World War I? For example, in 1902 William Woodley and others discussed the inclement weather during May and up to mid June, in which there were hailstones and it was as cold as March. The observation is even made in the *British Bee Journal* that in 1902 the weather was even worse than it had been in 1888.

It's also ironic that CBPV seems to be more prevalent in commercial apiaries, which in Britain are often comparable in size to Mr. Woodley's bee farm. Although studies are careful at this stage not to scapegoat commercial beekeepers, a question that arises is whether or not CBPV is linked in any way with queens imported by bee farmers. In defence of William Woodley, he worked with (and indeed preferred) the native Black bee, so it would be unfair to suggest that he might have contributed to Isle of Wight disease as a bee farmer. In those days (as now) the majority of imported queens ended up in hobbyists' hives, and these might have been part of the problem because of the larger colonies they produced. Clearly, the larger the colony the more crowded it would be in both a skep or a frame hive if confined for any length of time.

The question can be asked: did CBPV disease arrive in Britain with imported stock, possibly decades before Isle of Wight disease first broke out? Or were colonies more susceptible to CBPV because foreign queens were not adapted to variants of the virus to which they were exposed in Britain? The same question about imported queens can and should be asked today, and is another potential threat to the future of beekeeping and our native stocks. We worry about mites that we can see, but ought we also to worry about viruses we can't see, that might be carried by asymptomatic bees?

Regardless of whether or not Mr. Woodley was right or wrong about returning to skeps, or if he might have been partly responsible for Isle of Wight disease spreading because of his views as a commercial beekeeper, was he right to resist Bee Disease legislation? It's a question to which I keep returning. In some of his writing he seemed to be defending the poor cottage labourer with his outspoken criticism of *Bee Disease Bills*, but not always. In the January 30 edition, 1902 his *Notes By The Way* made the point that legislation in the US had not eliminated foul brood. With his telling phrase, '...the strong arm of the law' (13) he asserted, '...even if empowered by Act of Parliament to deal with foul brood, it would be probably many years before it was stamped out.' (14) Then he made a point that again suggested his fears for the poor cottager's livelihood, while also reinforcing his wariness about the force of the law: 'If compensation was given by the State for stocks of bees destroyed ''By Ordo'', many beekeepers would, no doubt, be willing to receive the inspectors...' (15)

I'm sure his concern about the destruction of hives and stocks, 'By Order' of, '...the strong arm of the law' without compensation were as much a threat to his own business as to the cottage labourer. In fairness, however, it's unlikely that legislation for managing foul brood would have been sufficient, without significant amendments, to tackle Isle of Wight disease too. Perhaps Mr. Woodley's observation was a good one – that legislation in the US had not made much difference there. Would it have made much difference to Isle of Wight disease either, which was much less understood than foul brood? At least for foul brood there were treatments that seemed to work, but no one could agree what was effective against Isle of Wight disease. Even today we're only just beginning to identify antivirals that might be effective against the CBPV virus, but as yet there have been no field trials. However much we might continue speculating about Mr. Woodley's  assessment and judgement of the situation beekeepers faced a century ago, we can say that legislation probably wouldn't have stopped Isle of Wight disease reaching a crisis. The problem for Mr. Woodley at the time, as a lone voice, was that he seemed to offer no alternative, and most people felt that something had to be done. It's very natural that, faced with a problem, that most people want to do something rather than nothing. There are times, however, when doing nothing is actually the right thing to do.

What about other factors that were involved: the movement of swarms and honey around Britain; hives neglected during the Great War, and escaping swarms; beekeepers who didn't belong to BKAs and would never have been inspected or have reported any disease; the importation and cross-breeding of foreign bees; the use of pesticides and synthetic fertiliser; the diminished growing of clover for animal feed; and probably a whole raft of other complex and interrelated issues? I'm inclined to think Mr. Woodley was at least correct in his sceptical view that the legislation proposed by draft Bills would not have been enough on its own to tackle the crisis. I can also see that his objections, on their own, were not enough either, and that by offering no solutions, Mr. Woodley had become, in many people's minds, part of the problem.

# MINDING THE BEES

## Chapter 14
### Pay Close Attention

'STOCKS, Straw Skeps.
10s each. 3 early swarms out of same, 12s 6d.
Free rail.
BROWN, Withington, Shrewsbury.' (1)

*British Bee Journal 1904*

Two and a half miles east of Beedon, in the village of Compton, Mr. G. W. Dyer and his wife were known to the Woodleys. In his *Notes By The Way* for 8 July 1909 Mr. Woodley reported on a great number of his swarms that season, adding that, 'Mr Dyer of Compton – had five the same afternoon...' (2) George Dyer was a plate layer who lived with his wife in a railway cottage at the Compton crossing, on the Newbury to Didcot section of the railway that ran to Southampton. While Mr. Dyer repaired the track (which he had also helped to lay), working 6.00 am to 6.00 pm, his wife, Sarah, worked as gatekeeper at the crossing.

A few years earlier Mr. Dyer was mentioned in the Journal in an article that referred to him as, 'Our friend Mr. Dyer...' (3) We learn from this piece that Mr. Dyer had bought his first swarm in 1893, having made his hives from old boxes. He had no beekeeping knowledge or literature, but eventually picked up the *Guide Book*. As the old cottage labourers had done, to make skeps, he used his winter evenings to make new hives, usually a couple a year. On average he took thirty pounds of honey a year from a hive.

Mr. Dyer represented a new type of beekeeper and a different kind of cottager, both of which emerged in the nineteenth century, and he helps us to appreciate that the world at that time, as now, is always more nuanced than we might think. It's easy to get into the mindset that imagines all rural cottagers at the beginning of the twentieth century were agricultural labourers of the kind that Mr.Woodley defended, such as in his *Notes By The Way* of 1 May 1913: 'The frame hive is out of their reach. How can a cottager on 11s or 12s per week invest in hives and other appliances for modern beekeeping?' (4)

Mr. and Mrs. Dyer moved into the cottage attached to the Compton crossing of the Didcot,

Newbury and Southampton line that had opened in 1882. Sarah was a local who knew the Woodleys and their beekeeping. Perhaps it was even Sarah who had encouraged her husband to take up beekeeping? At Compton they had a large garden with the house, in an isolated, rural location, with security of tenure, unlike the agricultural workers William Woodley observed changing employer annually; if there was work available and he was not among the many who moved to the towns and cities at that time. Making his hives on winter evenings, Mr. Dyer became accomplished, eventually winning national beekeeping medals for his honey, and becoming President of the Hampshire Beekeepers' Association. He also wrote 'Notes From Mid-Hampshire' in the *British Bee Journal* in later years.

In fact, from researching the years blighted by Isle of Wight disease in the editions of the British Bee Journal, it seems to me that the world of beekeeping at that time was much more nuanced altogether than many books would have us believe. Firstly, beekeepers didn't divide at that time neatly between skeppists and frame hive hobbyists, or rural cottagers and BBKA members. In between there were beekeepers who seemed to have used both traditional and modern methods side by side. Among these were many who experienced losses from Isle of Wight disease from both skeps and frame hives. There are reports also of wild colonies that were affected as badly as managed stocks, and managed stocks of foreign races that were just as affected by Isle of Wight disease as native Black bees were. Equally, there were beekeepers of native and foreign bees right through the worst years of the blight who reported no losses, and this doesn't seem to have had anything to do with the kind of hives they were using.

Mr. Woodley is just as complex, the more I discover about him. Just as it's remained almost impossible to fully explain what caused Isle of Wight disease, it's very difficult to pin down with any precision one single objection he had to Bee Disease legislation. It would be over-simplifying him, I think, to suggest that he was motivated purely by the plight of the agricultural cottager or to suggest that, contrastingly, he embraced everything about the Industrial Revolution. If you put him into any of these categories you end up with a man who appears to be a complete contradiction.

This doesn't mean he wasn't in some ways a contradiction. For example, in 1906 he bought a *Benz Velo* motor car, invented in 1885 by Carl Benz. The model became the first mass produced motor vehicle and participated in the first automobile race – the 1894 Paris to Rouen Rally. Interestingly, Benz (like Woodley) was fascinated by technology, and also fixed clocks and watches in his early career. The car bought by Mr. Woodley in 1906 (registration BW37), was bought from his daughter, Elizabeth Goodman, by the Science Museum in 1912 and is now in the *National Motor Museum* at Beaulieu, Hampshire.

Despite a fascination with technology, and teaching himself watchmaking from the *English Mechanic* (buying every edition from start to finish) he seems to have had some regrets about mechanisation. In writing about him in his obituary, his friend, Mr. C. Heap noted: '...he said the character of the farming has changed in recent years...in the old days the click of the cutting machine was not heard in the fields. Now the machine goes into a large field in the morning and the clover is all down by night-fall, whereas the swish of the scythe went on for days, giving the bees longer time to work on the bloom.' (5)

He would also have embraced the railway as an aid to his business. He wrote about using Newbury Station, while his swarms and honey sections going north would have gone via the Newbury to Didcot railway, right past Mr. Dyer's cottage. One of his adverts for, 'Ye Olde Englishe Bee' (6), priced 15s, in October 1907 assured the buyer that they were sent free by rail. (6)

Clearly, Mr. Woodley didn't embrace every novelty or innovation of the modern age, even in his beekeeping. He explained, for example, in his *Notes By The Way* of 18 December 1902 that he was dismissive of the American shook swarm method, preferring natural swarming, which he viewed as superior. Mr. Heap observed the same point about him in Woodley's obituary, '… his habit of thought saved him from the error of imagining that methods and practices that suit America are equally suitable for this country.' (7)

I suspect Mr. Woodley had a similar relationship with progress and technology to our own. We want our cheap, mass produced food, but we don't want to pay high prices for organic local produce. Consequently, we're locked into a system of monoculture farming and chemical spraying. We like the idea of re-wilding, but we're nervous about leaving large swathes of valuable land wild and uncultivated when we also want to reduce imports of food that we can easily grow ourselves. We want the convenience of the motor car, but we know the dangers of the pollution it causes. We want to use better forms of fuel, but we need electricity. As beekeepers too, we want our honey, but we think we can only have it alongside the chemical treatments and imported bees we're convinced are necessary in modern beekeeping.

William Woodley, like us in the early twenty-first century, straddled two worlds that were held in tension; one declining, the other gaining momentum, and there were cost-benefits that came with the Industrial Revolution, as we know, just as there are at the start of the Fourth Industrial Revolution. History teaches us that new laws were needed to protect workers from the worst terrors and abuses of the growing factory system. We know that small boys went up chimneys and sometimes never came down. We know that cities and towns not only improved the transportation and markets for manufactured goods, but that they also produced poverty, squalor and disease. We also know that people thought the steam train would turn sheep black in the countryside, but that didn't happen. Mr. Woodley's world of beekeeping in Beedon parish is a microcosm of this complicated relationship between progress and tradition. Significant too was the fact that he had come from a rural tradition and humble beginnings, which he could never forget, living in the middle of a hamlet all his life.

His beekeeping probably had the same uneasy relationship with progress that we have in our own times. Like us, he could not despise the advantages of industrialisation and Capitalism. Yet it must have been obvious to him (as it is to anyone) that before the middle of the nineteenth century beekeeping had remained unchanged for centuries; and that one of the changes that did not seem merely coincidental was the increase in bee diseases in proportion to the increased use of modern frame hives. I am quite sure that his call to return to skeps at the end of his life was a consequence of a very deep, '...habit of thought...' (8) about where beekeeping had come to and where it might be going.

Neither was Mr. Woodley being alarmist about the demise of the British Black bee. It is evident from many letters and articles in the war years of the *British Bee Journal* that there was a serious crisis that had implications not only for beekeepers but for the entire nation in the midst of food shortages caused by the Great War. Charles Heap pointed out that bees do more than just make honey, and that they are also important pollinators. He related in 1912 his visits to Worcester and Berkshire the previous autumn and that the lack of bees by then was evident from the fruitless appearance of trees in the orchards and village gardens. On some trees, his correspondence informs us, there were hardly any apples at all; and this was before the war had even begun, and in the middle of Isle of Wight disease, before it had done its worst.

I wonder to what extent the context of World War also shaped Mr. Woodley's views and the vehemence and anger of those who opposed him. As Isle of Wight disease escalated to its zenith in 1915-16 many people had lost much more than their bees: in 1915-16 there was the costly campaign of Gallipoli; Jutland in 1916, costing fourteen ships and six thousand men, despite victory over the German fleet; Verdun in 1916, the longest and costliest battle of the entire war, not to mention the carnage of the Somme in the same year, followed by the battle of Ypres in 1917. It was a time of national loss and crisis, when people's instincts were to pull together and to protect. It was a war that affected populations at home too, especially through food shortages. Though sweeteners are considered a necessity today, imagine sugar in short supply, and diminishing sources of honey - its substitute. The war certainly added bitterness to the bee crisis, and Mr. Woodley's views about legislation would have further soured the mood among beekeepers who wanted to pull together and tackle food security and the future of the native bee population.

Our times are not dissimilar. After Brexit and Covid, with war in Ukraine, economic crisis, strikes, rising prices and our own gaps in the supermarket shelves, we are living in a time of many crises. These have soured people's moods; they have disillusioned people about institutions, such as the monarchy, the church and even our political parties. We worry about food and water security and the environment. In the media we hear the constant apocalyptic doom and gloom of environmental crisis, global warming and climate change; political and economic instability, fear of escalating conflicts and the threat of nuclear war. We worry about an unelected global elite with ambitions of becoming a quasi world government with its post-human agenda to micromanage our lives in the name of saving the planet. Imagine if anyone were to deny the reality of this social, cultural and historic context; they would be treated like a holocaust denier. I suspect this is an analogy for how Mr. Woodley's views were regarded.

It can be difficult to assess someone's motives, views and actions, even when you know their personality, but we have to infer much about Mr. Woodley in order to make any attempt at understanding his beliefs and ideas about the state of beekeeping towards the end of his life. His use of the word scourged to describe people's attacks on him, and the blame he suffered for views that some believed had aided the spread of Isle of Wight disease, suggests this hatred of him caused him considerable emotional suffering. A letter giving notice to give up his tenancy of the bee garden at Stanmore in 1917 might also indicate that by then he had at least considered giving up beekeeping altogether: 'PS: I am sorry to give up the garden but all my bees have died with the I of W disease at Stanmore and nearly all my stock at Worlds End. I don't think there is a stock alive in Beedon Parish.' (9)

What we do know of him, from his bee farm, and his motor car, is that he was a pioneer, and a key trait of pioneers is perseverance. It's a watchmaker's main characteristic too, so we should hardly be surprised that earlier in 1917 he wrote again in a brief appearance of his column, 'When perseverance fails the swan sinks.' (10) He added that he was,'...not chastened by being wiped out, or nearly so, twice.' (11)

To those who might have wished Mr. Woodley had felt more punished by his personal experience of Isle of Wight disease, this was surely a defiant shot across their bows and a suggestion that his views about Bee Disease legislation were unchanged. I'm tempted to think, however, that he was an early victim in the end of what we now call the cancel-culture. The *Editorial* of the *British Bee Journal* explained his dwindling contributions to the publication in a different way,'...until through advancing years and his many other interests, he was unable to write so much.' (12) This seems consistent with C. Heap's other point in the obituary – about Mr Woodley's restocking,'...he had begun to work it up again, without any intentions after 1917, however, of going so extensively as formerly into the business.' (13)

This is confirmed by the last advert he ever placed in the journal, for various pieces of equipment he was selling off, suggesting that he had down-sized his business. Then on 13 January 1923 he sold Hilltop Cottages at Stanmore to Arthur Thomas Lloyd of Lockinge, for £240, while an advert appeared in the journal for the sale of his apiary. Right up to 1922, however, he was still advertising section honey and swarms for sale, the final advert appearing in June 1923, advertising section racks, dividers, clearer boards and porter escapes, for 3s and 6d.

Maybe at the end even his motives for semi-retirement were mixed. I'm sure at 78 he probably needed a rest, but I can't help supposing that he might have looked back at his former enterprise with the same words he had used in 1902 to describe his use of swarm decoys - that sometimes our best laid schemes fail.

On the subject of swarm decoys, or lures, my bait hives have now been out in position for a week, and there have been signs of scout bee activity around three of them. One of them is a twenty-five litre bucket hanging in a tree by the car park, about six feet up, and I notice a single bee visiting when I look a couple of days after hanging it up. Another box hive, on top of a storage unit beside the 1918 building, has had a constant stream of bees visiting when the weather has permitted, which probably means this site and / or its box is getting more votes from the scouts. This apparent preference might result in a swarm arrival, but at this stage nothing is guaranteed.

William Woodley's swarm decoys were different because his purpose was different. While I want wild or escaped swarms to settle in what they see as new homes, Mr. Woodley wanted his own swarms to settle nearby where he could collect them before they escaped. He described, '...bee bobs' (14) as worn quilts, waxed and propolised, which he hung in trees. He also placed *bavins* (bundles) of brushwood leaning against a forked stick rubbed with beeswax. According to Mr. Woodley, ' These contrivances attract some swarms every year.' (15) This implies that they sometimes lured swarms other than his own, but it is unclear if this was his intention or just a consequence of providing places for his own swarms to settle.

Looking at the weather at the moment, I'm not optimistic about seeing swarms before the end of May. Temperatures in the second week of the month are only just nudging fourteen degrees, and it's still showery and unsettled; this is the pattern for the next week at least, with temperatures improving only after the middle of the month. My beekeeping friend down the road has informed me that he has no idea how his own stocks are doing because the weather has not allowed him to open his hives. I'm thinking that is probably the case with many beekeepers, and I anticipate many of them being caught out when swarming eventually happens. If more beekeepers miss their swarms, this means there will be more swarms around for the taking, and that's good news for me. On the other hand, it's difficult to predict when swarming might begin when you can't inspect the hives and when the weather is unsettled.

Everything in the garden is at least a month behind owing to the depressed temperatures. Despite this, much of the natural world has continued, such as bird nesting. I've already seen juvenile blackbirds from this season's first broods, and I've seen a brood of robins leave the nest. The meadow, however, has hardly begun to grow. Going by my strongest stock that is building up with plenty of brood, the bees are catching up at last, as dry weather has permitted some foraging for early pollen and nectar.

All this considered, I think if a window of settled, warmer weather arrives at the end of May, there will be a sudden explosion of swarming; some will not be expected by beekeepers, having been delayed by the inclement weather. The rest will depend on the kind of June we get. The last time we had this kind of spring weather was 2012, as I recall, and that summer was a complete washout. I'm wondering if, at the end of this season I might be quoting Mr. Woodley, 'Sometimes our best laid schemes fail.' (16)

This need to adapt ourselves to the vagaries of the weather and a particular season are essential beekeeping skills. Like the watchmaker, we have to observe the smallest details of the mechanism and solve problems as they emerge. We have to diagnose where parts are rubbing, even in the smallest way, or where a mote of dust has entered the cogs. I think this is the reason I'm puzzling about how accurately Mr. Woodley read the developing situation of Isle of Wight disease in the context of World War I, the public mood at that time, and the increasing calls for Bee Disease legislation. It seems to me that, all in all, he failed to see the minute connections between these parts. The issue of legislation wasn't just about foul brood any more, or whether or not the cottager could transition from skeps to frame hives, for whatever reason. The debate had shifted and his arguments had failed to develop or to come up with an alternative to legislation. In the same way today I venture to suggest that the issues of locally adapted bees and treatment-free beekeeping aren't purely environmental, but are related to our own political, cultural and historical context.

Concern about the difficulty cottagers faced adopting the expensive frame hive date back in the *British Bee Journal* to at least 1873, as far as I can tell. In that year's publications there is a story of a vicar in 1868 who set up and ran a club for cottagers, to help them buy their first frame hive on an instalment plan. By World War I most of these cottagers were fighting at the Western Front or in the Dardanelles and people were more concerned about whether the nation could afford the loss of so many men to the carnage of war than whether someone might be able to afford a modern bee hive. Food production had also been affected, by war and by the declining number of pollinating bees. The debate about skeps and modern hives was by then more about the survival of our native bees, food security and public morale.

William Woodley's call to return to skeps in these later years was framed within this context. Was this a cynical way of shifting his ground and saving face for earlier misjudgements; or perhaps it was an attempt to move the debate away from the issue of legislation? Did he attempt to reestablish the only common ground with his adversaries, which was that the native Black bee was now in grave danger of extinction, and that something had to be done; not legislation, or more BBKA experts, but a return to skeps? It was the same old position, but using a different argument. I must admit, for all my admiration for Mr. Woodley, this remedy to save the native Black bee does seem like something of a *bee-bob* in the trees. It's not surprising that it didn't attract much interest or support.

I often recall the three beekeepers who arrived at Douai with the bees when I restarted the apiary, and how quickly they began arguing over the advice they were each trying to give me. That was my introduction to beekeeping, and I've come to a decision that I don't want to argue about beekeeping; I want to enjoy keeping bees. Very quickly I also realised in the early days of my new hobby that many beekeepers spend a great deal of their time worrying – mostly about varroa mites. I don't want to spend all my time worrying. If it's all about varroa mites and worrying, I'd rather not keep bees at all.

The bees living in our buildings for decades had no one worrying about them. Colonies living in old trees and buildings have no one pouring chemicals into their nests, but they are surviving,despite stories of environmental doom. They've been adapting for survival for millions of years. Foul brood came and the fittest colonies survived; Isle of Wight disease came, but some bees survived. In the US, before varroa was a problem, they had tracheal mites through the 1980s, but the bees survived and the problem was replaced by varroa mites. Foul brood went away by itself, as it did in Britain. We now know that bees are solving the varroa problem too – and they've worked it out without our interference. It seems to me that the bees are telling us what to do, and that minding them is more than taking care of them – it's also about paying attention. Minding the bees, for me, begins with paying close attention.

The French philosopher, Blaise Pascal, was right when he said that all of humanity's problems stem from our inability to sit alone quietly in a room. Sitting alone in front of a hive is the next best thing, only a lot easier. We begin to pay close attention, not only to the bees, but to ourselves. Our secular culture calls it mindfulness, and it's good for us on every level of our being. It happens also to be good for the bees, if we begin to pay close attention to what's happening to them.

*Minding the bees*, if nothing else, begins with our paying this close attention. *Mind* is from the Middle English *minde*, which itself comes from the Old English *gemynd*, referring specifically to the faculty of memory. To *mind* means, in this sense, to *call to mind* or *to remember*. To *mind*, broadened from the Old English definition, means also *to notice* or *become aware*, as in *paying close attention*, or *to heed*. We also use the word, meaning *to be careful* or *to take care of* or *to look after*. Furthermore, it means *to be concerned about* – hence the phrase *never mind*, meaning *don't trouble yourself or be concerned*.

I chose the title for this book for all these reason, beginning with the obvious one - that William Woodley was inspired to keep bees probably by that formative experience of minding the bees for his great aunt at Stanmore. More than this, however, his story and the story of how we got to where we are in modern beekeeping, means that we have to remember his story and the story of what happened to Britain's bees a century ago. That, in turn, should make us notice and pay attention to the similarities with what is happening to them and to ourselves now. It is my hope that in doing so we might take heed and be more careful about what we do in the future. And I believe that we *should* trouble ourselves and be concerned because if we look after the bees they will, in turn, look after us.

Never mind then about *telling the bees*; we do well to listen to what the bees are *telling us*. Since I began minding the bees they've helped me to notice that a century ago humanity shifted from an era of great change to a change of era, and that the same thing happened in beekeeping. Coincidentally, both bees and humans went through a plague at the same time, and a century later we've been through another recent plague, and coincidentally, so have the bees. We are also perhaps at the start of a change of era, for beekeeping and for us.

# MINDING THE BEES

## Chapter 15
### Long Overdue

'STRONG healthy stocks
of English black Bees, on wired frames. 25s -
WHEATON, Exton, Topsham, Devon.' (1)

*British Bee Journal 1912*

It's neither one thing nor another so far this month. It's as cold as winter still, most of the time, with heavy rain and sunny intervals, though the temperature is still struggling to reach fourteen degrees. Despite the cold, inclement conditions, there is still a sense of the spring's indomitable progress. New leaves have transformed the countryside with a light mantle of May's distinctive fresh young greens. Dandelion clocks fleck the lawns and meadow, and the last pink apple blossom is hanging on, despite a deluge of rain over the last few days; hail yesterday (17 May) and thunder crashing and rumbling through dripping leaf canopies from mid afternoon until about 7.30 pm.

Today begins with bright sunshine. After Matins and Lauds I walk in the grounds. As though to emphasise the strangeness of the season, a hare appears at the meadow gate as I reach the 1918 building. I stop walking and stand, motionless, as it approaches. It's relaxed and seems unaware of me, see-sawing slowly past me about thirty feet away. Goodness, hares are huge! Only after passing me does the hare realise what I am and changes gear, racing away. It's the closest I've ever been to a live hare. Usually I see these elusive animals across the car park early in the morning from the Abbey Church, or through the meadow's long grass on summer evenings at dusk. Rarely do I catch more than a glimpse of a tail retreating at speed, or those huge, dark-tipped ears twitching above the grass.

Eric the pheasant comes to be fed. He worked out a year or so ago that by staying in the Abbey grounds he can avoid the local shoots. Over the past week his mate has finally disappeared and I'm assuming she's now nesting somewhere nearby – unless she's fallen victim to something. Our chickens are waterlogged and muddy, their enclosure looking worse at the moment than it's looked all winter. I decide to let them out onto the grass, to free range for the morning – respite from conditions that have made them look bedraggled and miserable.

It's strangely compelling, but reassuring, to find a record of similar weather conditions back in 1902. May that year was as cold as March, with hailstones leaving the ground white, the inclement, cold conditions lasting right into the first half of June. By 15 May the *British Bee Journal* lamented that there had been hardly a working day for the bees during May. There were warnings of starving bees as late as 22 May, with an expert from the Essex and Suffolk *BKA* (W.A. Withycombe) warning of starving bees as late as 29 May. Even as late as 16 June Mr. Woodley wrote that it had been, '...sunless and wet as November, and cold as March...' (2)

In the same year natural swarms were reported to have been few in number, with drone brood destroyed by bees, thereby removing one of their incentives for swarming. Others, such as J. Rymer of Levisham, Yorkshire, observed on 31 May that where bees had swarmed they had left empty comb behind. By June Mr. Woodley was writing in the *British Bee Journal*, 'Swarming is late this year, but a few lots came off from straw skeps in the valley of Pang between Saturday and Monday last.' (3) There had, however, been some swarming at the end of May, such as the two excellent swarms from Miss. W. Curtis' apiary, of Kingsclere, Nightingale Lane, Balham.

Here, in 2021, I've heard news of another swarm on 13 May up in York, which closed some Council offices, proving that even in this weather the bees can surprise us. Our bees at Douai Abbey have surprised me as well. Taking advantage of a sunny morning (18 May) I decide to check that the stocks have sufficient candy, given that more rain and cool weather are forecast for at least the next week before any sign of warmer, more settled conditions seem likely. I am heartened by what I find.

The hive that had the virgin queen at the start of the month has somehow managed to get going, with the queen now mated and laying. There are a couple of palm-sized patches of sealed brood on the frames. The hive that I suspected had laying workers (later deduced to be more likely a drone-laying queen) have also produced an increase of worker brood and eggs at various stages, which means the stock is queen-right. I'm now certain that this queen was superseded last autumn and that I haven't seen her because she's unmarked and probably very dark. It's also been too cold to go through every frame looking for her, for fear I might chill the little brood that is appearing. The queen-right stock with the non-laying queen, to which I donated a test frame with eggs, has raised three open queen cells, each charged with royal jelly and a young larva. As I suspected, they wanted the chance to supersede the old queen, which they are now doing. Finally, the strongest hive continues to produce solid frames of brood and a good number of drones, the latter being a good sign that the season is improving for the stocks.

What has surprised me most is that a queen has been mated this month, despite the constant rain, frosts (even in the first week of May) and depressed temperatures. It's possible that she was mated in the vicinity of the apiary rather than at a Drone Congregation Area, but either way she's managed to mate. It's good news then, as all the stocks seem to be going in the right direction at last when a couple of weeks ago I was concerned that I might be reduced to a single stock. Hopefully, by the time fine weather arrives in June I will be able to split these stocks into nucleus colonies, making increase from the survivor stocks. Much will still depend on June's weather.

I mustn't lose hope. Back in 1902, after an almost identical start to the season, there was a heatwave in the second half of June, with Mr. Woodley reporting swarming, '*...ample to meet all wants*' *(4)* within ten days of the cold weather. No such luck here though. Instead, I've received notification from my Beekeeping Association of a zoom discussion group this Friday, prompted by the strange and challenging season and the difficulties it is causing for beekeepers. Although my problems seem to be resolving, I think it would be helpful to attend the meeting, to compare notes and to learn more about how this start to the season is affecting other beekeepers in the area.

It strikes me, however, in the light of this spring especially, that every year beekeepers encounter different situations. No two springs are the same, while every beekeeper's situation will also differ, according to their hive histories and what they did or did not do last season and through the winter. Moreover, no beekeeper is right about everything. Like gardeners who learn by making mistakes in gardens that differ in many ways even from the garden next door, we beekeepers also learn by making mistakes. Such differences exist in the bee garden as in any garden. Mistakes are an important part of learning, whether it's gardening or apiculture, and they can be your best teachers. It's certainly a truism in beekeeping that the person who never made a mistake never made anything.

None of us have all the answers, and we all have slightly different experiences. I'm wary of anyone (especially self-appointed Technocrats) who claim to have all the answers to the world's problems. With this in mind, it strikes me, the more I reflect on the crisis of Isle Of Wight disease and everything I've read about it in the *British Bee Journal*, the more convinced I am that neither Mr. Woodley nor the editors, nor even the vast majority of beekeepers at that time were completely right about issues such as foul brood and Isle of Wight disease, skeps, frame hives and legislation. Views on these controversies tended to polarise, as we've seen with issues of our own day, such as Brexit, or Covid vaccines, with each side becoming more entrenched in their views. In reality, just as Brexiteers and Remainers each had some fairly good arguments for their positions, so too beekeepers a hundred years ago all had some plausible arguments to support their own positions regarding the bee crisis of their time. So too, beekeepers today who advocate treatment-free management of their stocks and use locally adapted bees, and those who favour a particular strain or subspecies and import foreign queens, or manage *varroa* with chemicals, all have plausible arguments in their defence. Nothing, it seems to me in the beekeeping world, can be reduced to simple either / or arguments, not least of all because beekeepers all have very different situations and experiences each year compared with other beekeepers, so discussions are never based on completely 'like for like' comparisons. An additional and inevitable factor is also quite simply that some beekeepers are better than others, either from greater experience, innate ability or both.

Perhaps an exasperated G. B. writing in the *British Bee Journal* on 8 January 1920, understood this: 'What is wrong with bee-keeping, or, rather with bee-keepers? We have had teacup storms about rival antiseptics, skeps, the price of bees, the standard frame, and now legislation, the last of which can only be settled, if we are to take some writers seriously, by burning the Editors on a pyre of the JOURNAL. What is the matter?' (5)

That year the journal invited readers to express their views on the topic of legislation to prevent *bee diseases*, and the variety of responses throws up some very interesting material for discussion. What's striking about these responses is that beekeepers at that time don't necessarily fit neatly into one camp or the other. A decade or two earlier those who resisted legislation, like William Woodley, were presented as pro skep-beekeeping too, while it seemed that frame hive apiarists on the whole were pro legislation as part of the progress desired in the craft. By 1920 those simple categories (if in fact they were ever real) don't seem as distinct. The issue of skeps seems almost by then to have disappeared, the concern instead being whether or not legislation could halt the spread of Isle of Wight disease. It's clear, moreover, that by 1920 the old definition of *cottager* had also all but gone, with a new generation (like Mr. and Mrs. Dyer of Compton) having other types of work than agricultural, and often having frame hives as well. Unlike the old cottagers who kept skeps to supplement their low income from agricultural work, these new apiarists, like the Dyers, didn't necessarily need the income from their bees to pay their rent. They were a new breed – the hobbyist beekeeper.

Throughout the 1921 editions of the *British Bee Journal* the debate on Bee Disease legislation continued, the editors calling for beekeepers not only to write in but to write to their MPs to support the Bill. The editors' bias became increasingly evident in their responses to the various letters, with footnotes similar to those in 1912 they ran to counter Mr. Woodley's opposition to the proposed Bill. Those who supported legislation have no editorial footnotes, but those who opposed, such as J. Sowrey in October 1921 were cut down by the editors, '... we have no intention of starting a discussion for the benefit of anyone who has apparently learned nothing from those of the past...' (6) The editors also reminded readers in November 1921 that, '...ordinarily the Editors' footnote to a contributor's communication is, I believe, to be considered final.' (7)

J. Thompson of Dalbeattie, wanted a much more sweeping Bill before he would support it – one that banned foreign bee imports, which he called, '...foreign rubbish' (8) and the, '... root of all the troubles now connected with beekeeping.' (9) It is unusual, but interesting to hear his voice on this subject, and the link he suggested between diseases such as *Isle of Wight* and imported queens: 'The Bill I want must prohibit all importations of foreign bees and queens for the next ten years, and that all foreign bees and their crosses found in Scotland be destroyed.' (10) To which the editors added, '...we are not prepared to go the pace of our correspondent.' (11) This, it seemed, was an attempt to shut down that line of argument. It's easy to miss too, but J. Thompson referred to native Black bees in Scotland that had survived the Isle of Wight disease – further evidence that Britain's native stocks had not been completely destroyed, as some claimed.

Throughout the *British Bee Journal*, over many years, it's apparent that there were indeed many beekeepers who never lost their *native* bees, either to foul brood or Isle of Wight disease, right up to 1921 when the crisis was drawing to an end. These beekeepers had kept bees for thirty, forty, or more than fifty years without ever introducing new stock. As J. Truman wrote, 'The modern man yearly purchases queens, nuclei or driven bees, just to keep his stocks from extinction. Why should the latter, who cannot keep his own bees, try and rule the farmer, whose only fault is that of having prevented the race from extinction?' (12)

There are a number of interesting issues here. Firstly, these beekeepers (including Truman) who, '..,prevented the race from extinction' (13) were working with the native Black bees, which provides clear evidence right up to 1920 that there were native bees that had survived the crisis. Secondly, the writer alludes to the notion that there were (and still are) good beekeepers and bad beekeepers. J. Truman did not believe that the reason beekeepers did not lose their bees was just good fortune; he mentioned a local expert half a mile away from him who had lost his bees and had to restock repeatedly after 1916.

The same writer resented the confrontational, even bullying tone, of some who advocated legislation. He objected to the idea that such people would have been permitted, with or without his being present, to examine his bees, especially if such a person were a so-called *Expert* like the one half a mile away who didn't seem qualified to preach to successful beekeepers, like J. Truman who had never lost their bees: 'A few weeks ago this beauty informed me: "Just you wait until that Bee Bill passes, then me and a bobby will see your stocks when we want, and may Gawd help you." What does a Leicester Bee-keeper say to this?' (14)

The rather belligerent attitude of *this beauty* seems to have been a characteristic attitude of some pro legislation beekeepers, reminding us today of similar belligerence we have witnessed on both sides of the Brexit issue. A. E. Stanley, for example, received a scathing response from the *Editors* for his irresponsible and aggressive suggestion in the Journal that beekeepers with neighbours who had diseased bees should visit their neighbour's hives at night and set fire to them.

Others, such as Mrs. Mabel Goodacre, were equally scathing about the Government's actions during the parallel crisis of the Great War, and on this basis didn't support legislation, because she claimed the Government had: '...failed in every thing they have touched upon – meat, butter, milk, cheese – and now the beekeeper is to be harassed so as to get rid of the English honey.' (15) These people, however, are contradicted by *Lover of Bees* who, as an inspector of apiaries in Lancashire, noted that he had not come across anyone against legislation and felt that, '...it is long overdue.' (16) Chas Little, on the other hand, who lost all his bees, was one of those in favour of legislation. Surprisingly, William G. Wells was prepared to accept and welcome legislation, too, despite an experience from which you'd expect him to oppose:

'...I have been asked to remove lead from many buildings in this neighbourhood, when the bees have become a nuisance to the owner or tenant, and so far as my knowledge extends, I have never seen any sign of disease. As a matter of fact, there are two stocks of bees, one beneath a lead gutter, and one beneath a slate roof, that have been there for the past fifteen years, and to my knowledge have never shown any sign of disease although the whole of bees numbering many stocks kept in the gardens belonging to the house have been destroyed with the 'Isle of Wight' pest.' (17)

Not only does this evidence contradict Brother Adam's assertion that no native Black bees survived after 1915, but it once again suggests the possibility that frame hives or managed bees, as opposed to wild bees, were somehow more susceptible to Isle of Wight disease. Was this due to inexperienced or poor beekeeping? Was it something to do with frame hives? Were wild bees somehow less likely to become affected, such that A. J. Brown's question was irrelevant, that: 'If we have inspectors coming round, what about colonies that are in hollow trees and roofs and buildings? If bees diseases are so infectious, these would be impossible to deal with them.' (18) A. J. Brown, a cottage beekeeper who had kept bees for thirty years, had experienced foul brood and Isle of Wight disease, yet was still against legislation. Further to his point about wild colonies, he pointed out that inspectors and a government body would need to be paid.

D. Davis was also against legislation for a variety of reasons. Firstly, he argued that Isle of Wight disease's infectious nature was unproven. Out of ten hives he had only one perishing from the disease, while nine close by were unaffected, '...and the same kind of bees.' (19) Secondly, he observed, that even in areas where there had been no bees for five years, reintroduction had led to a reemergence of the disease, whereas there were other areas where the plague was unknown.

He argued that legislation would end attempts to cure; rather than beekeepers continuing to struggle to find a way of saving their bees, inspectors would simply destroy affected colonies using officials who would need to be paid. The Board of Agriculture experts had by then failed to discover any suspected microbe in the laboratory, so a cure was more likely to be found by a beekeepers than by anyone else. What was needed was more research and study, to establish the cause of the disease, which would be a more effective use of money. Local *BKAs and* bye laws would surely be more effective (and cheaper) than national legislation. A counter argument was made by the editors that: 'The cause of leprosy is not yet known, therefore we suppose our correspondent would suggest that until the cause is found lepers should not be treated or segregated.' (20) Lastly, there were others who simply maintained that where legislative laws were in force they had resulted in great benefits to beekeepers.

From this small sample of responses to the *British Bee Journal's* invitation, what emerges is a very mixed picture of individual beekeepers' responses, based on very varied personal experiences of bee diseases and keeping bees. There were polarised views about whether *Isle of Wight disease* was even transmissible. The same polarisation presented a variety of views about how a cure might be found. Beekeepers were just as divided by their experience of losing bees or never having lost any, with no consensus that wild bees or managed bees were more or less susceptible, or that wild bees might even be unaffected. That's beekeeping for you (or beekeepers); it was ever thus!

It seems to me that the key to understanding what exactly caused Isle of Wight disease is somewhere within this very varied experience of beekeepers; which is that there was no single cause, even if a microbe or a virus were involved somehow in the phenomenon. In my view, there were many variables operating – beekeeper levels of skill and experience, weather conditions, the Great War, the unavailability of sugar, questionable products used in an attempt to cure affected stocks. Perhaps there were even viruses and pathogens at work, together with pests such as tracheal mites and the movement of stocks by people selling swarms. Even the sale and movement of honey might have played a part, as had the importation of foreign bees for over half a century, which would have also led to the hybridisation of stocks in certain areas, possibly weakening their resistance to diseases.

In other words, when an experienced beekeeper, such as Brother Adam, diagnosed his bees in 1916 with tracheal mites (acarine) we can only believe that his judgement was correct regarding his own bees. That was his particular experience, and we can trust his diagnosis based on his own experience because he was a highly skilled beekeeper. Equally, beekeepers who were at that time more experienced than Brother Adam, who had maybe thirty to fifty years experience of beekeeping, were probably no less correct in their assessment that experience and skill were not inconsequential factors in the spread of the strange plague. Let's not forget that in 1915-16 Brother Adam was a mere novice at beekeeping and didn't take over the apiary at Buckfast until 1919. This, however, does not mean that someone like William Woodley, who lost his bees in two successive years during the Great War, despite having kept bees for decades, was not in some ways more right or more wrong than Brother Adam in his assessment of the crisis. Perhaps, because there were so many variables in play, both beekeepers were correct, as were the others who weighed in with their correspondence in the *British Bee Journal*. Perhaps where they were incorrect was in generalising from their particular experiences, which were dependent on a unique set of variables in a particular location and at a particular time.

&#10086;&#10087;

Monastic life doesn't allow much opportunity to attend local BBKA meetings, as they are usually in the evenings or at weekends when I have other commitments. On 2 May, however, with the Abbott's permission, I attend an evening zoom meeting with about thirty members of the Newbury and District BKA. It's clear that most people know each other, from the online greetings and small talk at the start of the meeting. I'm the outsider, but I'm not here to make small talk. I've decided, in view of the unusual season and its challenges in the apiary, that it would be useful to compare notes with other hobbyists in the area, many of whom are more experienced than I am. I might learn something valuable.

Everyone agrees that this is the worst season since 2012, which was the worst honey harvest for fifty years. There is a consensus that there will be no spring honey this year and that no one should really expect a harvest this summer either. Everyone is feeding their bees; some with syrup, while others (like me) prefer fondant. Only one person seems to have attempted any splits, and he tells us that they're not strong and that he'll probably end up uniting them with other stocks, to strengthen them. Another member has Chronic Bee Paralysis Virus in one hive (which makes me sit up), and another has had an inspector visit to assess an outbreak of Sac Brood. There is a general agreement that inspections are necessary at the moment, but to go into the hives only briefly, to assess food stores and possible swarm preparations. Interestingly, there are a number of members who confirm my expectation that when the weather improves (hopefully next week) there will be a sudden explosion of swarming.

I expected more disagreements, but the only one centres on whether or not to give hungry bees frames of honey from another hive. Most say they would do this, but the lady who has CBPV in a colony says that she wouldn't do it because of the possibility of spreading the disease; not surprising, given that her bees have CBPV. In general there is agreement about not using imported queens, the value of locally adapted bees, that wild bees are extremely resistant and tend to thrive; that inspections are necessary at the moment, and that honey is not the most important reason for keeping bees, despite the fact that most of the old beekeeping books give advice based on the premise that honey production is invariably the common goal of all beekeepers.

What's clear is that most of the beekeepers in the meeting keep bees for a much wider variety of reasons than perhaps beekeepers did in the past, or a hundred years ago. A general enjoyment of bees and an interest in keeping them healthy seems to be the most important consideration for most at the meeting. At least two members seem to have native Black bees. From a single meeting it's impossible to assess the attitudes of all beekeepers everywhere, but I am reassured that I don't feel at all out of step with the consensus in the meeting that the welfare of our bees is more important than any other agenda for keeping them. I'm greatly encouraged and heartened by this.

Whether or not I catch any swarms this season, I need to think now about planning to make increase from the stocks I have. If I don't catch swarms, I'll still need to take enough stocks into winter to be able to lose some and to have at least a few survivor stocks again next year. I would normally try to make increase in May, which means the first mated queens are laying by mid June. This gives them time to build up before the autumn. I certainly don't want to leave it until July, as the nucs will be weak in August when wasps appear, and wasps will rob out and destroy a weak nuc, wasting all the work the beekeeper has done and destroying the chances of making new stocks. It's a gamble this year, because I only have a few stocks and I will need to split them into nucleus hives – one nuc for each new queen, but I don't have many stocks to work with this year. Fortunately, I can afford to weaken them by splitting them, as I don't expect a honey crop this season. My focus now is on increasing my stocks. If I begin the process this month, I can hope for new queens by the time the weather brightens up in June – at least that's the plan!

Raising new queens is something I have done for some years, and it's an important area of treatment-free beekeeping. While there are things I plan not to do, this is one thing I have to do. As well as catching swarms that might have the genetics for developing varroa resistance, I need to raise queens from my survivor queens. Every year I do this from survivor stocks, the closer I come to having stocks that also have a wider immunity to a whole range of diseases as well as varroa. If there are beneficial epigenetic modifications occurring in my survivor queens, they might not yet be expressed in the survivor colonies, but can be inherited and expressed in the daughter queens' colonies. In other words, my breeder queens probably already have the genetics to develop the range of traits needed to survive varroa and other diseases, but these might be expressed in the next generation.

If you want to be a treatment-free beekeeper and you are interested in developing locally adapted bees it's going to be important to raise your own queens. This area of beekeeping can be daunting for new beekeepers especially, and it can appear to be a rather complicated process, but it really isn't. Some people might make the argument that if you do raise your own queens there's no way you can guarantee their genetics because they will be randomly mated. One alleged consequence of this is that you could end up with a regression in your stocks' temperaments, while in some cases you might even end up with aggressive bees that you are unable to handle safely. That can take all the enjoyment out of beekeeping. Another argument that might be made against raising your own queens is that they won't be good quality, especially if they are raised under the emergency impulse.

When I began writing about this process I had been treatment-free for three years, but I am now a good few years in, and I have had several years of raising my own queens. I have not had any significant problems with the temperament of the bees in our apiary, and I don't think there has been any deterioration, despite some colonies being more docile than others. The only really unpleasant colony I've had were the bees I first took when they swarmed from Kelly's Folly in my first year of beekeeping, but I think they were probably not that bad. They appeared unmanageable because I had been keeping Buckfasts which were incredibly docile, and by comparison the swarm made me nervous because I lacked experience handling *sassy* bees. I think in relative terms they could have been described as temperamental, but not in general terms. As for the many queens I have raised (an average of a dozen to twenty new queens every season) none of their colonies have given me any problems with temperament. Occasionally there is a colony that runs more on the comb or is slightly more defensive than I would like, but I don't select from them for my next batch of queens. In this way I am selecting for docility. Several years down the line and I have mostly docile bees. There are some stocks I can handle without gloves, and I could probably handle them without a veil, though I tend to regard my bee suit and veil in the same way as the seat belt in the car – it's not an option, but a sensible precaution. As with any bees (even Buckfasts, in my experience) if the beekeeper does something stupid and careless, or if the weather conditions are unfavourable, they will let you know their displeasure. My own experience over several years then is that locally adapted bees do not necessarily become aggressive. You're more likely to produce aggressive bees (F2 aggression) from crossing certain races, or even in a first cross, so if you start off raising your own queens from Italian or Carniolan stock, for example, the random crossing you get in the first year might well result in a deterioration of temperament. That is why I would suggest starting to raise locally adapted queens from a captured swarm that is more likely to have diluted genetics of any particular race.

When it comes to methods of raising queens some beekeepers are sniffy about queens raised under the emergency impulse, claiming that these queens are inferior. I disagree, having had some very fine emergency queens. The argument (which has some wisdom) against emergency queens is that the larvae are not fed as well as queens started under the supersedure or swarm impulse. This can be true, if the bees only have the option of raising queens from older larvae, rather than having the availability of eggs and day-old larvae. It seems plausible that a larva fed the required diet to raise a queen for half the time would not be as good as a larva raised from an egg that is fed longer on royal jelly. On the other hand, if there are eggs available, the bees will use these for raising emergency queens, which will be fed for the full number of days required to raise a queen. The emergency impulse is nature's insurance policy to get a queenless colony out of trouble, so we have to give it some credit as a strategy that has evolved because it works. Occasionally, I admit, the bees will raise an emergency queen  which they don't like for some reason, as though she is just required to kick-start them, after which they sack her and raise another queen. This new queen generally satisfies them.

There are a variety of methods for raising queens that are quite simple. You don't have to go to all the trouble that you sometimes find in books and videos where you see beekeepers raising frames of thirty or more queens cells from very complicated procedures using lots of specialist equipment. Most of us don't need that many queens, and there are simpler methods for producing smaller quantities of them without having to make cell-starter colonies or having to learn complicated techniques like grafting.

Once I discovered a couple of key principles about how bees work, it unlocked for me how to prevent swarms, manage swarm preparations and raise queens. Let's begin, just for fun, by imagining the hive as a simple watch mechanism. You could call the queen the mainspring, because that's really what drives everything. Then you have the brood (larvae at various stages) and eggs, which we could call the movement or engine of the hive. Covering this are the young bees (nurse bees) up to three weeks of age, who, along with various housekeeping jobs, keep the larvae fed and warm. We might see these as a battery. Lastly, there are the flying or foraging bees that interface with the outside world, like the hands of the watch. Like the hands of a watch, they are also indicators that something within the hive is not working quite right or that the mechanism has broken down.

This analogy works in so far as any analogy approximates to what it is describing, but the point I am making is that there are parts or components to a hive, as there are in a watch, and they all depend on each other. If you take something away, the watch will not behave in the same way. Remove its hands and it still works, but it doesn't tell the time; remove the mainspring and nothing works; take away the battery and everything stops. With this model in mind, if we remove the queen from the rest of the colony it will have an effect. Similarly, if we remove the foraging bees, this will alter how the hive is working. If we remove the brood and nurse bees, the workings will be affected in a different way. By knowing this, we can manipulate the colony with the beekeeper's sleight of hand and prevent swarming, or encourage the building of queen cells. For example, without a queen the colony will not swarm, so to stop swarming you can simply start by removing the queen. Similarly, a swarm needs a high number of flying foragers, so if you can remove them, leaving the queen and the brood and nurse bees, you will short-circuit the swarm impulse. The last option is to remove the brood and most of the nurse bees, as a swarm will always want to leave behind enough brood and nurse bees to continue raising a new queen and a viable colony.

The easiest way to raise a queen, which doubles up as a swarm prevention technique, is to do a walk-away split. This will raise you either one queen in the split or a few nucleus colonies, if you cut out the emergency cells and distribute them between a number of nucs. All you do is remove half your brood frames (with brood at all stages and eggs) and put them in a new hive or a nucleus box with the attached bees. Fill up the gaps in the parent hive with fresh frames of foundation or drawn out comb. You don't even have to know which box contains your queen after the manipulation. It doesn't matter. One of the boxes will be queenless, but as long as you've given them a frame of eggs and young larvae they will raise a queen under the emergency impulse, producing perhaps a few decent queen cells, only one of which will reign after she has killed her sister princesses. The box that still contains the queen will be weakened. It will have all its original foragers, but the nurse bees and brood will be significantly reduced and the colony will not feel affluent enough to swarm.

Another method is to make an artificial swarm by removing all but a couple of brood frames and the queen and distributing them between two, three or four nucleus boxes. The parent hive will still have its foragers and will continue making a honey crop, but the stock will have been weakened to the point that it won't be prosperous enough to swarm. The nucs will each also raise you a new queen. Don't worry if you have a few emergency cells in a nuc; the first queen to emerge will kill off the competition.

If you prefer to raise queens from the swarm impulse (as Mr. Woodley said he did too), as long as you get the timing right you can do this, before your bees abscond. Swarm cells are numerous, so you could make up a good number of nucs from one stock that makes swarm preparations. The key thing here is not to panic, but to let nature do your work for you. The first thing you need to do when you see a colony starting a large number of swarm cells is to stay calm and see this as an opportunity. Secondly, they won't swarm until the first swarm cell is sealed, which is usually about day eight from the egg and the beginning of the cell. If the weather is bad for a few days after the cells are sealed this can delay swarming further, but assume the prime swarm with the old queen will plan to depart as soon as the first cell is sealed. Don't be caught out here and count from the time you first see unsealed cells, because they will already be up to four or five days old by then, and if you leave them a further week your colony will have swarmed by the time you next look.

You can make a split into a number of nucs, donating a frame with its queen cells to each nuc, with a couple of frames of brood and an extra frame of bees shaken in. Knock down all but a couple of cells in each nuc, so each new stock has only a couple of queen cells. Alternatively, if you want to keep the parent hive strong and make a honey crop, remove the queen and make up a nuc for her with some frames from her colony (check that these don't have any swarm cells on them). In the parent hive, remove all but a couple of queen cells and fill up the gaps with frames of foundation. The hive will allow one of the cells to produce their next queen and won't swarm. It will continue making honey, as it will have all its foragers.

There are many more methods of raising queens and making increase, but I decide this year to use the *Alley method*, as it's good for raising a small number of queens. I don't want twenty queens because I have insufficient stocks to make up that number of nucs, but the Alley method will give me several good queens – about the number of stocks I can probably make.

To start with, towards the end of May, I make up a nuc of six frames of capped brood with attached nurse bees and some stores of pollen and honey from the survivor stocks. This is my *cell-raiser* colony which will raise queen cells when the conditions are right. If each stock allows me two frames, this won't deplete any of the hives too much and will give me a strong enough cell-raiser to produce decent queen cells. The stocks are obliging and I manage to fill a nuc box with four frames of capped brood and nurse bees, with a frame of stores at each end. I move the cell-raiser away from the other hives. Any foraging bees on the frames I've moved will return to the parent hives, leaving me the nurse bees who will stay put.

Having set up the cell-raiser,  I need to leave it queenless for at least a week. In that time the capped brood will also start hatching, increasing the population, by which time some of the older nurse bees will have become foragers and will start to bring back pollen and nectar stores. Meanwhile I must make a decision about which queen in the parent hives I will select to be my breeder queen. I will use her eggs to donate to the cell-raiser in a week's time, from which it will raise my queen cells.

Timing is everything in this process. A day before I want to start the queen rearing, I will need to do two things. First, I will have to donate a frame of drawn, empty comb to the breeder queen's hive in which she will lay the eggs to donate to the cell-raiser a day or two later. Next, I will need to check through every frame of the cell-raiser to see if they have raised any emergency cells from eggs or young larvae that were donated when I set up the cell-raiser. If they have started queen cells, I'll need to remove them, so that they get to work raising queens just from the frame of eggs donated from my breeder queen.

If it all goes to plan I'll have several new queens in a month's time, and by the end of June they should all be laying in new colonies. If it all goes to plan.... but that's not always guaranteed in beekeeping. If it doesn't work, I will simply have to begin again. Beekeepers, like watchmakers, are used to starting again. I once heard of a monk who used to start his day by saying, *Today I begin again*. Certainly in monastic life and in beekeeping I've often learnt the best lessons from having to start all over again.

I console myself – what's the worst that can happen? I could learn something.

The problem is that unless we all learn something from the worst mistakes of the past, we will be doomed to repeating them. For example, while it is impossible to prove any theory that either the importation or cross-breeding of races caused or helped cause Isle of Wight disease (by importing the disease with stock or by compromising the bees' immunity to it and to other diseases such as foul brood), no one can deny that importing bees and our obsession with increasing honey production resulted in the emergence of the *Varroa destructor* mite in Western honeybees.

There is, moreover, another alarming example of the possible consequences of cross-breeding and importing. It is the case of the *Africanised* honeybee in America. In 1956 Dr. Warwick Kerr imported bees from East Africa, of the race *Apis Mellifera scutellata*, with the intention of hybridising the bees with European stock, to produce a honeybee for the subtropics that would increase honey production in South America. In the 2005 film *The Monk and the Honeybee*, Brother Adam can be seen in Tanzania encountering *scutellata*, which he evaluated as having the worst temperament he had ever encountered among any race of bee. The bees were crossed in Brazil with *A m ligustica* and *A m iberiensis*, but twenty-six swarms escaped in 1957. They spread rapidly throughout South America, reaching North America by 1985 and Texas in 1990. By 2004, having continued hybridising with US stock (which are all European bees), the Africanised bees were recognised as highly invasive, with a reputation for sometimes usurping European bees' hives and killing the queens, killing pets and, on some occasions, killing people. The Africanised bee has now earned itself the moniker *Killer Bee* - though perhaps unfairly. It's not that they go looking for trouble, but if troubled they are extremely defensive, stinging in very large numbers and even following a threat up to a mile from the hive.

The Africanised bee has become controversial because it is invasive and highly defensive – sometimes even dangerous, and yet it is also extremely productive and hardy, with higher resistance than European bees to varroa, chalkbrood and Colony Collapse Disorder. The drones are known to be stronger than European drones, however, and outperform them in mating with European queens, which means the Africanised bee eventually has the potential to become the dominant bee as it spreads further throughout North America. If Americans want a hardy, disease-resistant and highly productive bee, the Africanised bee could be received as a blessing, if a mixed one. Others regard it as an invasive problem. Either way, it demonstrates how hard it is to contain our experiments with the honeybee, and how easy it is for those experiments to get badly out of control. Unfortunately, we do have a propensity for making a mess when we mess with nature, while when we've messed with the honeybee, to perfect it in the name of domestication and commercialisation, time and again it hasn't ended well.

# MINDING THE BEES

## Chapter 16
### Sweet Isolation

'QUEENS – English are undoubtedly best.
See cover advertisement last week.
By return post, 2s, 8d., guaranteed fertile.
CHARTER, Tattingstone, Ipswich.'(1)

**British Bee Journal 1907**

According to *The Isle of Wight Beekeepers Association: A Short History*, Harry Cooper of Thorley, Isle of Wight, carried until his death a deep resentment about the way he had been portrayed as an ignorant scaremonger when he first raised the alarm about a strange new disease on the island in 1906. Born in the 1880s, a farmer's son, he was not school-educated, as his father thought he should work on the farm. Instead, he was taught at his mother's knee, later becoming a Sunday school teacher, an expression of his reputation as a community-spirited man. He became area secretary, then secretary of the Hampshire and Isle of Wight BKA at the time the epidemic known as Isle of Wight disease broke out. He was the owner of fifty-seven stocks, all of which he had lost by 1907.

Harry Cooper would have been aware, as pointed out in Mr. John Silver's report in the *British Bee Journal* in 1907, that the first symptoms of the mysterious malady had first appeared in the summer of 1904, in the south of the island, at Brook, Brighstone, Wroxall and Shanklin. It was not until 1906 that it spread to the centre and north of the island. In a short time nearly 50% of the stocks on the Isle of Wight had succumbed.

In February 1906 he wrote to the *British Bee Journal* and to the Press, warning of a strange new disease on the island. His report was titled 'Bee Paralysis: is the cause known?' (2) As a leading beekeeper in his BKA, and as one of the first to experience the ravages of the apparent epidemic, his initiative was responsible and laudable, with fears among many in his BKA members that the strange disease could have a devastating effect if it reached the mainland.

Harry Cooper was not to know that from its first appearance the epidemic would cause great confusion. According to a paper by Brother Adam in 1968, Dr. W. Malden of the

Pathological Laboratory, Cambridge University, first observed the disease in the autumn of 1904, just south of Newport, on the Isle of Wight. By 1905 he reported that it had appeared in the neighbouring villages. In February 1906 H. M. Cooper, a secretary of the Hants and Isle of Wight BKA, wrote an article in the *British Bee Journal*, entitled 'Bee- Paralysis', in which he described the symptoms of a disease he believed to be highly infectious: 'I feel quite sure that it is infectious, as a colony in my apiary noted for their robbing propensities, got at a diseased hive one day...and about three weeks later were crawling on the grass in front of the hive by the dozen...' (3) Mr Cooper described the affected bees as sometimes having swollen abdomens, wings twisted back as though dislocated, and bees crawling, unable to fly.

According to the *British Bee Journal* in May 1905, Mr Cooper's hives had performed well that year, averaging sixty pounds of honey per hive for a number of seasons. 1905, though a dry season, was reported as averaging twenty-three to sixty-seven pounds of honey per colony, depending on the location in Britain. By July 1906 the 'Editorial' clearly felt that there was a good deal of panic in the air that they needed to calm, referring to the story as a scare (4) and that it had, '...not yet died out'. (5) They referred to various press articles that had come to them, alluding to scaremongering by the Press. There was also an implicit criticism of the Secretary of the Hants and Isle of Wight BKA for his apparent silence: 'The curious part of the matter is that our good friend Mr. E. H. Bellairs, the active and indefatigable hon. Secretary of the Hants and Isle of Wight BKA for many years past, has not said a word in print on the subject, while one of his local secretaries; Mr Cooper...has, so to speak, made himself famous by letters to the Press; in which he refers to bee paralysis as ''a new and highly infectious bee disease'' .' (6)

The editors dismissed the idea that Isle of Wight disease was Bee Paralysis, on the basis that the latter was unknown in Britain and only emerged in warmer climates. In an effort to respond in some helpful way, they reprinted information on Bee Paralysis from the authoritative American publication *A. B. C. of Bee- Culture* which described the symptoms as swollen abdomens, the black, greasy appearance of the bees, and shaking or trembling of affected bees.

Mr Cooper did not appear to be making progress. Perhaps because of lack of support or action from Mr. Bellairs and the *British Bee Journal*, he continued his stories to the Press, which the editors of the Journal quoted on 19 July 1906, clearly alarmed by his suggestion that the disease: '...had not yet appeared on the mainland' and that, '...if it got a footing there it would probably mean the ruin of the bee industry so far as England is concerned....quite half of the bees kept in the Isle of Wight are now dead.' (7)

The editors' language reveals that from the start they did not trust the information circulating, or the Press; and that this was one of a series of bee scares, '...which have had a ''turn'' ' in the newspapers.' (8) In fact, it's evident that they regarded the Press as irresponsible and sensationalist: 'This is stated to be "a new and highly infectious disease", which we find a so generally solid leading journal as the Standard giving prominence to in type as large as one expects to see when notifying something of national importance.' (9)

The fact that they also reported contact from representatives of, '...several leading papers' (10) (including the *Standard)*, asking for, '...reliable information on the subject' (11) proved their dismissive response to the first reports of a disease.

By May 1907 the situation was serious enough for the Board of Agriculture to appoint Dr. A. D. Imms to investigate the bee losses on the Isle of Wight. Some weeks later, with Dr. Imms' report pending, Mr. E. H. Taylor of Welwyn, alarmed by the situation, showed letters (received from Isle of Wight beekeepers) to Mr. John Silver of Croydon Grove, Croydon. After a conversation between the two, Mr. Silver visited the Isle of Wight, riding on his bicycle from Croydon. For three days, working from dawn until late at night, he travelled around the island, interviewing beekeepers and looking at their stocks.

He spoke with thirty beekeepers who between them had owned three hundred and twenty-six hives, of which twenty-nine stocks were still alive. Fourteen of these were not expected to survive. What he found surprised him, especially as he had expected to find carelessness and inexperience, but fourteeen beekeepers who had lost typically between ten and twenty stocks each were all very experienced in the craft.

Reverend Leslie Morris of Brook, for example, had thirty years experience of keeping bees, but had lost twenty-eight stocks.

Reverend John Vicars of Colbourne had lost sixteen stocks.

Misses Gibson of Porchfield, prize winners at honey shows, had lost twenty-three stocks and some of the finest bees on the island.

Mr J.W. Cooper of Shanklin had only three stocks left.

Harry Cooper of Thorley had lost fifty-seven colonies,

while Mr. Tyman of Newbridge, a skeppist with forty years experience, had lost fourteen pots.

Even stranger, the strongest stocks seemed to fail first, often having made a good harvest. It was reported to Mr. Silver that the affected bees usually had swollen abdomens which were full of yellow matter, found to be pollen. The island beekeepers were unanimous about one thing in sharing their experiences of the disease with Mr. Silver, asserting that the disease was, ' ...highly infectious.' (12)

By 6 June when Mr. Silver's report appeared in the *British Bee Journal,* confirming the extent of the outbreak on the island, the editors decided at last to reprint part of the annual report of the Hants and Isle of Wight BKA that had appeared in the national newspapers, under the title *Bee-epidemic: Isle of Wight Scourge, (13)* alongside another article called *Bee-epidemic in the Isle of Wight*. It was the first recognition that, '...a new – or, at all events, not understood – disease among bees has spread east and west, north and south, through the island.' (14) The 1907 Editorial, however, referred to, '...reports of a sensationalist character' (15) from the previous year, reiterating their assessment that the disease was bee-paralysis. By then the story had reached, 'Even so serious a paper as the Daily Telegraph, which published on May 21 last, the following extract from the annual report of the Hants Bee keepers' Association.' (16)

By that time they had acknowledged that bees were becoming extinct on the island, but insisted again that the disease was bee paralysis and was known. The report extract from the Hants beekeepers, however, identified the disease as new, and contained an indirect defence of Harry Cooper and others whose efforts through the newspapers to warn of an impending crisis were unfairly criticised by bee journals and some of the Press. By that time, they reported, 50% of the island's stocks had been lost, most of those affected belonging to poor labourers. Their language pulled no punches and their warning remains as clear and uncompromising for our own times as it was when Harry Cooper first raised the alarm: 'To the bee expert it is as terror-inspiring as anthrax to the cattle-man, and should it spread over England, as it has done over the Isle of Wight, there will be no need of bee-societies, for there will be no bees.' (17)

Though at pains not to minimise the seriousness of the disease on the one hand, the Beekeeping Association offered some reassurance with the other, stating that they did not believe the disease to be paralysis at all, but one frequently confused with it. On the continent it was known as *mal de mais* in France, or *Maikrankheist* to Germans, owing to its appearance in May and June when an early spell of warm weather is followed by cold, wet or foggy days. They recalled 1853 and 1855 when, after cold springs, epidemics of this malady raged through France and neighbouring countries. In 1865 it was so severe in several parts of Northern France that hive numbers there were reduced to a third, or even a quarter in some places.

In 1865 Dr. E. Assmuss had identified that disease's cause as a micro-organism, and attributed its outbreak to improper food and pollen damaged by frosts. The Hants Beekeepers' Association, however, seemed reassured by this apparent naming of the beast and offered hope to beekeepers on the mainland, 'Whatever it may be, beekeepers in England need not be alarmed, as it does not necessarily follow that the disease would become epidemic...' (18) They ended the article referring to to the Board of Agriculture's Inspector, Mr. A. D. Imms MSc and his pending report, though they suggested that the Board had indicated ahead of publication of the report that the disease seemed to affect the bees' digestive system. There followed some description of how the bees' abdomens, distended with congested pollen that could not be passed, put pressure on the abdominal air sacs of the tracheal system, interfering with its function. Unable to expand these air sacs with air, the bees were unable to fly. Coupled with their increased weight, this accounted, they added, for the crawling, flightless bees seen during the epidemic on the Isle of Wight. Death, the report explained, followed from blood poisoning as toxins accumulated in the intestines of affected bees.

When Dr. Imms' report was published in 1907 it was inconclusive, but he ruled out bee paralysis. In a lecture given in 1909 Dr. Malden referred to the outbreak as, '...the new disease, unknown till five years ago.' (19) By then the epidemic had reached Berkshire, Hertfordshire, Hampshire, Sussex and parts of Essex.

Already we can see the areas of disagreement emerging over the strange phenomenon as it spread, some arguing that beekeeper ignorance was a key factor, though there were ample examples to the contrary. Others argued that the disease was not infectious, though there was persuasive evidence that it was. Some scientists thought one thing; others disagreed.

What about weather conditions? In fact, from the evidence of weather reports in the *British Bee Journal*, there was no sequence of unfavourable seasons. Between 1900 - 1920 seasons were no less favourable than previous ones, but were, on the whole, better. 1905 was dry, but yields of honey were reported as good, averaging twenty-three to sixty-seven pounds per colony. 1906 was another good season, with average yields of sixty to one hundred and twenty-nine pounds. 1907 was a bad season everywhere, followed by a good 1908, then a dry but good 1909 and a poor 1910. 1911 was extremely good, but 1912 was poor, followed by a series of good seasons. It doesn't seem plausible from this evidence to suggest that the *disease* was obviously linked to bad seasons or poor weather. In fact, the *British Bee Journal* often reported instances of stocks dying after they had yielded a good or even excellent crop, while strong and productive colonies were also often the first to collapse.

According to Brother Adam, writing in 1968, beekeeping methods at the time did not seem to be linked with the plague either, and in his view most beekeepers at that time did not lack skill or knowledge. At a time of transition between primitive and modern methods, some have suggested that skep beekeeping contributed to the problem, while others maintain that skep stocks were largely unaffected. According to Brother Adam: 'Indeed, the disease spelled the doom of beekeeping in skeps; in the village of Buckfast (which was quite small then) there were a number of cottagers with skeps. However, all native bees in the area died in the winter of 1915-16, whether they were in skeps or modern hives.' (20)

In 1909 Professor Enoch Zander identified the problem as Spring Dwindling, or *May Disease* , another form of nosema, caused by the fungus *Nosema apis* or *Nosema ceranae*. He found that the microsporidian was found abundantly in mature bees' intestines. In his view this accounted for the death of thousands of colonies annually. He classified the germ as a protozoan (a single celled animal) that was parasitic. The disease caused by the infestation of this parasite is dysentery, with symptoms of crawling, reduced brood production, a dwindling population and greasy-looking bees. Although *Nosema apis* is thought to have affected the *European* honeybee for centuries, *Nosema ceranae* is thought to have originated in the Eastern honeybee *Apis Mellifera cerana*. *Ceranae* is particularly virulent, destroying colonies in only eight days.

Professor Rennie (1865 – 1928) waded in and completely disagreed. Rennie, a Scottish scientist of Aberdeen University, had first worked on a cure for diabetes, later lecturing in parasitology at the University. He was also a leading scientist at the North of Scotland College of Agriculture at Craibstone, where he lectured in agricultural zoology. In 1921, while searching for the cause of Isle of Wight disease, it was his colleague, Bruce White, who discovered a mite called *Acarapis woodii*, or tracheal mites, following their experiments.

By 1911, however, Professor Zander and other scientists engaged to the Board of Agriculture had concluded that *Nosema apis* was the cause of the epidemic on the Isle of Wight. This didn't seem right, however, as nosema losses were normally only in spring, with dysentery often present, but no mass-crawling of bees, except in June and July. In the Isle of Wight epidemic the queen also seemed to survive, though with nosema she was usually the first victim. Despite this, by 1912-13 the Board of Agriculture agreed with Professor Zander's diagnosis.

It was Dr. J. Anderson in 1916 who first observed that nosema and Isle of Wight disease were not identical. In fact, some stocks suffered with both nosema and Isle of Wight disease simultaneously, some with foul brood, sac brood and chalk brood as well, according to Brother Adam who concluded, 'Multiple infections were probably the rule rather than the exception.' (21)

Rennie, on the other hand, stated in 1919:

'In a stock infected with *Nosema apis* the behaviour of the bees has in our experience been in striking contrast to that of the members of a colony afflicted with the condition known as "Isle of Wight" disease...loss of flight power has not been found to be a characteristic of Nosema infection until the insect is actually dying. In the Isle of Wight disease it is usual for this symptom to appear a considerable time before death, if the bees are prevented from sacrificing themselves by crawling and subsequent death from exposure. We have not observed Nosema infected bees to loiter in large numbers about the doorway or to gather in clusters on the ground as Isle of Wight crawlers do.' (22)

Moreover, Rennie had observed a high frequency of queen losses with *nosema*, contrasting with many queen survivors of the Isle of Wight disease.

If this were true, and acarine was the real cause of the epidemic, where did the tracheal mite originate? According to Brother Adam's 1968 paper, there was a widely held assumption that the mite had been present in stocks for centuries, or more. If this was so, then it was reasonable to speculate that the British *native* Black bee had suddenly, for some reason, become more vulnerable, or nature would have exterminated the subspecies long ago. He also made the persuasive argument that if *Acarapis woodii* had been common in Europe from earliest times, it would also have shown up in other continents where every other new bee disease had appeared through imported stocks from Europe. Another possibility he proposed was that it was brought into these countries but then died out because of the climatic differences, before later reemerging: 'The most plausible explanation is that it was brought into this country in some way or another, from some place not yet identified.' (23) If this was correct, the smoking gun was the importation of foreign queens.

In conclusion, Brother Adam did not blame unfavourable climatic conditions for bringing about the Isle of Wight disease. He believed acarine was responsible, as the main symptoms of the mite were identical to what appeared as Isle of Wight disease. Severe losses, in his view, were brought about when susceptible stocks came into contact with the mite, and not all strains were equally susceptible to *Acarapis woodii*. As with other diseases, stocks had varying degrees of resistance and so did the different subspecies of bee, which is why Buckfast's native Black bees were affected but not their Italian stocks. Brother Adam pointed out too that acarine was still found in susceptible stocks such as Carniolans and Black bees (presumably he meant all European Dark bees, and not just native British bees which he believed had been entirely wiped out by 1915).

Thus far then, we have convincing arguments and scientific evidence that both types of *nosema* were present in infected stocks together with the Tracheal mite (*Acarapis woodii*). There were no consistently bad seasons for the duration of the disease, but there were the usual vagaries of our northern climate in particular years and seasons, as indeed has happened the year I write this. Other diseases were also present, as a consequence – dysentery being one of the more obviously identifiable. Rather than simple disagreement between scientists about the overall underlying causes of Isle of Wight disease, there seems to be agreement that identifiable agents of the disease were in some way involved. The disagreements centre upon which single cause was ultimately responsible for the deaths of so many stocks. In Brother Adam's opinion it was *Acarapis woodii*: 'But, judged on the basis of what I observed at first-hand, there is no doubt whatever in my own mind that the trouble known as I of W disease was the primary cause of the epidemic, that it did the killing, and well nigh swept the country clear of bees.' (24) And Isle of Wight disease, he maintained, was acarine.

What about all the attempted cures during the years when the epidemic raged? We know from the *British Bee Journal* that many people wrote in, claiming to have found something that worked. People were desperate and were willing to try anything, especially if it seemed to have worked for someone else. On the other hand, how much were these efforts merely stabbing in the dark, and to what extent did they contribute to greater losses? Brother Adam, I suspect, was probably right when he wrote, '...I am convinced that the methods and practices in use had no more to do with the wholesale loss of colonies then than at the present time.' (25) On the other hand, he also made the sensible observation that some remedies must have offered some benefit, or they would not have caught on: 'Some of the remedies applied may have done more harm than good, but not obviously so, or they would not have been widely used. The addition of a small quantity of salt added to the syrup – considered beneficial ...a hundred years before – may well be poisonous to bees in the laboratory, but this does not prove that it is so in their normal environment, when they are at liberty to fly.' (26)

It's worth considering how heavy losses might have been in the centuries of primitive beekeeping before the turn of the twentieth century. Beekeepers would have had very little control over the welfare of stocks housed in simple skeps, and for reasons such as bad weather and poor seasons there must have been bad years when many stocks died. Given the lack of communication or ways of collating such losses in more primitive times, there are no records that this ever happened – which doesn't mean it never did. There would have been years when bees starved or had depleted stores, just as wild colonies do. If bee parasites and diseases, such as bacterial, fungal and viral infections, have lived alongside bees probably since recorded time, there must surely have been episodes in history when epidemics such as Isle of Wight disease broke out, perhaps on a smaller scale, as bee stocks were less likely to be moved large distances centuries ago. On the other hand, given that people in past centuries were more dependant on bees than they were even at the time of Isle of Wight disease, would an epidemic have been recorded by someone, somewhere?

It was Mr. R.W. Frow of Wickenby, Lancashire, who was the first to find a cure for the strange disease, according to Brother Adam, though fully effective acarides were not developed until 1952. When the disease appeared in Switzerland in 1922 the Bee Department of Lebfeld Institute became the main centre of research, under the direction of Dr. O. Morganthaler. He discovered that resistance to the disease seemed to be governed by age, effective when bees are five days old and when *Acarapis woodii* mites are found at the base of the wings. As the disease continued abroad, so too did research, with the prevailing view that the only solution seemed to be to develop resistance to tracheal mites, as Brother Adam was doing.

From the start, Isle of Wight disease caused confusion, even among scientists. Some identified it as paralysis, observing the symptoms of black, hairless and shiny bees, even though these symptoms did not present in every case. Others maintained that the ailment was *May pest*, though this was known to happen only in May and early June, and its symptoms included soiled hives and crawling bees. *May pest*, however, was very rare in Britain. Others thought nosema was to blame, and others thought the smoking gun was acarine, while some thought that other factors were involved such as the wide variety of methods used to treat the problem, many of which were ineffective and even harmful. Not everyone agreed either that the disease was even infectious, while some, such as Brother Adam, had convincing evidence, from their own experience, that it was. He cited two stocks of bees that had travelled from the south of England to the Outer Hebrides on 28 September 1909. By the following January they showed signs of Isle of Wight disease, despite being six miles apart. In a similar way the disease entered the Isle of Man, and Ireland in 1912.

In 1964 Dr. Leslie Bailey, a major figure in bee pathology, published his paper, *Isle of Wight Disease: The Origin and Significance of the Myth*, in which he proposed that the designation Isle of Wight disease included, '...several maladies having analogous superficial symptoms.' (27)

After service in the Royal Air Force he had joined the Bee Department of Rothamstead Experimental Station in 1951, to work on bee diseases. His open-minded approach made him question many accepted ideas about the Isle of Wight disease that had by then found their way into books on apiculture. When Rothamstead became one of the first places to purchase an electron microscope, he employed it to study bee viruses. In 1963 he finally identified and described two viruses, the first of many to be isolated at Rothamstead over the next two decades.

Eventually he came to the view that the description of bee colonies as either healthy or diseased was simplistic. Many colonies coexisted with multiple pathogens without presenting any symptoms. Bailey argued that under certain circumstances or combinations of situations these pathogens could become harmful, with different pests and pathogens interacting. He went on to demonstrate that Isle of Wight disease had been caused by Chronic Bee Paralysis Virus (CBPV) which was unknown at the time, combined with weather conditions, which had confined bees in dense colonies with insufficient food. According to Bailey, Brother Adam and those who believed that acarine had caused the virtual extinction of the native British Bee were wrong. Bailey explained: 'Why Acarapis woodii became so firmly established as the cause of I of W disease is hard to understand...Perhaps  some thought it was the last adult bee parasite that would be found...This ignores the possibilities of other pathogens, especially of bacteria and viruses, which we now know to exist and cause disease with symptoms resembling those reported to be the I of W disease.' (28)

Firstly, Bailey identified ( as we have seen) that the myth of Isle of Wight disease was, '...promulgated by sensational but uninformative articles, which I have read, in the Standard, a now defunct London morning paper, and in several provincial newspapers.' (29)  Bailey observed that the distended abdomens of affected bees were normal for bees confined to the hive and was therefore not a symptom particular to Isle of Wight disease. He argued that the intestines of healthy bees confined to the hive closely resembled the intestines of diseased bees. Malden, he pointed out, had examined the anatomy and trachea of affected bees at the beginning of the emergence of the disease, but had only found more bacteria in the bees' guts; there was no evidence, however, that these bacteria were pathogenic. Moreover, in 1922 Bullamore pointed out that bees prevented from flying sometimes showed symptoms similar to Isle of Wight disease, such as crawling. Bailey argued too that due to lack of available sugar during the war, starvation was often an added factor accounting for stock losses. Similarly, feeding was an issue because some beekeepers removed pollen after an official report that bees were short of nitrogen because they were full of pollen. For years afterwards some beekeepers would therefore have removed a vital source of protein from their bees. Still others applied remedies, including salt, Izal, sour milk, phenol and formalin, while poisoning from the use of agricultural sprays would doubtless have compounded the problem. With regard to *Acarapis woodii* Bailey argued that, '...flying workers were frequently more heavily parasitised than were bees of the same stock which were unable to fly.' (30)

William Herrod Hempsall wrote in 1937 that *Acarapis woodii* had spread from the Isle of Wight to Europe after 1918, but in the same account claimed that it had infected the honeybee for centuries across a number of countries. Bailey thought it more likely that the latter had been the case, '...for several thousand millenia.' (31)

In short, Bailey's assessment was that in poor seasons of dearth the mite reemerged, the result of abnormal circumstances, and that bees kept by beekeepers are, by definition, not in normal circumstances: '...I suspect that the I of W disease was assumed to be the cause of all the losses from which there was no obvious explanation at the time. In this sense it was truly a myth.' (32)

Brother Adam observed tracheal mites in Buckfast's bees in 1915 and it was reasonable for him to assume that they were the cause of all losses everywhere between 1904-18. Indeed, they might have played a major part in the loss of Buckfast's native Black bees, though other additional factors might have been involved. That does not mean, however, that *Acarapis woodii* was to blame for every lost stock which could not otherwise be explained. *Acarapis woodii*, Bailey believed, was scapegoated, and that lack of knowledge about the disease allowed it to become a crisis. For other beekeepers their losses might have had nothing to do with tracheal mites; their bees could have been lost through starvation, poisoning (in the name of treatment), nosema, weather conditions leading to confinement, together with a myriad of unknown viruses we now know are always present in a hive. All each beekeeper could go on were the observable data with regard to their own bees, which probably differed, according to which variables were most prominent in their own stocks.

As for William Woodley, when expressing his sympathy with beekeepers who had been affected before he was, he referred to the problem as a plague; and in one of his *Notes By The Way* even suggested that proposed legislation might increase its transmission on the feet of inspectors travelling from place to place. He saw the problem as a new and separate disease, unknown and highly transmissible.

What do we make then of his eventual suggestion that beekeepers should have returned to skeps? Was that a proposed solution to what he saw as a new disease that had arisen from misapplied modern methods of beekeeping, or a reaction to modern methods because he had begun to believe that they might have actually caused the 'plague'? It is understandable that someone like Mr. Woodley might have looked at the developing situation between 1904-18 and that it might have seemed obvious to him that there was a connection between modern methods and this new 'disease'. From such an assumption the most logical conclusion would have been that nothing comparable had ever happened to bees (as far as we know) in many centuries of skep beekeeping; therefore, it would have seemed sensible to return to skeps, at least until the problem died out. Mr. Woodley, we recall, also observed that he had never seen *foul brood* in a skep, so he had a prejudiced opinion, based on his own experience, that bees were, de facto, healthier in skeps.

What difference, however, might legislation have made? Firstly, any legislation introduced for *foul brood* would have needed amendments to cover Isle of Wight disease as well. That aside, it's hard to see what contribution it might have made when no one could agree about the nature of the problem called Isle of Wight 'disease' or its cure. In defence of Mr. Woodley's opposition to *Bee Disease Prevention Bills*, one might argue that legislation, had it existed, would have been a useless instrument with which to solve the problem. At best it would have been stabbing in the dark and might occasionally have saved some stocks. On the other hand, it might also have resulted in the loss of many more, if only by a policy of stock destruction by inspectors simply because of the proximity of healthy bees to affected stocks, and on the mistaken assumption that in every case the affliction was invariably contagious. Some of these bees that might, under legislative powers, have been destroyed by inspectors ultimately became important survivor stocks.

How much of Mr. Woodley's opposition to legislation was petulant opinion is a moot point. His opposition over the years shifted its focus between different arguments to justify the same position, almost as though he held an intransigent opinion for which he was grasping at the best justification. On the other hand, he might also have had many reasons to oppose the proposed legislation. There is, however, some interesting evidence as far back as 1902, before Isle of Wight disease, that Mr. Woodley's opinions could isolate him and that they had already aroused the anger of his readers, which might throw some light on his character. In 1902 he wrote something on the subject of honey shows that angered Rev. Donald Moore of Clifton: 'There are some who, if they cannot have everything their own way, would kill a show. Conspicuous among these I cannot help thinking is Mr. Woodley. He seems to live upon objections.' (33)

Rev. Moore explained how he went to great efforts at a honey show to introduce something new, which was met with great approval by most, but that the objections of a few had put him off repeating the effort. He criticised William Woodley for his dismissive and uninformed views about the use of honey in cosmetics:

'For instance,"in my opinion,"he says (I should think in this matter he enjoys sweet isolation),"no English honey has ever been used in "the manufacture of such things "as cosmetics, soaps, hair-restorers, and such like "articles. What a confession! The horizon of Mr. Woodley's knowledge is manifestly a very near one. With sublime innocence he fails to realise that possibly there may be uses for bee-produce of which he has never heard or conceived.' (34)

Rev. Moore continued, 'I know for a fact that honey is used in the manufacture of many of these articles to which Mr.Woodley takes exception.' (35)

This letter is fascinating because it suggests something about Mr. Woodley's character and seems to prefigure his eventual isolated response and opinions to *Bee Disease Bills*. When Rev. Moore referred at the start of his letter to the, '...petulant objections' (36) of those who caused contention at honey shows, I wonder if he made an astute observation about Mr. Woodley's character? Scathingly, Rev. Moore implied that Mr. Woodley's motives were purely professional self-interest, and that: 'An ounce of knowledge is worth a pound of "opinion." ' 'His remarks with reference to "trophies" could never be made by one who was by nature and not profession a bee-keeper.' (37) Did William Woodley live upon objections? Was he simply petulant and opinionated, but ultimately lacking in knowledge? Is there early evidence in this letter of an emerging hubris that was eventually his undoing over Bee Disease legislation?

Similarly, in 1904, his notes in the *British Bee Journal* were dismissive of new methods of queen rearing from America. Articles began to appear about small mating boxes in which a queen could be mated and brought into lay with only a few hundred bees, which also relied upon raising large numbers of queen cells which could be cut out and donated to these mating boxes. Mr. Woodley did not believe these modern methods could compare with the traditional methods of working with nature: '... but I doubt if we shall improve on Nature's plan. I have cut out of a hive to-day, which swarmed on June 4, five or six as fine queen-cells as one could wish for - cells which contain queens reared under the swarming impulse in a colony on twelve frames. These are good enough for me, and queens such as these have for these twenty years been the backbone of my profitable bee-keeping.' (38)

From readers' reports of their own experience of imported queens and home-bred queens of other races, there is no doubt that queens with many good traits were being raised by these new methods. Mr. Woodley's dismissive opinion of them, which was not evidenced by personal experience, seems rather unfair, and dare I suggest – even petulant: 'The results produced by the purchase of two queens for the same sum may prove dear at any price, as we lose a season trying to find out the good qualities which they do not possess.' (39)

If William Woodley was an argumentative and petulant man (and he might have been!), perhaps we ought to be grateful that these traits might have contributed in some way to the obstruction of Bee Disease legislation. There is an argument that such legislation might actually have brought about the extinction of the *native* British bee. A defence of William Woodley's position could be made on this basis by suggesting that his opposition to legislation actually saved many stocks that would otherwise have been destroyed. Those like him, who dared to stand against the weight of opposition that held them accountable for the nation's severe losses, could arguably be viewed a century later as inadvertent saviours of our native Black bee.

# MINDING THE BEES

## Chapter 17
### Hitting Our Stride

'FOREIGN BEES AND QUEENS
ABBOTT BROS. Southall, London
BENTON F. Munich, Germany
SIMMINS S. Rottingdean, near Brighton' (1)

*British Bee Journal 1886*

On 26 May the rain stops tapping the nettles and clattering through the great green oaks and the sun is shining again. The delicate transience of May is revealed as it should be, the wood pigeons cooing from trees heavy with blossom, pale lilac cuckoo flowers delicate among the green, rising sward of the meadow, and hares flopping through dappled sunlight in the field across the lane. At 7.30 am on 27 May I'm watching a hare across the lane and listening to the joyful chorus of morning birdsong.

I notice flower buds on the limes – nature's green version of those little metal jacks we played in our childhood. The buds tell me that the main nectar flow is only a couple of weeks away. Pollen is coming into the hives too, so there's forage somewhere, but this is now the beginning of what beekeepers call *the June gap* and we haven't even had a spring nectar flow this season.

It's wonderful to be able to watch the gentle movements to and from a hive on a June afternoon. There's something peaceful and restful and altogether timeless about a beehive in the shade of a tree, surrounded by long grass in an early summer afternoon. A male tawny *woo-woos* at 2.15 pm in the garden. A shriek follows from a female.

When the next sound happens, the moment is as close to a dialect of the transcendent as the natural world can speak:

*cuckoo...cuckoo !*

From the line of trees across the car park, only a hundred yards away, there follow two or three minutes of unmitigated bliss as the cuckoo rolls out its announcement of early summer.

It's late to be hearing it though. Usually they've stopped calling by now. Are they late, I wonder, and this is the late start of early summer, or does this signal the close of a spring we've missed altogether? Who cares? For now I stand joyfully immersed in the sound, the way wine drinkers savour a rare vintage. Last year I heard a cuckoo only once, briefly, in the distance. This jubilant celebration of itself at such close quarters, if it be the only encounter I have for another year, more than compensates for all the time spent listening and missing our favourite spring visitor.

Four out of five of my swarm boxes receive regular visits from scout bees as soon as the weather warms up. I re-bait them with a couple of drops of lemongrass oil on a cotton bud pushed into the entrance of each hive. The bees have found my lures. The question is: are they voting for or against these possible choices of a new home? It's one thing to have visits by scouts, but another to see the frequency and the number of visits increase. If the latter happens, more bees are being recruited to a particular site, increasing its chance of selection by the imminent swarm.

One box in particular seems to have more visits than the others, especially on 26th. On the 27th it's warmer, but visits are less frequent. My guess is that a swarm has issued somewhere, but my bait hives were not selected. Not this time. But there are still some visits on 27th, and it's only the start of the swarm season, so there's every possibility of another swarm arriving.

The same day, with the temperature reaching nineteen degrees at last, it's warm enough for me to make proper hive inspections without having to rush or worry about chilling the brood. Smoker in hand, I begin opening each hive, in turn. The hive that I thought had laying workers or a drone-layer, then a suspected supersedure queen who was hiding, finally reveals a queen in all her dark glory. She's unmarked, so she's a supersedure queen after all. She's laying and has a couple of full frames of brood. I guess she was late starting because she might have some *Carnica* DNA.

The strong hive continues building up, and in the third hive the honey-coloured queen, who superseded the winter queen in April, also shows herself and is laying well. That's three stocks that are queen-right with brood. Then there's the polyhive with the marked queen who doesn't seem to be laying yet. A couple of weeks ago I donated some brood and eggs in case they wanted to supersede the queen or they were queenless. I look for emergency cells or supersedure cells, but there are neither. Then I find the queen and she looks healthy enough, but there's still no brood or eggs. There isn't even any drone brood, so she isn't a drone-layer; she simply isn't a layer at all. I'm mystified by what's going on with this stock.

Yet this is one of the absorbing fascinations of the craft. I enjoy identifying the problem and resolving it, or knowing when to allow the bees to take control. I step back and look at the situation like an horologist presented with a faulty watch. Firstly, the queen is probably all right, or she wouldn't be alive. I gave the stock a test frame a couple of weeks ago and that was their opportunity to raise a queen if they aren't happy with this one, but they didn't. If the frames are old and black she might not lay, but she has some fresh comb that's still yellow. I surmise that she's probably fertile because she produced winter bees last autumn that took the colony through the winter. If she's infertile or failing, I'd at least see drone brood with its distinctive dome over each cell, or drone brood scattered among worker brood, so that can't be the problem. Honey-bound frames can also stop queens laying due to a shortage of available cells for eggs, but at this stage of the new season the colony's stores are low, so there's plenty of space in the brood frames for her to lay. Lack of pollen can prevent laying too, but they've had pollen candy for months, which they could have used, as have the other stocks. Is she late starting because she has *Carnica* lineage (and late laying is a Carniolan trait)? That was the case with the supersedure queen that I've just seen for the first time. If so, what's the difference between the two queens – that the supersedure queen is laying while this one isn't?

...nurse bees? If a queen doesn't have sufficient nurse bees, which are usually the younger bees, she won't lay because there's no chance of her brood being fed and kept warm. At this time of year nurse bees are the youngest bees – up to three weeks old, and the only ones would be from the frame I donated some weeks ago. By now they'll be foragers. I recall that I gave the supersedure queen's colony a frame of sealed brood, more than once, not only to boost dwindling numbers of bees, but also to suppress what I thought might be laying workers. I conclude that the queen doesn't have enough young bees and that I should donate a frame of sealed brood. Within a week to ten days that brood should emerge and might trigger the queen into laying. I'm also transferring the stock to a nucleus box, to reduce the available space and to cram the bees together. That will help keep the brood warm.

The next day is one of those cloudy, still days when cow parsley is the only thing twitching in the hedgerows as cars pass along the lane. As I prepare the nucleus box with the right combination of newer and older frames, adding some honey frames stored from colonies that died out in the winter, I make an irritating discovery: a few silvery moths, about an inch long, flutter to the floor from the stored fames; then I find a few stored frames with cobwebs across the comb peppered with black dots of moth faeces. Lastly, I find a number of inch-long yellow-white caterpillars: wax moth larvae.

Wax moths come in two species, Greater and Lesser, but neither is a lesser nuisance. I seem to have an infestation of Greater wax moth *(Galleria mellonella)*, otherwise known as the honeycomb moth. I've kept an eye on the stored frames of honey throughout the winter, watching for signs of this pest, but I've taken my eyes off the ball with the other empty frames. Eggs are laid in the previous season, so there have probably been hibernating moths or dormant eggs in the boxes all winter. In particular they favour the oldest, darkest wax in the brood box, which are the frames the beekeeper has to change every few years, in the interests of biosecurity, so the moths are nature's way of cleaning up after the bees.

It's the caterpillars that really do the damage. They introduced themselves to me in my first year of beekeeping when I lost my first hive. As the colony dwindled away and weakened ( I didn't realise at that stage how easily I could have rescued it) wax moths took over, destroying frames of wax and even chomping channels into the wood of the brood box.

I expect this spring they've taken hold in the spare brood boxes that I'd usually have outside by now; colony losses in the winter and spring, and the late start making splits, has meant spare equipment with stored frames have been sitting around when they would normally be in use outside. I'll need to go through every box as soon as possible, killing moths and removing the caterpillars and cocoons. At first glance, it looks like a light infestation, but I'll have to act quickly if I want to avoid a complete mess. Although they prefer the oldest wax that the beekeeper cuts out or removes anyway, I've known them to attack cleaner wax and even honey frames, despite what the books tell you.

It's a quick job to move the stock with the non-laying queen into a nucleus box, to which I add a frame of sealed brood from another stock. I'm hoping the warmer conditions, the pheromones from the brood and the prospect of nurse bees will stimulate the queen to lay – assuming she's all right. I also add a couple of newly-drawn frames of empty comb, as queens prefer laying in fresh comb. The box with the transferred stock is put on the stand, in the same position as the original hive. Lastly, I give them a feed of sugar syrup to stimulate a nectar flow, as queens are stimulated to increase their laying during a nectar flow.

Every year the bees seem to be a little ahead of me, always stretching me to understand them better. This is what I most enjoy about beekeeping; you never quite know what they might do next, you never know everything, and every year something new happens and they seem to break all their rules. Take, for example, this year's queens: I have an emergency queen that was produced at the end of April and was somehow mated during a cold May. She's small – much smaller than the other queens, and yet she's laying well at the moment and her colony seems happy with her. Sometimes they will almost immediately supersede a new emergency queen, as though they only have a temporary use for her, but apparently not with this queen. Then there's the queen in the hive which I supposed was the strongest stock all winter. She hasn't laid at all, though I've donated brood and eggs, hoping these might set her off, but she's done nothing. Usually the colony would remove a failing queen or one not properly mated, but this queen appears to have a charmed life, despite her apparent uselessness. They've had every chance to get rid of her, but she keeps going, a lovely big, healthy-looking queen who looks beautiful and suitable in every way. It's a mystery to me, but I have to trust that the bees generally know what they're doing, and that if I leave well alone they usually put things right in their own time. Some beekeepers might have pinched her out of existence by now, but anomalies interest me; I might learn something new by watching what happens next.

What happens next though is that rain returns, and I must confess to feeling despondent about my prospects of catching a swarm. Recovery in the apiary is slow, adding to my frustration. Usually by mid June I'd have at least a couple of new queens mated in my first nucs, but rain interrupts this and foraging, confining the stocks, which I fear will knock back brood production even further.

I make a couple of new nucs with brood and eggs and a couple of frames of nurse bees from the survivor hives, hoping to steal a march and that they'll have raised a new queen in time for June's weather to brighten up for mating. Every day I glance at my swarm boxes as I pass by on my way to feed the poultry, but there's no sign of scout activity at the moment.

A week into June I return to the cell-raiser colony that I set up a week ago. It's now full of bees and looking strong, as some of the capped brood has emerged. I lift out each frame, shake off the bees and check them for queen cups or capped queen cells, but luckily they haven't raised any, as there were no eggs or larvae young enough on the frames I moved from the donor hives. That makes life easier. A day or two earlier I donated a brood frame of drawn, empty comb to my breeder queen's hive. By now it is full of eggs and ready to donate to the cell-raiser.

In the shade I begin to prepare the Alley method of raising queen cells. I cut a strip of comb full of eggs from the donor frame – about the length of a frame and an inch wide. I attach this to an empty frame with thin wire, orientating the cells straight downwards. I use my hive tool to dent the comb at regular intervals along its length, so that I crush cells each side of the cells from which I want the bees to raise queen cups. This distributes the queen cells more evenly and stops the bees making too many in one place that can't be cut out without damaging the queen cell beside it.

Raising queens is fascinating and is also an example of epigenetics at work. Any fertilised egg can become either a worker or a queen; they are genetically virtually the same. There are no different or additional genes in the bee genome that express the specific characteristics of a queen, and yet the queen and the workers are physiologically completely different. How can this be? With epigenetics we have to remember that the DNA remains unchanged, but certain genes along the genome can be either switched on or off. In the case of making queens it isn't the environment that switches on the genes that express for a queen, but diet – specifically royal jelly, a secretion from the nurse bees' glands in the hypopharynx of the head. Workers and drones are only fed royal jelly for about three days and are then fed bee-bread, made from honey and pollen. It seems to be the pollen that creates what's known as *DNA methylation*, or the silencing of genes that express for a queen. Queens, on the other hand, are fed exclusively on royal jelly, which results in less methylation of the DNA, allowing expression (switching on) of the genes that create a queen; the larger, longer body and legs, ovaries that can produce eggs and the creation of a pheromone known as queen substance, together with a life-span of up to five years.

*Summer in the apiary*

I remove a frame from the nuc box from which most of the workers have emerged, making a space for the donor frame, which I slot in, positioning it in the centre of the box. The queenless cell-raiser colony is now desperate to raise a queen and will begin drawing out these cells into acorn-like cups in which they will begin to feed the larvae, turning them into queens. In eight days the colony should have raised at least several fat, sealed queen cells from this donor frame. I can then cut out these cells and donate them to other queenless nucs which have been prepared to receive a queen cell.

Ten days later there are five decent sealed queen cells on the donor frame; not quite as many as I'd hope for, but they should allow me to increase my stocks. The cell-raiser is strong enough now that I can split it into three three-frame nucs if I add some frames from the other stocks. I manage to find enough frames to make up five nucs, including the cell-raiser, and cut out the queen cells, one by one, as carefully as I can so as not to damage them. I keep them vertical and avoid any shaking that might dislodge the larval queen inside. I put each queen cell into an orange plastic protector that looks like a toy basketball net. This protects the cell from any bees that might decide to tear it down, allowing the new queen to emerge from the opening at the bottom. The protector is inserted between two frames, held in place by the top bars each side of it. I am careful not to squash any of the donated cells as I move the frames back together just tight enough to grip the cells. A couple of the nucs are a bit light on bees so I take a few frames from the full hives, checking first that the queen is not on them, and shake the nurse bees into the nucs to top them up. The virgin queens from these nucs should hatch in about six days. Another week after that they will begin mating flights and be settling down to lay by the third week of June. I now have to hope the weather is fine enough by then that my new queens will be properly mated.

 

Midsummer's Day, perversely,  is colder than the winter solstice six months ago, reminding me that the season for prime swarms calms down considerably towards the end of June and that swarms taken after this tend to be smaller casts. One of the elderly brethren reminds me that Biddy never touched a swarm in July. It's a reminder that my chances of bagging a swarm this season are diminishing with each passing day, with my hopes hanging on the possibility that the swarm season will be pushed into July along with everything else because of the late spring and delayed summer.

It was a midsummer evening at Highgrove with my sister a few years ago that was the beginning of my slow conversion as a beekeeper. It had been a disappointing season and by Midsummer's Day I'd already lost a couple of swarms and my hopes of a decent honey harvest. Out of the blue came the invitation to Highgrove gardens, from an investment company looking to attract business from religious houses. The Abbot gave me the two tickets and I invited my sister.

It was a warm, pleasant afternoon when we arrived at Highgrove for refreshments before a guided tour of the gardens. By the time we reached the chickens and the bees it was early evening. We stopped while the guide explained to us that Prince Charles (as he was then) kept Anglesey bees. I'd never heard of *Anglesey* bees, but something about the name caught my imagination and my curiosity. My Buckfast queens (but for one in her final year) had by then been superseded and my hives had become swarmy with mongrels and I hadn't yet worked out how to control them. I'd been thinking of replacing the queens, and was tempted to buy more *Buckfasts*. Then I heard mention of *Anglesey bees* and I was interested.

It's well known that King Charles is passionate about caring for the environment and our nation's heritage. In 2018, just after I visited Highgrove gardens, he was interviewed on BB2's *Gardeners' World* from the Highgrove Estate about the issue of biosecurity and the threat to our native plants and trees from foreign pests and diseases coming into the country with imported stock. The presenter, Adam Frost, referred to a *DEFRA* list of 1000 pests and diseases, to which 100 new ones are added every year. In particular the then Prince of Wales recalled Dutch Elm Disease and how it had hit Gloucestershire especially hard, and expressed his concern about the current threat to native oaks and that gardens like his own at Highgrove could end up a wasteland unless policy on quarantining were improved.

Back at the Abbey I did my homework and discovered that the Anglesey bee isn't actually a race, but a strain of the native Black bee (*Apis Mellifera mellifera*) bred from survivor stock. The Isle of Anglesey, North Wales, is a remote location, providing an ideal environment to develop pure stocks of this strain of locally adapted Welsh bees by its own Beekeeping Association. They strongly discourage the importing or spread of non-local bees on the island, such as Buckfasts and Carniolans which might introduce diseases and pests and F2 aggression from cross-breeding.

This was the moment my eyes were opened to the issue of imported bees and the existence of our own native Black bee. It was the first time I'd heard of locally adapted bees too. Earlier that year I'd requeened my first captured swarm of very dark bees and the blackest, most beautiful queen I'd ever seen, that came from Kelly's Folly. They were rather difficult to work with compared with my other stocks, so the queen had to go. As I discovered more about native and locally adapted bees, however, I began to kick myself that I got rid of that lovely black queen.

Questions were forming in my mind: how were wild bees surviving in our buildings, as they seemed to have done at least since the 1980s, according to the older monks of Douai Abbey, without *varroa* treatments? Why were they so dark, compared to the *Buckfast* bees with which I'd relaunched the apiary. It seemed to me that they were either Black bees or were somehow reverting through natural selection to characteristics resembling those of the *native Black* bee. Perhaps, I speculated, it was for the same reason that the bees in our apiary were also becoming darker as the original Buckfast strain became diluted through natural breeding.

*Kelly's Folly -front view*

According to Brother Adam: 'The present day dark bees originate from importations made in 1919 and subsequent years in an effort to restock the country. These importations came mainly from Holland and France and the bees brought in belonged to the same group of races as our former native variety.' (2) The question that preoccupied me was whether or not the bees I had caught were related to the stocks Brother Adam wrote about that replaced those lost to the Isle of Wight disease up to 1919? Alternatively, (and this was more controversial) were they related to the original native bee - a separate subspecies of the European dark bee? If they were, then Brother Adam's claim that the original native Black bee was swept away by Isle of Wight disease was incorrect.

For months I became absorbed, researching the Isle of Wight disease in archived editions of the *British Bee Journal*. I read editions from as far back as the 1880s, including articles and letters on foul brood and Bee Disease Bills, right up to 1904 when Isle of Wight disease first emerged. This is where I found William Woodley's fortnightly *Notes By The Way* and observed how he was drawn into the debate on legislation and the long running argument between the traditional straw skep and modern methods of beekeeping.

The world of apiculture a century ago began to reveal itself and I became intrigued by parallels with the beekeeping issues of our own time. It was an era of debate about the merits of modern scientific methods against the traditional craft in areas such as the best race of bee, cross-breeding, genetics, and the most suitable way to keep bees. In letters and articles they discussed pests and diseases and the extent to which beekeeping needed regulation and legislation. They debated the interests of rural cottagers who kept bees to pay the rent, against others as varied as the hobbyist clergyman, the commercial queen-breeder and the businessman like William Woodley of Beedon.

⚇

I go out to the apiary with a couple of empty nucleus boxes, hoping to make a couple more splits from the survivor stocks, but am frustrated to find that the queens have slowed down their laying again. Then I realise why – it's now the June gap, between the spring flow (when the stocks were prevented from foraging) and the main nectar flow. I know this from the lack of flowers and the fact that the limes around the meadow are still not flowering. In a good year the limes would flower about now. Without pollen or nectar coming in, the queens slow down laying, or stop altogether. If some of my queens have *Carnie* traits (as I suspect) a brood break during a dearth is another of their typical behaviours.

Of course, it's also obvious by now that, having missed a spring crop of honey, we'll also miss the summer honey. The stocks have to get going by late April to build up enough to have sufficient foragers for the main nectar flow. This year the stocks just haven't had time to build up. Any small amount of honey they make will be theirs and not for the taking.

25 June – sultry and cloudy after rain overnight. I look up as I'm walking to the chickens to collect eggs after lunch. A pair of swifts flutter over. Sometimes I like to take a quick look at the apiary on a day like this, just to see that the bees are busy and to check that badgers and foxes haven't knocked over any hives.

Clover is starting to flower and the first tiny cream stars are opening in the limes. Ground elder flowers around the chicken enclosure fence are covered with honeybees. Bracken among the grass in the apiary has reached its full two metre stature.

On the edge of the apiary site I stop beneath the sweet chestnut tree and peer through the long grass towards the nuc boxes I left on their stands a few days ago. They're white boxes and so what I see next is unmistakable – a swarm bearding the entrance of one of the boxes.

It can't have issued from one of our stocks because none of them have been making swarm preparations or are in any shape to be able to. In any case, they're all too weak now after splitting. This is an alien swarm, for sure.

My heart jumps a beat with excitement. This is typical of the bees; they never do anything quite the way you plan; they choose the wrong time, the wrong place and the wrong way to do so many things. Ignoring all my carefully positioned swarm boxes that I've gone to so much trouble to site and prepare, this swarm has chosen a nucleus hive rather than a full hive body, that's only a foot off the ground, right next to the other hives, that I abandoned there because I couldn't make a split and I couldn't be bothered carrying the box back inside. The magic ingredient, it seems, is that it's full of frames of drawn comb.

I rush inside to put on my bee suit and gloves, bringing out with me a full hive body with frames. Most of the bees have crammed themselves into the six-frame nuc and there are still bees waiting in dark, gooey clumps, to enter their new nest site. I set up the hive (also white) and transfer the frames of bees from the nuc. There's no sign of the queen as I glance across each frame while transferring them, so I guess she's probably unmarked, but by the size of the swarm I'd also say it's probably a prime swarm, which means there's a good chance it has arrived with the old queen (prime swarms leave with the old queen).. and if she's unmarked, that could mean she's a wild queen. That's exactly what I want.

If I catch a swarm and put them into a hive on a site they haven't chosen I usually lock them in for a day or two by stuffing the entrance with grass, and I give them a frame of brood; bees won't leave brood, so this persuades them not to abscond. In this case they've chosen the site, so a full hive body on the same site can only be an added attraction.

All is well the next day. The swarm stays put when I remove the grass plugging the entrance.

On 26th, before supper, the bursar calls me to his office, saying that bees are coming in. My responses are equally of elation and a sinking feeling; we've been here before, I'm thinking. A couple of years ago wild honeybees took up residence in the roof above the bursar's office, accessing via air bricks above the windows. At the time I pleaded for clemency, but a death sentence was executed and the air bricks were blocked.

I rush outside and look up, hopefully. Bees are swinging around the air bricks, one over each of the two air bricks above the office. I'm elated at the sight of them. They're back! They haven't chosen my swarm boxes, but in this location I know where they are and they will be unmanaged, in which case if they survive and swarm next season I can take a swarm from them, as I've done before with swarms issuing from our buildings. What's more, the longer they survive, the more they become locally adapted wild bees and the more likely they are to have developed resistance to varroa and immunity to other diseases.

*Bees in the air brick*

Now all I have to do is work out how they're getting in, to keep the bursar on side. He joins me in the office after Compline (Night prayer) and we look up in the area corresponding to the suspected ingress and the location of the air brick outside. Clusters of exhausted bees are lying in the stone well of the window-ledge inside the office, but they haven't come in through any window gaps, as far as I can see. They've exhausted themselves against the windows in a desperate bid for freedom when the windows were closed. Looking up, to the level of the air brick on the outside wall, there's a narrow gap about the size of bee space between the wall and the wooden ceiling; that's the obvious point of entry. I suggest a line of sealant along the gap, which should stop them bothering us. In fact, when bees first arrive at a new site, it takes them a while to orientate themselves, so I suspect that some bees came in accidentally through the windows as they explored the new area, then the windows were closed, trapping them inside. I think only a small number came back through the gap near the ceiling, but once they work out where they are I doubt they'll continue coming in.

It's another example of the bees doing something 'out in left field', but I'm delighted that they've returned to the same nest site a couple of years after they were exterminated. Firstly, I know where they are, and I can always look out for a swarm to take from them next year.

Secondly, if they survive the winter, they'll be survivor stock that has been unmanaged for at least a year, which might suggest they are already survivor stock from the wild. These are exactly the kind of bees I'm interested in catching.

Things are working out interestingly, though looking increasingly like Plan B rather than what I originally intended. What else was I expecting, I ask myself: doesn't this happen every year in the apiary? Nothing ever goes quite to plan with beekeeping, but that's precisely what makes it endlessly fascinating. It's another reminder that honeybees are always wild creatures, and that we'll never really be completely in control of their behaviour or their genetics.

More surprises follow on 1 July. It's a cloudy, sultry afternoon, ideal for swarming. The bursar calls me after lunch; the bees above his office and mine are bearding the air brick. What's going on? I thought they'd just arrived this season, so I wasn't expecting swarming; or is it another swarm entering the roof area through the same entrance point, as the area within is large enough to contain more than one nest? When we had work done on the roof of the 1923 building some years ago there were a dozen or so old nests in the roof space, though I never worked out how many might have been occupied at the same time.

I hurry to put on my bee suit and pick up a nucleus box, but when I return the bees are moving back inside. I still can't work out if they're a new swarm arriving or the same colony that was already there. It's not common, but bees can sometimes swarm and return to the nest – perhaps the queen isn't quite ready? If this scenario is the explanation, then the swarm must have arrived last year and overwintered here, and hasn't actually arrived this season.

After twenty minutes there are many bees inspecting the nuc, which has three old frames and is baited with lemongrass oil. I place it on a green wheelie bin under the windows. I calculate that the swarm is from a colony already behind the air brick and that they've aborted swarming today, but will probably swarm again tomorrow. Once again, I'm not in control of events and I haven't really any clear idea what's going on. I love every moment of the enigma and suspense of the unfolding drama.

Later, another scenario occurs to me - that the bees I've seen entering and leaving the air brick over the last few days were only scouts, and that the swarm has actually arrived today. That would explain the dead bees exploring the bursar's office. Whatever the situation, the fact is that we have bees in our buildings again and that's good news for developing locally adapted bees. Once a colony establishes in the wild it can survive perhaps a few years; or, exceptionally, for over a decade or more, each year producing swarms of survivor bees that are an increasingly adapted and distinct ecotype.

Friday 2 July: white, cloudy skies, sultry and still; typical swarm weather. I stand under the limes, looking up at their creamy clusters of tiny flowers. I hear the persistent low hum of bees foraging. All morning I work in the parish office, the nuc that was left outside my window yesterday buzzing at last with a steady stream of bees. Are they robbers from the swarm above the window, or are they scouts from there, or from somewhere else? After lunch there's no sign of swarming from the air brick above the office, but a hunch persuades me not to move the box.

At 5.00 pm I look again and the whole front of the box is bearded with bees. It's a swarm. Have they issued from the air bricks after what seemed an aborted attempt yesterday? Or are they new bees from elsewhere? Ultimately, it doesn't matter. The fact is that I've bagged my second swarm. I'm elated.

<hr/>

July has arrived. A swarm in July usually isn't worth a fly, the old country adage reminds me, as does the advice from some of the brethren that Biddy was never interested in swarms this late. Usually by July swarms are casts, which are smaller swarms, issuing after the prime swarm has gone. There might be several or more new queens that emerge in the week or so following the departure of the prime swarm, but each cast has fewer bees and less chance of survival. If it's a reasonably sized cast, at least you get a young queen from the current season. Either way, a swarm this late won't do much this year, but none of the stocks are going to perform well anyway, given the late start they've had and the strange weather and seasons. In any case, I'm not interested primarily in honey this year, but in locally adapted bees, and the swarms I've taken so far this year might turn out to be exactly the kind of bees I want.

In relative terms, however, the two swarms I've collected are not actually late this season. In a usual year they would be, but everything is about three or four weeks late this year. Purple loosestrife and crocosmia, to name but two perennials usually flowering by now, still haven't hit their stride. That means, in terms of the swarm season, it's more like early to mid June rather than early July. The size of the two swarms suggests that they are prime swarms. That's good news because it means I might have another two or three weeks of the season in which I can bag one or two more decent swarms.

After Vespers I put on my bee suit and shake the new swarm into a nucleus box, closing them in for twenty-four hours. As I've moved them, there's a chance they might abscond (I know from past experience). In the morning I carry them to the apiary, the box purring like an engine. Later in the day I open their entrance disc and give them a frame of capped brood from another colony – an added attraction to their new home which should persuade them to stay.

It's always prudent to quarantine swarms, just in case they are carrying diseases. I put them on a site some distance from the other stocks, so there can be no drifting between the swarms and the Douai stocks. The process of monitoring will take a year or so before I can integrate them with the other stocks. I'll be looking at the brood as it appears, checking for signs of sac brood or foul brood, or deformed wing virus in the workers. One test will be their survival through the winter. If they come from treated stock, the varroa load this autumn will compromise them in the winter months and I could lose them, but this is part of the strategy for selecting and making increase from survivor stocks. If they are wild bees, it's possible that they are already locally adapted and have developed some resistance, in which case they will survive.

Thoughts of a decent after-cast persuade me to leave the swarm box under the office window. I set up the box with two frames of old comb and bait the entrance with a cottonbud soaked in lemongrass oil. After lunch I notice a steady stream of bees visiting the box. Interestingly, they're not coming from the air brick above, but are flying east to west, from the direction of the monastery garden, along the flight path between the side of the monks' refectory and the wall between the monastery and Abbey gardens. I've named it *Swarm Alley*. This is exactly what happened before the last swarm settled on the box, the numbers gradually increasing as scouts recruit more scouts to assess the site. If the numbers start increasing it's a good indication that the site is moving higher in the colony's election process. Eventually so many scouts will be visiting, and each visit becoming a vote, that one particular site is elected. Could my box be elected a second time in the same position?

What's clear is that *Swarm Alley* seems to be a hot spot for swarms. In the past I've caught swarms in this area. My first one appeared outside the cottages, directly in line with Swarm Alley, although the bees moved to another location, where I took them. Another year I bagged a swarm from the same area – the hedge-lined path leading to the cottages. Yet another landed on a small tree to the right of the cottages. And pre-dating the current swarm in the air brick there was another colony in there a couple of years ago.

The first week of July is mainly cloudy, with showers. Despite this, there are signs of progress with the stocks in the apiary. More of this year's nucs are showing signs of mated queens who are now laying. With the captured swarms this takes the stocks almost back to the number I took into last winter. Capped brood appears in the two swarms as well, and seems to be a decent pattern, with no signs of disease, though I still haven't found the queens, which are obviously unmarked.  Hopefully, I still have time to raise another couple of nucs this season, which means the queens will be laying by mid August, allowing plenty of time for the stocks to build up and prepare for winter.

One evening, after Compline, having put the poultry to bed, I take Banjo, my pet white Call duck, for a swim on the pond. It's a cool evening with flies swinging over the black, simmering surface of the pond. While she swims, preens and flutters her wet wings, I sit and think about my apiary plans.

As it's the main nectar flow, I hoped I might take a small amount of honey from the first swarm, as it was large enough to fill a full hive body. Unfortunately, when I inspect the super, it is light. It's also been too cold and wet for sustained foraging. In good weather a decent stock can fill a super in a week, but not this year. Plan B is that I split the swarm stock into three, which would give me two new stocks and their queens a month from now. Splitting this stock would also be an effective *varroa* management, reducing the mite load in the nucs during the weeks until new brood appears. In the original queen's colony the mite load will also crash as she loses most of her brood to the splits.

After a busy morning in the parish office I take an hour off to split the swarm into two nucleus boxes. It's a sunny, hot afternoon at last, though showers are still forecast for the next week or so. The bees are in a good mood, making it easy to split them. I take a couple of frames of brood and eggs, with attached nurse bees, and a frame of stores for each box, leaving some brood and eggs in the parent hive. I still haven't seen the queen, so I don't know which box she might now be in after the splits have been made. It doesn't matter, as long as each box has eggs so that there is the possibility of raising a new queen. I fill up the boxes with drawn comb and give the nucs a slab of fondant. In a week's time when I look again queen cells should have formed in one of the boxes, confirming that the swarm queen is absent. The other box, in which the swarm queen remains, will not have built any swarm cells.

While I'm in the apiary I give the other nucs more fondant, as they're likely to be confined again when the rain returns. Looking through one of the nucs that has been raising a queen, I see eggs, which means she is now mated and laying. Generally, looking over the stocks this year I'm satisfied that all the queens are producing a good laying pattern and that none appear to have any problems. The brood patterns are fairly solid, with a good number of healthy larvae being raised; a pepperpot pattern can indicate the early stages of queen failure and supersedure, but all looks well at the moment. After a terrible winter and a shaky start to this season, the apiary is in recovery mode at last and I can breathe a sigh of relief.

❧❧❧

A heatwave arrives in the third week of July, with a succession of days in which the temperature hits thirty-plus degrees Celsius. It's humid after recent rain. Our gooseberry and blackcurrant crops have never looked better, but have to be harvested by the community in blazing afternoon heat. There's still plenty of clover in the grass, and flowers in the garden that should have been over by now are suddenly starting to bloom – even late lupins. It's a rule of thumb that the lupin flowers indicate the duration of the swarm season, and that when they stop flowering the swarm season ends – an unreliable indicator this year.

After splitting the first of the captured swarms, the super on the parent hive continues to gain weight. Foragers return to the parent hive after the split, leaving only nurse bees with the nucleus; this means the swarm stock is still sufficiently populated by workers of foraging age to make a small crop of honey, though I doubt I can take any from them this year. One of the survivor hives is also looking very full of bees at the moment, with plenty of brood and with frames dripping honey as I carry out an inspection. It's a strong enough stock that I can split it in half, which gives me another stock from the winter survivors, and a new queen which they will raise with the survivor traits of the original stock – hopefully along with its other desirable traits of docility, a good laying pattern and winter-hardiness.

It's very satisfying on these long days of full summer to see how the apiary has recovered from the winter and the poor spring. Approaching the apiary, I notice the meadow bleached to hay in the scorching heat and loud with whirring grasshoppers. The sound reminds me of the summer days of my childhood when I cycled everywhere. I enjoyed free wheeling my bicycle and listening to the grasshopper whir of its wheels. Singing meadows and bicycles along lazy summer lanes are the soundscape of high summer. And now, in the apiary, at the height of long, drowsy summer, suddenly I'm free wheeling at last with the bees after the uphill struggle of the last few months. The stocks are now back to where they were at this time last year. From three survivor stocks I'm back up to twenty, including the nucs and this season's captured swarms.

The next tasks are to find and mark the year's new queens and to monitor each queen's performance into the autumn. I'm looking for good laying, which means full frames of healthy brood, and plenty of them, and a good laying pattern. I'm looking for docility in the colony and bees that look likely to make a decent crop of honey next season. This is a job for cooler weather though. In thirty degrees  my wet hair is plastered to my head under a heavy bee suit and my shirt is stuck to my back. In these extreme temperatures even my hands sweat, the water dripping from my gloves when I take them off. For now, while the heatwave blazes, all I have to do is allow the bees to make the most of the weather and the available forage.

23 July brings another surprise. In the cool early morning after Lauds I wander outside. I walk towards the car park and check the two bucket hives in the trees. As I approach, I see movement at the entrance of one of the buckets. On closer inspection, the bucket is slightly tilted, suggesting a weight inside. The entrance hole is full of bees. Another swarm!

I admit that I'd almost given up on these two bucket lures, though bees have visited them in small numbers, on and off, throughout the season. This is a late swarm though, another indication that the swarm season has come much later this year, along with everything else.

As dusk falls and all the foragers return to the bucket, I take it out of the tree where it's hung by a piece of rope. Before lowering it I put a piece of black tape across the entrance, to keep the occupants inside. As gently as possible, I carry the bucket the short distance to the apiary and place it on a stand, weighing it down with a brick. Rain is forecast for the weekend, so it's unlikely I'll be able to move them into a nuc box until early next week. I'm concerned that they might go hungry in the meantime. Swarms leave the old nest full of honey, which fuels the move and comb building at the new site, but it's unlikely they have sufficient stores or comb, having had only a couple of days at most to forage from their new home.

If I catch no further swarms this season I'll be content with my captured stocks this year. I don't know yet if they're all locally adapted, wild bees, but there's every chance that some of them might be, in a rural area like this. It will be interesting to see how they fare this winter and next season. As for next year, I now know that at least two of my chosen locations for taking swarms have paid off and are worth repeating. I admit that I was starting to have doubts too about the bucket lures in the trees and had almost decided to take them down. Now I know that it's worth putting them in the trees again next year.

I take stock of the common features of my successful lures: none have been more than six feet off the ground, though the first swarm chose the air brick which is at least twelve feet up; not all have contained old comb, but all were painted with old wax and propolis, and the most successful have been re-baited with lemongrass oil; none have been the size of a full hive body, but have had capacities comparable to a nucleus box; and I've identified Swarm Alley as an attractive location for swarms, making it a good location for swarm boxes next season.

After a rainy weekend the 26th is sultry and cloudy most of the day. It's ideal swarm weather, following rain. In the second bucket hanging in the trees by the car park I notice a couple of bees visiting early in the morning. Numbers increase through the day, and by dusk there are a dozen or so bees guarding the entrance. My guess is that they've chosen this site and are sitting it out overnight before the swarm arrives tomorrow.

In the apiary I manage to inspect the other bucket lure that was occupied a few days ago. Lifting off the lid, I find they have drawn several beautiful white combs attached to the lid (note to myself - making it harder to release!)The swarm looks quite large too; hardly a cast, which is a further indication that I'm getting prime swarms and that this is the main swarm season this year, despite being late July.

I decide not to disrupt their building at the moment as they seem happy enough in the bucket and have plenty of room. At some stage I'll need to do a 'cut-out' and transfer the combs to frames, but I don't have the equipment with me today. In any case time is short today. I've spent the morning on parish admin and emails and I have to prepare to cant at Vespers, so this inspection has to be a quick one.

Next morning the scouts are still hanging around the entrance of the bucket lure, though by mid morning there are only a few remaining. Where have they all gone? Perhaps they've returned to the parent colony to assess the situation. Alternatively, the colony's scouts have elected a better site and the bees I watched yesterday at this lure have been recruited to vote for another site.

I recall that Professor Thomas Seeley has discovered that new nest site scouts do many complex things, including initiating the signal for swarming. These scouts' behaviours include a whole series of tasks that involve the most incredible abilities to communicate and to operate together. First they have to sense when the colony is ready to swarm. Then they have to determine that conditions, such as temperature and weather, are right. They have to assess a suitable interim site nearby, where the swarm will cluster after leaving the old nest. Then they are responsible for initiating the swarm. When this happens worker bees can be seen making piper-signals that warm up the other workers. After this they begin to run through the hive making *buzz-runs*, exciting the other bees into a swarm impulse. If this set of behaviours isn't enough to astound us, they also mark or guard the prospective new site and guide the swarm there. All this is done by only a few hundred bees.

⟨⟨⟨⟩⟩⟩

Another showery, cooler day follows and suddenly there's no sign of guard bees at the bucket lure. Where have they all gone? Perhaps the swarming that seemed imminent two days ago has been called off for now and the bees have returned to the nest to sit out the inclement weather. Bees won't swarm on a day like this. Tomorrow, however, the forecast is for sunny, warm weather, and a fine day after rain often triggers swarming.

It strikes me that catching swarms this way is very similar to fishing. You set your lure with ground bait and you watch. Then you get some nibbling and you become focused on the entrance to the lure, the way an angler watches his float or the tip of his rod. It's just as addictive too. To paraphrase Walton, I envy no one but him, and him only, that catches more swarms than I do. (3)

As the morning warms up, I visit the bucket lure again on my way to feed the poultry. A single bee is motionless beside the entrance. Is this one of the same scouts that visited and staked-out the lure a couple of days ago? Or is it a scout from another colony? Either way, all it takes is one interested bee and the chances are that others will be recruited if the weather remains fine. The fishing float has twitched, but will another swarm bite?

By late afternoon the lure has been abandoned. Though sunny, it's blustery and I doubt that any swarm will favour such a day to move house. We're more or less at the end of the swarm season now and the forecast for the next week is rain. It's starting to look as though the year's brief swarm season is over and that my chances are narrowing of taking another swarm to end the season.

My nucs have not disappointed me though. By the end of June there are five new queens mated and laying and building up their colonies. My timing has worked and I have increased my stocks back to a level that the Douai apiary feels less vulnerable. The swarm season, however brief, has not disappointed me either! Given that the season began so badly and has been the worst season for at least a decade, I haven't done too badly after all. It's another lesson not to worry. If I can pull back from a season like this, why worry? I have survivor bees and I've made increase from them. I've raised new queens whose survivor genetics could increase my chances of having yet more survivor stocks next year. And I've now successfully captured swarms, some of which might carry the adaptive genetics I want in the Douai apiary. These swarms have reassured me that even if the worst happens and I lose all my stocks, I know I can catch more bees. All in all, I'm well on my way with treatment-free beekeeping; the bees are surviving, I'm raising new queens, splitting my survivor stocks and am catching swarms. What's the worst that can happen; I continue learning something new.

Raising your own queens is not only very satisfying, but you know their full history. If I order a queen from a breeder or from a supplier who imports them, I know nothing about the way the breeder raises them, how long they are banked and how long they are travelling. I know nothing of the huge number of environmental variables that can stress a queen and trigger unwanted epigenetic modifications.

There is now a lot of evidence of epigenetic modification in other species, such as rats and mice and worms caused by stress and environmental changes. Not only can environmental influences and diet switch on or off certain genes in an organism, but those that are switched on in the gametes can be passed on and can remain heritable for many generations. Think of all the possible environmental triggers that can potentially stress a queen coming from a breeder, even before she arrives at her final destination: temperature fluctuations matter; it matters whether or not the queen was instrumentally inseminated, as this requires clamping her into a vice under a bright light; it matters what age she was when she was taken out of the colony that reared her and banked her for sale. How long she was banked and the conditions and length of time in which she travelled are equally important factors. Was she handled roughly at any time, from the cutting out of queen cells and donating them to a nucleus, to banking her? What additional dangers was she exposed to during transit, such as damage to her limbs, or extremes of temperature?

If these variables can become triggers for epigenetic modifications today, I wonder what the situation was like a century or more ago, when queens were bred  as far away as Syria and Cyprus and took weeks by sea to arrive in Britain? It seems more than likely to me that many of these queens would have handed on epigenetic changes, either to progeny in the apiary in which they headed a colony, or to progeny resulting from cross-breeding with native queens or queens from other races. Did some of these modifications switch off immunity to certain diseases such as European foul brood, or switch on a susceptibility to new and emerging diseases, such as Isle of Wight disease?

# MINDING THE BEES

## Chapter 18

### Disorientation

'STRONG healthy stocks
Dutch Bees in skeps for sale,
delivery March, April, May 1922 -
DAVIDSON, Forest Road, Burton-on-Trent' (1)


*British Bee Journal 1921*

By mid July the nectar flow has slowed down. There's still blackberry blossoming in the jaded, mildewed hedgerows where cow parsley has gone to seed, while among the singing grass of the bleaching meadow bright yellow rugs of Birdsfoot trefoil *(Lotus corniculatus)* continue flowering, red buds ready to offer more nectar-rich blooms for the bees. I wander across the meadow, dried pods of Yellow rattle rustling underfoot and shedding black seeds.

My favourite meadow flower is an unassuming little jewel that is almost hidden in the long grass, just a few inches high, though it holds its own admirably. *(Pilosella aurantiaca)*, Foxes and Cubs, the Orange hawkbit, is a European alpine, its red buds (cubs) and ember-orange, smouldering flowers (foxes) could plausibly be the dandelions of Hades, which perhaps gave them the alternative name of Devil's paintbrushes. I love their jewel-like intensity, glowing in understated clumps at the base of the taller perennials. If they were two feet high they might seem immodest, but their diminutive stature lends them a certain humility and grace that endears. Perhaps that's why I like to call them by the less familiar name of Golden mouse ears.

When I'm not inspecting the hives, there are still bees to observe elsewhere. In the monastery garden the pond's water level has dropped this month, exposing the base of the yellow flag irises now in full bloom, and leaving the green, algal mud of the pond margins high but far from dry. Foraging honeybees arrive at the wet, slimy margins of the pond to collect water. In hot weather they use water to spread across the combs, cooling the hive, while other bees fan their wings, creating ventilation. There are scores of them landing at the wet mud as I watch.

It's an opportunity also to look for other insect life visiting the pond. There are over a dozen pairs of Common blue damselflies *(Enallagma cythigerum)* mating low over the simmering water, hovering like strange alien craft from a science-fiction movie. A Banded demoiselle (*Calopteryx splendens*) flutters across the pond, bright blue with black, butterfly-like wings. Then, hawking back and forth across the water, a pea-green and yellow female Southern hawker dragonfly (*Aeshna cyanea*); the males have blue and green bodies. Hawkers are the fastest of the dragonflies. Resting on the flag irises at the side of the pond is yet another species of dragonfly. It has a brown head and thorax and a stumpy, wide abdomen that looks as if it's been dipped in white paint with a very small touch of blue - the Broadbodied chaser (*Libullula depressa*); males have the blue abdomen.

We have an unusual, mysterious visitor at the end of the lavender bed in the guest garden - a Hummingbird hawk moth (*Macroglossum stellatarum*), a day-flying moth that flits between the flowers, its body at a vertical axis like a little hummingbird; brown thorax and fore-wings, orange hind-wings and a black and white, spotted abdomen, its long proboscis mimicking a hummingbird's bill. They are more common in Mediterranean countries, most of the UK populations being migrants from Southern France. They do breed here, but can't survive a British winter unless it's very mild, but there is evidence of some return migration to Southern Europe in the autumn. A remarkable fact about them, apart from their unusual and exotic appearance, is that they will often return to the same flower bed every day at about the same time. I have now seen this one twice at the end of the lavender bed just before Vespers. Hummingbird hawk moths obviously have some memory.

At supper one evening I watch the birds in the refectory garden. A nuthatch spirals around the trunk of the old oak tree, always bigger than you remember them. Birds are flying in and out of the thick growth of the wisteria on the wall opposite. Wood pigeons are nesting there. A small bird flutters out, thuds into one of the cloister windows and drops to the ground. I watch... no movement. Many birds are killed against our windows and glass cloisters. Most are pigeons, which are big enough not always to be killed outright, often leaving a grim, forensic image on the glass, recording the moment of impact. Last month three Lesser spotted woodpeckers banged into the same window overlooking the monks' refectory. On one occasion several years ago I was sitting in the garden with the Abbot when one thudded against the window and dropped dead below.

As supper ends Benjamin asks me to identify a bird. It's hopping along the rill in the refectory garden. It's a chaffinch. I was reading recently in Helen McDonald's book *H is for Hawk* about a scientist called Thorpe who experimented with chaffinches in the 1950s. He reared young finches in isolation, to understand how they learn to sing. He found that in their sound-proofed cages there was only a short window of time in which they could reproduce the complex song of adult chaffinches. Beyond this window, if they didn't hear adult chaffinches singing they never managed to produce the song accurately themselves.

After supper I go out and look for the latest casualty of our windows. As I approach, I see a tiny bird on its back - not moving. I almost give up on it, but I want to know its species, so I move closer. It's legs are moving slightly. I pick it up and turn it over. It's a juvenile greenfinch. It's eyes are wide open. I think it's just stunned. For several minutes I hold it, cupped in the warmth of my hands, while it glances up at me, panting a little from an open beak. I observe the distinct green area of feathers on its rump, between its wings. A little sadly I think of all the forlorn finches in small cages who never learnt to sing. Then, unexpectedly, with a flutter... up it goes into the oak tree. I smile as it disappears into the foliage. I think briefly of all the birds that have been through my hands ever since I was a fascinated boy standing by the apple tree at home with a crashed rook perched on my arm. Something in me yearned to be feral from that moment. The earliest experiences of life *teach us who were are.*

My own experiences in life have taught me that I'm a survivor too. A freak bacterial infection in my heart half a lifetime ago and lifesaving heart surgery on two occasions is enough explanation. My default is survival. Nature's is too, which is why I am an eternal optimist about the bees; Foul brood at the turn of the century went away; Isle of Wight disease went away; nature returns to abandoned engineering and heals the messes we might make. Whatever the prophets of doom come back at me with, the objective reality is that due to increased $CO_2$ in the atmosphere the planet is fifteen percent greener than it was in 2000. Some of these areas are amongst the driest on the planet. $CO_2$ is a plant fertiliser so that might actually help balance the current deficit in ammonia fertiliser, at least in the short term. Another survival story is that the native Black bee survived in Britain a century ago; and another is that our locally adapted bees are showing signs of coming through varroa.

One hot afternoon I am reminded that the bees are still mysterious creatures that are always just a hop away from a hole in a tree. In the apiary I observe something common, but strange happening at one of the hives. Hundreds of bees have been coming out and lining up in rows above and below the entrance and are doing something rather peculiar that no one fully understands. It's called wash-boarding. Like some insect version of line dancing, they are rocking rhythmically on the spot, anchored by their middle and hind legs. Their heads are down, abdomens in the air, and their front legs appear to be working at something. The behaviour goes on for hours, but it doesn't change the surface on which they are gathered, so it doesn't seem to be linked to cleaning or polishing behaviour. Are they ventilating the hive? Who knows? It's another reminder that we really don't know everything about these tiny, complex creatures that we've kept for thousands of years. We micromanage them, we throw chemicals at them, but we now have more diseases and losses of both managed and wild honeybee colonies than those simple, illiterate, uneducated peasants of hundreds of years ago who managed to keep bees successfully. Like Covid, wash-boarding humbles us in the right way, challenging us to wonder at what we still don't understand. The limitations of our knowledge stop us in a silent moment of awe in the face of the universe's vast mysteries.

It's been a poor summer for honey, but I knew in May that the stocks would not be strong enough to make a good harvest, owing to the late spring and the slow expansion of the bees. Despite this, they've made a little honey for themselves, though not enough for me to take any this year. But that's beekeeping for you, and it's not all about the honey, to be honest. There's so much more to beekeeping.

It was the same in 1933, when the Abbey's first stock only produced thirty pounds of honey and the monks left it for the bees. I shall also leave this year's modest yield on the hives. Like those monks, my enthusiasm is not dampened, though we have never achieved their early aspirations in apiculture, which the 1934 *Douai Magazine* records as a target of one hundred hives, 'As time goes on, it is hoped to establish this new industry on a firm foundation that the production of honey may eventually become a notable feature of the Abbey.' (2)

Beekeepers, like farmers, have to take the rough with the smooth. I'm reminded of this when I stumble in the *British Bee Journal* across an article from 5 May 1898. It's part of a series they ran called *HOMES OF THE HONEYBEE – THE APIARIES OF OUR READERS*, and at first I think I'm reading about William Woodley's apiary, except that the apiary mentioned was in Caversham, Berkshire. The article features William's cousin, Mr. A.D. Woodley – the one who had introduced him to beekeeping. It informs the reader that Mr. A. D. Woodley's first year in beekeeping (1879) proved to be a failure, as it was a poor honey season, giving him no return on his outlay. He had started beekeeping after reading an article in *Chambers' Journal* in February 1879, followed by a text book on beekeeping, where he found the instructions to build his first frame hive. Nevertheless, the article assures the reader that all had not been lost and that Mr. Woodley had gained valuable experience and skill during that first year.

It would seem from the article that beekeeping was Mr. A. D. Woodley's great consolation amidst a rather troubled life. He had married in September 1880, but shortly after the birth of their first born his wife suffered some kind of mental crisis which put her in a Reading asylum. She was still there, sixteen years later, in 1898 when the article appeared. Mr. Woodley is quoted: 'I hardly know what would have happened to me but for my "hobby" which since then has formed my chief pleasure. Truly stranded on the threshold of matrimony, my bees have proved a great source of happiness, and I have benefited in health and purse thereby.' (3)

Born 26 December 1852, A.D. Woodley was educated at the village school and put to work at the age of fourteen as a grocer, like his cousin William. A decade later he was working at the offices of Messrs. Huntley & Palmer of Reading, where he was still employed in 1898. In 1882 he joined the Berks & Bucks BKA, becoming an expert, assistant and honorary secretary, with connections to the Berks BKA. In his active years in the Association he made several tours through Berkshire and Surrey, giving demonstrations and lectures in village apiaries and stimulating interest in the craft. He was no doubt proud to make known his early association with his famous cousin, 'I can claim to have practically introduced my now well-known cousin, Wm. Woodley, to modern bee-keeping, having made him his first bar-frame hive at this time...' (4)

A. D. Woodley was a survivor. The mention of his wife and the tragedy overshadowing their marriage puts a lump in my throat, in the same way that I was deeply affected by reading the news of Annie's death and the effect it clearly had on William. But it also reinforces a strong conviction I have formed that beekeeping is about much more than honey, just as gardening is more than eating home grown produce. Both are processes and ends in themselves rather than simply a means to an end. Both are a great pleasure and a source of happiness.

We saw this with the surge of interest in gardening during the pandemic – especially that first lockdown, assisted by a beautiful spring and a fine summer. It was more than people being bored or confined indoors and looking for activities or distractions. It was more than people worrying about food shortages or wanting some fresh air. It was perhaps something deep in our national psyche that yearns for the simple life that affirms our deepest sense of freedom. It was a reconnection with nature that we are always in danger of losing as technology advances and the pace of modern urban life accelerates. Perhaps ever since those Victorian cottage labourers left villages like Beedon for the towns and cities, something in us has wanted to return and to pick up something of their ancient cottager economy. With their departure perhaps we even lost the dialect of transcendent realities. Now people are home baking bread again; they are sewing and making cakes; they are queuing up on waiting lists for town allotments; they are interested in keeping bees.

For the first time I understand poignantly something of Virgil's elegiac tone in his 30 BC *Georgics*, a four-book treatise on smallholding and beekeeping. His call for the peace of agricultural crafts after a time of civic unrest and the shocking death of Caesar strikes a note with our own recent experience in the midst of the battlefield of a pandemic, civil protests, clenched fists and the toppling symbols of heritage and history. Virgil saw the passing of an era, and so it seems do we.

In 1942, during World War II, the *Douai Magazine* expressed the monks' own experience of this phenomenon: 'When the war ends we shall have to make use of our newly won and hardly acquired skill and experience. Our farm work, together with our keeping of bees and hens, are beginning to make us feel quite rustic.' (5)

Virgil's *Georgics* celebrated a way of life in which the ancient beekeeper (*Book IV*) knew who he was and from where he had come, having been a cultured, independent farmer whose craft was once considered noble. Under the Roman Empire the beekeeper's importance declined in favour of the owner of the bees, reducing for centuries afterwards the noble poet-beekeeper to the status of a lowly slave – lower than the swineherd. Even if we modern beekeepers do not articulate Virgil's epic poetry at our craft, I am sure that today we recognise in ourselves while standing in our apiaries the same joyous feeling he conveyed. I am sure we find in our apiaries an optimism among the bees for a society that, like Virgil's, seems too often on a determined path to tear down and topple.

Perhaps it was also because those lockdowns were the biggest crisis we'd experienced since the Second World War that something of the war spirit returned. My grandfather was typical of many who grew vegetables and kept chickens in the back garden during those years. We reconnected with that lost world, epitomised by the black and white British Pathe Film *In Your Garden* (1944), featuring every gardener's friend, Mr. Middleton.

Mr. Middleton was a tall, lanky, middle aged, bald man in a suit, with 1940s circular, dark-rimmed spectacles. His memory now inhabits the world of YouTube, recalling a lost era of black and white Pathe films narrated with ebullient, clipped accents; the gardening equivalent of Dame Vera Lynn. With his matter-of-fact, slightly stilted speech that would have also appealed to the working man, he had an air of polite, quiet formality, interrupted only a little by the incongruity of his long legs stepping over a picket fence (by invitation) to inspect a fellow gardener's handiwork. I watch him congratulate an amateur gardener as they stroll past his vegetable patch observing the businesslike appearance of his orderly rows of cabbages. In the background an orchestra plays an upbeat soundtrack. You wouldn't know it had been made during the war.

Mr. Cecil Henry Middleton galvanised the nation's *Dig For Victory* campaign in 1939 on the BBC's *World Service*, but began his broadcasting career in 1934 with afternoon wireless gardening broadcasts called *In Your Garden*. The first celebrity gardener, he paved the way for Percy Thrower, Geoff Hamilton, Alan Titchmarsh and Monty Don. Something of that world lived on in my own childhood. I would pass the Metroland gardens of elderly gardeners as I walked to the shops or home from school – old *make-do-and-mend* gardeners who dug the ground while wearing a tie and a tweed jacket even on a summer day; businesslike, avuncular fellows who mentioned the war quite a lot, who stowed their meticulously cared for and oiled gardening tools and sipped their hot tea in corrugated Anderson shelters left over from the Blitz.

It was more than a possible food crisis in 2020 that sent us on our knees, sowing seeds. It was something to do with the simplicity and the freedom of another time we longed to reclaim. Covid also made us aware that we are not in control of everything in an increasingly disorientating world, and we instinctively went down on our knees to hold the earth. When the world shakes, what else is there left to hold on to? Getting our hands in the soil helped us hold on and orientate ourselves. More importantly, we sowed seeds of hope.

We find those vestiges of hope and freedom in an orderly vegetable garden or a burgeoning cottage garden - a world that hasn't changed since that Pathe film was made in 1944. Gardens are freedom, life and continuity. Cabbages and carrots still grow in neat rows where weeds have to be hoed, as they did when Mr. Middleton spoke to the nation in black and white. Lupins and foxgloves and alliums still adorn the borders in late May, even when the Chelsea Flower Show has been cancelled. We can still return to our gardens with some sense of returning to the old, familiar world before Covid. Our gardens are a symbol and an expression of the vestiges of our small, local communities in which everything that is meaningful and valuable about human life is affirmed in the face of what sometimes seems to be a vision seeking to delete in us of all that is worthwhile in the human experience.

If anything of that withered world survives, it is because it is rooted deeply in the freedom and the life-affirming refuge of an English garden. It was rooted in me too, having spent a lot of my childhood in a suburban garden. That root grew into a great love of the outdoors and of the natural world. To this day I have a strong desire to spend part of my day outside where I can see the sky, hear the birds and be surprised by anything else that would otherwise pass by unnoticed. Not that my childhood days were entirely halcyon along the lines of Tennyson's *Eleanore*,

'*The yellow-banded bees,*

*Thro' half-open lattices*

*Coming in the scented breeze,*

*Fed thee, a child, lying alone*' (6)

Many people realised in that year of lockdowns that we don't want supermarket-standardised vegetables and polished fruit. Perfection? No thank you! We want carrots with soil on them, that look like funny pairs of trousers; we want parsnips resembling gurning codgers that make us laugh. We want downy feathers stuck to our eggs, still warm from sitting hens. We want parts of our country to remain wild and free and pollination to belong solely to the bees and other pollinators. As for the perfect bee – my idea of the perfect bee is the one that got A.D. Woodley out of bed in the morning when he was stranded on the threshold of matrimony; the one that kept William Woodley focused when Annie died; the same bee that helped me get back to the basics of life in the midst of a pandemic. The perfect bee is always first and foremost the source of any beekeeper's happiness and health. The perfect bee embodies survival and life.

Is this mere sentimental nostalgia for the past? *Nostalgia* – from the Greek *nostos*, meaning *returning home*, and *algos*, meaning *pain / ache*. Nostalgia is more than a sentimental longing. It's a deep desire to return home to the inner freedom to be ourselves. Knowing who we are and where we are orientates us. My musings then are more than what might appear to be nostalgia for an era that has gone. It's more complex than that. I realised this as I progressed with my research into the many changes that have happened in beekeeping over the last one hundred and fifty years, which has begun to orientate and liberate my own beekeeping.

Knowing what happened to apiculture in Victorian times and the direction it took during and after Isle of Wight disease and throughout the twentieth century has given me a stronger sense of how we came to where we are now and the kind of beekeeper I want to be. Far from mere nostalgia, looking back to the past reorientates us in the present, and that can set us free from the mistakes of the past.

Below the surface of this nostalgia there is a deep desire in us for freedom from today's chaos. There is chaos everywhere. In recent years this kind of nostalgia has occasionally appeared in a television drama, such as the BBC award-winning comedy series *Detectorists*, about a couple of middle aged detectorists who belong to the Danebury Metal Detecting Club in the fictional village of Danebury, Essex. Lance and Andy are forty-something men whom life is passing by, whose hobby is metal detecting for treasure. As they plod across ploughed fields in search of the one find that might change their lives, their hobby at least momentarily disturbs the quiet tedium of modern life. If you dig below the surface, that fictional world is not about any life-changing event that sets the world alight, but a succession of lovely moments of pure freedom that are an antidote to modern life. Dig below the surface and its not dissimilar to the reasons people just like Lance and Andy belong to BBKAs or spend their spare time fishing or working on their allotments and gardens. Every episode of the drama begins with a little scene at ground-level; a ladybird walking up a blade of grass, a snail turning on a leaf, or an iridescent magpie in the morning sun. It's about history, conservation, a love of nature, the simple joy of sitting under an old oak and the quiet happiness of a friend's company. The treasure of these freedoms is always right under Lance and Andy's noses in the simplest, smallest things of life that are so small they are often easily unnoticed. The antidote to the utopian vision of a micromanaged world that has all the answers is at ground level in the truly human, the strong bonds of meaningful relations and the kinship of our small local communities.

For detectorists like Lance and Andy finding gold is the holy grail of all their endeavours. In one episode Lance finds a gold object. It ends up on display in the British Museum, but Lance's joy is short-lived, partly because he suspects a curse is at work when all kinds of things begin to go wrong. Lance ends up taking all his reward money to a coin shop and buying several gold coins (perhaps Golden Guineas), which he then buries in a field. Lance learns that there's more to his hobby than profit, and that he has to put back what he takes out. What really gets him out of bed in the morning to spend hours in a field isn't actually the gold. It's the time he spends in the outdoors with his best friend, where the modern world with all its disappointments ceases for a few hours to pass him by. Detectoring is their survival.

In another episode the two detectorists actually find the holy grail of all detecting – and this time it is literally the Holy Grail; but it isn't a gold chalice, as we might suppose, but a rather small, unprepossessing earthenware bowl. Neither of them want the bowl. The lesson here is that the real treasures of life are usually quite prosaic. The real treasures of life are so ordinary that we sometimes disregard them because we often don't value them.

There might be some beekeepers for whom a certain weight of honey per hive is a holy grail. Others might pursue, like beekeepers of the past, the holy grail of all races or, through selective breeding, aspire to creating the perfect bee - Sladen's Guinea Gold, Brother Adam's Buckfast strain, or a futuristic utopian Frankenbee. None of these get me out of bed in the morning. They remind me what we have taken out of the honeybee for the last century and a half. They recall for me the curse we put on ourselves by outraging nature with our meddling. I want to put something back, and for me the real treasures of the craft are the things most people might consider rather prosaic, but I notice them. It might be no more than a grasshopper on the side of a hive or the transient flash of a jay's wing in the trees, the distant call of a cuckoo or simply the sound of summer bees fully alive in a nectar flow. You can keep your Guinea Golds. My beekeeping is an earthenware bowl and I'm joyously happy with that.

Moreover, it's at ground level that a real difference is made to any of the world's problems, just as in the past it was the individual gardener and citizen, like my grandfather, who also dug for victory. In our war against the adversities that afflict modern beekeeping it's also going to be the humble beekeeper in their back garden and the small local BKA who do the spade work in solving the problems in apiculture. If it doesn't happen at ground level and we look to government, Big Business or Technocrats who promise us every solution through new technologies alone, we will end up cursing our beekeeping and ourselves.

Backyard beekeeping got William Woodley's cousin out of bed, just as metal detecting gets Lance and Andy out of bed, because it's not just about honey, any more than metal detecting is all about finding gold. For me every visit to the apiary begins with noticing an insect on a blade of grass, a bird perched in a tree and the joy of sunlight slanting through summer trees. I suspect, as with many beekeepers, every moment I stand at my hives disturbs a bit of the nonsense and chaos of the enslaving modern world. The sound of the hives on a warm afternoon decodes for me the very grammar of freedom and joy, and I'm at home again. Each visit to the apiary digs down into the real treasure of life – that I am here and it's good to be here; and that the world is good. The fact that it isn't any more complicated than that is exactly why it is so endlessly absorbing and so deeply rewarding. And that's freedom. That's the moment when life can not pass you by because you are fully alive to the moment.

I doubt much has changed in this regards for beekeepers. If we were to go back a century we would find the same themes in their lives – of meaning, joy, the search happiness. It's why I like t look at old photographs of those beekeepers. Old photographs from the infancy of photography don't so much capture the moment the way modern ones do; they capture something of the mystery of a person. I wonder if this is why William Woodley had a passing fascination with photography. Or was it that photography could be combined with his other obvious interest in people?

I've become fascinated by those old photographs. At first I passed over them and the articles 'Homes Of The Honeybee: The Apiaries Of Our Readers' in the *British Bee Journal*. Now I am spending more time looking at them. It's something to do with the long exposure time during which the subject had to remain still; concentrating the gaze into the lens; and the foggy images of generalised, blurred features conveying ghostly portraits of faces that haunt the viewer. Despite their indistinct characteristics, nevertheless, something of the presence of the person is captured in the image, enhanced by the years since the exposure was made.

They remind me that Isle of Wight disease was not just about bees or beekeeping; it was about beekeepers. These were the same people who struggled with bee disease in a time of rapid technological, social, cultural and economic change. These were the people whose lives were changing with the lives of their bees: Mr. Ford was a grocer shopkeeper from Wood Hayes, Wolverhampton, pictured with a boy dressed in black in 1900. In his spare time he was a church organist and choirmaster. And he kept bees. Among his favourite bees were Italians and English queens crossed with Italian drones. Mr. Balmbra of Alnwick, Northumbria looks tall and slim, dressed in a black suit and a flat cap, and sporting a dark Victorian moustache. Mr. Edgar S. Miller of Chippenham, Wiltshire, is dressed the same way. Mr. Crawford of St. Clement's, Jersey, Channel Islands, looks indistinct in the distance with his hives, but is dressed in overalls. Others, such as Mr. W. M. Clarke of Gateshead-on-Tyne, Northumbria wore a white shirt with the sleeves rolled up and a waistcoat. From nine years of age Mr. W. Loveday of Hatfield Heath, Harlow, Essex, had worked as an agricultural labourer. After getting a chill he developed rheumatism, followed by a spinal injury in 1894. Despite ill health, he managed to work with his bees and it is clear that they brought him much joy and consolation. A survivor! In Chaddesley, Kiddermnster, a country schoolmaster named Mr. Edward A. Millward was photographed with his eldest son. He told the readers that his happiest memory of beekeeping was a chat with the great W.B.C. (William Broughton Carr, co-editor of the BBJ and designer of the WBC hive in 1890). He recalled their genial chat about bees while eating bread and cheese with a drop of something unspecified. Mr. R. Mackinder of Newark, Nottinghamshire, suited, with his flat cap and big white beard leans on one of his hives. His interest began at a flower show where he had heard a lecture on bees. In their town bee-garden Mr. S. Powlson and his wife were photographed in 1902. He sports a hat like the one Van Gogh wore in a painting; his wife sits near the hives in a long, dark dress down to her ankles. On her lap she has some needlework, or lace-making perhaps? She took an interest in the bees and was well able to hive a swarm.

Many of these were the new urban beekeepers of the growing towns who had taken up the mantle from the old country skeppists; but what is clear is that beekeeping enhanced their lives and brought them the joy of an interest that the old skeppists would not have been able to enjoy in the same way. Some of them recalled the skeps kept by their father and the sulphur pits at harvest time, or had been given their first 'driven' swarm as rescue bees by an old village skeppist who would have put them to the sulphur pit.

Beekeeping is about such people. Minding the bees isn't only about looking after our bees; it's about remembering and celebrating the anonymous and forgotten people who for various reasons discovered the joy of beekeeping and whose lives were never quite the same afterwards in many wonderful ways. For some of them, minding the bees helped them survive the plagues and disasters that none of us can avoid for ever.

I'm not worried this year then if there's no honey from the bees. Keeping the bees is an end in itself now, to some extent, and honey is a welcome bonus, while minding the bees is as much about minding ourselves as well. It's the bees that are every bit as much minding me, such as when I'm in the apiary on a still summer evening as the day empties of sound and I look up and notice clouds as pink as a wood pigeon's breast. It's in those simple moments that I feel grateful and glad with life, when my whole heart rejoices with the old West Country saying, *I'm 'appy as a bee in a snumper (foxglove).*

During the pandemic I was conscious of how beekeeping took care of me, just as some people's gardens took care of them. At the time I wrote a poem about it:


**First Lockdown At The Hives**


I watch the bee-workers

outward-bound

this May morning,

heaving and hauling

away across the waves of purple meadow;

bee-songs and shanties-

flying folk-tunes

of the honey-mouthed hive,

while I wait through this long lockdown

for the strange world to re-open.

No social distancing,

their seams like seamen lined along a yard,

letting down their white sails of comb;

singing with bee shantymen

the rambling tales of long-haul lives.


I wait with them for the Wellerman to come,

the bees with their own strange plague

vectored by *Varroa's* viral bite -

bee-mite-bombs in their bully-hives.


But *Covid* has closed my world.


I envy the clean slate of their sky,

un-scrawled by vapour-trails.

I admire these colonies' busy crews

who can not read the times

but fly in the face of a pandemic;


while I am grounded here,

the wind sucked out of my sails.

Since then I have become more aware of how the world remains disorientated by that experience of the pandemic, and in many ways how we struggle to adjust to its repercussions. It was said at the time that we would not return to normal – only a new normal, and that seems to be the case. Tom Seeley, Professor of Biology at Cornell University, New York, has called honeybees ambassadors of nature, which is a very profound idea. What he means is that honeybees give us feedback on the natural world and how we're managing it, but they also foster our friendship with nature by nurturing our fascination, awe and wonder about the natural world. I think this is another reason why we must not consider honeybees to be domesticated; their intrinsic wildness is exactly why we must listen to what they are telling us through *Colony Collapse Disorder*, *varroa* and the various struggles they experience in the modern world. A cow in a field is not the same barometer of the environment.

Beekeeping gives us a way of connecting with nature through these special little ambassadors, making us more aware of the pulse of nature, the heartbeat of the natural cycle of the seasons, and changes to the environment. Bees also hook us, as they hooked William Woodley, or Mr. Mackinder at the flower show. It's probably one of the main reasons why we know more about the honeybee than about any other insect.

Professor Seeley relates in his *The Lives of Bees* a touching story about how he became hooked on bees as an eleven year old boy. One day he saw a swarm of bees heading for the knot-hole of a walnut tree. At first he crossed the road because he was afraid, but as he watched them he realised they were setting up home in the tree. He gradually overcame his fear and each day found that he could get a little closer and that he could watch the bees coming and going without upsetting them.  He described the experience as the moment bees impressed themselves on his life.

People have returned to the monastery after the pandemic, just as they always came, but with lives more broken, damaged and disorientated. They are my own barometer for the state of modern humanity. I speak to many people who visit Douai Abbey for whom life has been very difficult, especially since Covid. Many people have their own private agonies as well, just as A. D. Woodley and his famous cousin had. The human condition has not changed. They come to a place like Douai Abbey looking for some peace and gentleness in the midst of broken and damaged lives. They come looking for hope and they find a kind of hope here. They come to reorientate their lives and to find an inner freedom. They are attracted to the simplicity and order of a monastery, which is not unlike the order and a certain simplicity we find in beekeeping.

I can fully understand how A. D. Woodley found hope and consolation in his bees. I'm sure they were a distraction and in some way a healing from the agony of his personal circumstances. When Annie died I'm sure William also found comfort in the routine of his hive inspections and orderly work in his two apiaries. Beekeeping, like gardening and like so many crafts, is not only endlessly fascinating, but is deeply healing when we struggle to survive.

There is a serene order in the lives of bees. An afternoon spent in solitude with the bees, practising a craft that is centuries old, is endlessly reassuring in its simplicity and unchanging predictability. Bees occupy a binary world in which every individual has its ordered role and a specific identity. A bee's world is one of order within nature's immutable laws. There is a beautiful and simple logic to their lives; you know where you are with them. Their beautiful order and elegant operations contrast sharply with the chaos and irreverence of post modernism. They represent a continuity with the past in a world that is increasingly insecure, unpredictable and afraid because it is in danger of becoming post human, devoid of any recognition of what is special, ideal, unique and meaningful about human beings.

In 2020 a team of researchers called the phenomenon of Covid's global reduction of human activity an *anthropause*. When the world was in chaos and crisis I found serenity by sitting for long periods at the hives, simply watching. Some of the monks here tell me that Biddy, my predecessor in the Douai apiary, used to do the same. I couldn't help at that time but notice the striking contrast between the increasing activity of the bees and the corresponding reduction in human activity. It reached its most extreme expression in the swarm season when simultaneously we had the swarms and the *anthropause*.

When people go through crises they are often kept going by the things that not only distract them but that keep the world ordered and moving. At one difficult time in my life I found that simply putting my watch on in the morning set me up to face the day, but it doesn't matter if it's gardening or doing the housework or minding the bees, as long as they have the same effect. It was immensely satisfying then during lockdown for me to notice how the bees simply got on with life, and how nature wants to survive, even when the rest of life seems to be in crisis.

On a sunny day in April 2021, as the world opened up to something resembling normality, I found myself doing something I don't usually do near the bees: I sat close to the hives in my habit, without my bee suit or veil, just watching. It was something to do with having shielded for months in 2020, and wanting to feel again the hopeful sun on my skin and to open up with a world that was gradually opening up again. It was something to do with a relief that the worst of that time was over. It was something to do with the joy of spring and the joy of life. It was freedom. It was comparable to waking up from a heart operation on a spring day and seeing the sunlight through a daffodil in a vase, as bright as stained glass. I doubt that anyone will tell the bees about my inevitable passing, when it eventually happens, but that April day I wanted instead to tell the bees for myself that I was alive. I wrote another poem about that strange experience:

**After Lockdown**

Enough for now the freedom of this April afternoon,

safe beneath the flickering swifts,

to sit and watch, unshielded now,

these newly-woken hives.


*I have come to tell the bees I am alive!*


Reunited friends

after the long lockdown

survivors of those grave, dark months -

winter's furloughed workers are we all,

now touching and  talking again

in renewed intimacies

as small as bee-space.


I sit here, motionless,

barely distanced,

listening, and immune

to the murmured threat of stings.


A new sun's transmission of life

warms the uncovered skin

of my face and arms.

To touch this bare moment

is to breathe again clean air;

to hear once more

the resurrected cuckoo of the wood,

and to feel within myself

the very dance of deep life.


I greet their song -

the soft sound of promised meadows,

exposed to the full force

and shameless feeling

of my own season's naked joy;

that I have lived to see a world

where bees have flown again,


*that I have lived to tell the bees I am alive!*

During the Victorian age beekeeping advanced in leaps and bounds from primitive skeps to frame hives with all their modern apparatus and methods, together with the discovery of other races of honeybee, their complex biology and the new field of genetics applied to cross-breeding. A new era of beekeeping began, but it coincided with bee diseases on a scale that had never been known before and a mysterious plague that completely changed the face of British beekeeping. A century later, and exactly a hundred years since William Woodley died, humanity has shifted again from an era of great technological and social change to a new change of era. I believe that beekeeping is also entering a new era, with the emergence of new diseases and pests and another mysterious malady known as Colony Collapse Disorder. It is an era that promises new technological solutions to beekeeping problems, and new problems, and not all of them in the form of diseases.

Lewisburg, Pennsylvania 2006: bee farmer Dave Hackenberg's bees started disappearing. Literally. In the morning his colonies were fine, with a queen, brood and stores. By the afternoon the hives were virtually empty, with just the queen and a few bees. Colonies went out to forage and never came back. No one knew where they were going. Hackenberg lost most of his bees that year. Seventeen-thousand colonies were reduced to three-hundred. In 2007 Bret Adee, the biggest commercial beekeeper in the world, lost fifty-thousand hives, a staggering two-billion bees from his Lost Hills farm, California. Like Hackenberg's bees, they simply went missing. A new phenomenon had appeared. They called it Colony Collapse Disorder (CCD). To date it has spread to other countries, having the same effect it has had in the US. Defra has not yet recognised any cases in Britain, but I have read a statement by at least one scientist working in the field who disagrees with Defra.

Vast amounts of money have gone into research, to establish what might be responsible for CCD, but it remains a mystery. Like the Isle of Wight phenomenon, there seem to be many factors at work. Analyses of bees have shown that they are full of traces of large numbers of pesticides and sprays. Varroa is also a factor, no doubt, while moving commercial bees vast distances to the monoculture farming of California every spring probably also stresses and weakens the bees. Among the almond farms and other monoculture crops, a diet of pollen exclusively from one plant might result in nutritional deficiencies compared with bees that have access to a variety of forage plants. Some people have also blamed cell phones, though this has been dismissed. There simply isn't one identifiable cause for CCD, just as there was no single identifiable cause for the Isle of Wight losses.

One US scientist has pointed out that all commercial bees in North America are related to no more than five hundred breeder queens and that this narrowing of the gene pool might also be responsible for the susceptibility of commercial beekeeping to new diseases or phenomena such as CCD. This could be another warning about the dangers of the selective breeding of bees, and especially about the risks associated with inbreeding honeybees.

It's not all bad news though. On YouTube I find a couple of beekeepers in Snowdonia who seem to share my vision. One of them was in building work and found some bees living in buildings, which were infected with varroa, but were surviving. The pair began studying other wild colonies in trees and buildings and found that these also had varroa but were surviving. In the early 1980s they found an oak tree with a wild honeybee nest and began checking the tree after varroa emerged in Britain. They found that these bees were surviving too. They estimated that the nest survived over nineteen years. Having followed several tree colonies, they confirmed that the ages of these colonies were over six years and over five years, on average, with the nineteen (plus) colony being the exception. These estimates corroborate similar research by Professor Thomas Seeley in the US and his findings that on average a colony can exist continuously for up to several years. The evidence of a colony surviving over nineteen known seasons is even more interesting, however, because it confirms the claims made by my monastic community that colonies were living continuously in Kelly's Folly at the Abbey from the early to mid 80s, right up to 2017. This is comparable to the tree colony of nineteen seasons up in Snowdonia. More interestingly, this couple's studies are further evidence that the colonies in Kelly's Folly might already have been resistant to varroa when I caught my first swarm from the building several years ago.

In Wiltshire there is another interesting beekeeper. In 1995, after varroa had been around for a while, he found that one of his hives seemed to be immune to the mites. When he examined the mite drop under a microscope he found some of the mites had been chewed.

Had the bees chewed the mites? What was going on? He began to select queens from this hive and artificially inseminated them, to breed more resistant queens, while also releasing drones in the area, to flood the gene pool with resistant genes. The real mystery was why his bees were surviving when varroa also spread viruses through the colony. It's these viruses, such as Deformed Wing Syndrome, that seem to lead to colony collapse.

A leading expert in honeybee colony collapse, Dr. Declan Shroeder, analysed these bees, looking for two identified viruses, Type A and Type B. Type A was the Deformed Wing Syndrome virus associated with Colony Collapse Disorder. Type B was the less dangerous *Varroa destructor* virus. Strangely, Shroeder found that only Type B was present, and in large amounts. It was an astonishing discovery; the bees were using Type B to vaccinate themselves against Type A. As long as they carried the Type B virus, Type A could not enter their bodies. In other words, the bees and *Varroa destructor* mites had adapted to one another, which actually makes sense because parasites need their host species to remain alive.

Dr. David Heaf, a Welsh beekeeper, believes this co-evolving will eventually create a stable situation, and he hasn't treated his bees since 2009. Each year he suffers 24% losses of his colonies, which seems to be the accepted rate of loss among treatment-free beekeepers. On the other hand, there is growing evidence from surveys that treatment-free colony losses each winter are actually lower than for beekeepers who treat their hives.

In an edition of the *BBKA News* a couple of years ago there was an article by two beekeepers from Lleyn and Eifionydd Beekeepers' Association about treatment-free beekeeping. The writers were in their eleventh year of treatment-free beekeeping and claimed to have solved the problem of living with varroa mites. They knew then of 104 other beekeepers who had shared the same experience of not treating their colonies for varroa, and claimed that they knew of many others throughout the UK. Between 2010 – 2015 they collected data on 1, 573 colonies, of which 477 had been treated. Their survey revealed that the average winter colony losses for treated colonies was 19%, while for untreated colonies it was 13%. After several seasons treatment-free, none of their colonies had any problems living with varroa, while they were gaining increasing data on the long term survival of wild colonies in trees and buildings.

In Lleyn and Eifionydd BKA very few beekeepers treat their hives. In certain areas of the UK others have also stopped treating; one man in Swindon has been treatment-free for twenty-four years, while another beekeeper in the northeast has been treatment-free for seventeen years. It does not seem that different races, or methods of beekeeping, or hive designs make any difference to varroa-resistance. The one factor that seems consistent is the use of locally adapted bees.

Interestingly, recent genetic analysis shows that many Welsh bees have a high degree of genetic similarity to *Apis Mellifera mellifera*. This and the existence of wild colonies that have survived long term resonates with my own thoughts about the bees in Kelly's Folly at the Abbey. I think it very likely that in the same way these were locally adapted bees that had developed varroa resistance. If they also had a high degree of *Amm* genetics, this points strongly in support of the theory that the solution to varroa mites lies with local populations of bees that are genetically similar to *Amm*. There is, however, evidence from Professor Stephen Martin's work at Salford University that in many countries different subspecies are also showing the ability to develop resistance to varroa through natural selection. Indeed, Professor Thomas Seeley notes from his New York work that wild bees in that area are hybrids from *Apis mellifera, Ligustica, Carnica* and *Caucasica*.

According to an issue of the BBKA magazine in 2020, US colony losses from CCD were 43.7% in the previous year, with 22.2% being lost over the winter 2019-20, a decrease compared to the previous year. In the summer of 2019, however, 32% of managed colonies were lost, the highest summer rate of loss ever recorded. The dilemma commercial beekeepers face is that the cost of not treating would invariably result for a few years in even higher losses until their colonies begin to develop resistance to Type A viruses, and that's a high price to pay in the commercial world. Yet it means that US beekeepers are constantly running to keep up with the losses from CCD. The only solution, to break this vicious cycle, is to allow colonies time to develop immunity to Type A viruses, but the average of five years this takes is not feasible in the commercial world, while agriculture would also suffer from the smaller number of available colonies to pollinate crops in the interim until resistance has been established.

Each of our little actions and subjective choices all have huge repercussions that we might never see or be aware of, as illustrated by the beautiful story *The Man Who Planted Trees* published in 1953 by the French author Jean Giono. It tells the story of a solitary, heroic shepherd, Elzeard Bouffier, who spends his life single-handedly reforesting a desolate valley in the foothills of the Alps in the early twentieth century. The story is an allegory, the old man's acorns representing every human action, good and evil, which have far reaching consequences. At the end of the story the narrator looks back on the old man's accomplishment and finds in him something admirable in all humanity.

The story reminds us that we must all plant seeds of hope for the future, even if we feel that our single acorn hardly makes a difference. It is the same life-giving goal that Brother Adam exemplified (to give him credit) and it made him admirable too. He saw, in his own time and circumstances, as best he was able in 1915-16, that the British honeybee was in trouble, and his response was a lifetime of work planting the seeds of hope that something like Isle of Wight 'disease' could not happen again. There is really little difference between the work of Brother Adam of Buckfast and the fictional Elzeard Bouffier. Both represent those admirable and extraordinary people who dedicate their lives to a work that reminds us of the beauty and wonder of the world. In the same way that Elzeard Bouffier's heroic work moves the human spirit, Brother Adam's extraordinary life and work must also win our admiration. We should be thankful that we have people in real life, and not just fictional characters, who remind us, despite our capacity to make such a mess of the world, that humanity is intrinsically capable of greatness of spirit.

But the simpler world of Elzeard Bouffier is no longer with us, while Brother Adam's project to produce a superbee belongs, in my view, to another era. The goal of beekeeping ought no longer to be solely focused on increased honey production; today we need a bee that is able to flourish, despite the increasingly hostile environment of our modern world of pollution, pesticides, varroa mites and colony collapse. I wonder, if Brother Adam had his time again he might be breeding a very different kind of bee? He would probably be breeding a honeybee that is increasingly resistant to varroa, I have no doubt, perhaps with all the admirable traits of his Buckfast strain, but I wonder how his ideas might have changed, knowing that the old British bee has reemerged, and having reflected upon the detrimental effects of decades of varroa as the consequence of a century and a half of importing foreign bees. How would his work have changed in the light of new areas of science, particularly in the field of genetics and the new technology of genetic engineering?

I'm not altogether dismissive of everything he accomplished either. When tracheal mites reached the US in the 1980s some beekeepers imported Brother Adam's Buckfast strain because it had been originally bred for resistance to tracheal mites. I've heard American beekeepers who were immensely helped by the Buckfast bee at that time, whose livelihoods from beekeeping were saved by Brother Adam's work. On the other hand, varroa came along on the heels of tracheal mites and diverted everyone's attention, and eventually tracheal mites disappeared. Perhaps there's a lesson there for those, like me, who have at times obsessed about varroa. If we pay too much attention to something we can actually encourage it and make it harder to eradicate. Sometimes if we do nothing a problem will sort itself out.

I wonder what Brother Adam would have thought about genomics too - the study of the complete set of DNA of an organism? How would he have approached the science of transgenics and Precision Breeding– the introduction of a gene into the genome of another organism? Genomics is certainly useful in the laboratory, as it can help us understand the genetic basis for health and behaviour. Now that the honeybee genome has been decoded this can help scientists towards a better understanding of bee health. It could also take the guesswork out of bee breeding by revealing gene markers that make stocks resistant to diseases and pathogens, revealing how we can increase our bees' chances of surviving new diseases like CCD.

Transgenics, however, is a more controversial topic. The ability to build genetically modified organisms (GMOs) might seem, in certain circumstances, be a good thing. For example, what if you could build a transgenic mosquito with immunity to malaria? That would have the potential to relieve a great deal of human suffering. What if we could build a bee capable of defeating varra and CCD? That would certainly seem be a game-changer for commercial beekeepers in places like the US. Conversely, if transgenics were used in other situations, driven by political or economic ideology or greed for profit, its use could be completely immoral.

Aquarists can now buy florescent tiger barbs or zebra danios in some parts of the world, which have been genetically modified to glow yellow, green, red or purple. It started with research to create a fish that might glow when exposed to pollution. Scientists used transgenics to insert the gene for florescence from jellyfish and coral into zebra danios. Whether the experiment worked or not, they were quick to realise that there was a commercial opportunity in the market for garish aquarium tanks and in which garish tropical fish glow under blue light. The invention was patented, with breeding of the fish prohibited outside the company. Aquariums and accessories followed, cornering a market for the first commercialised GM pet fish - if you like that sort of thing.

No thank you. I prefer the natural look, although I've nothing against those who want pretty fish, and I can see the attraction of such an aquarium if you have children. By years of selective breeding fish-keepers have produced goldfish in a variety of colours (the native *Carassius auratus* is a bronze-coloured fish, like the Common carp), guppies in many shimmering colours, fancy budgerigars, canaries and mice, orange carrots, and numerous plant, vegetable and fruit cultivars for the gardener. Some might argue that if transgenics gets us to the same ornamental or productive outcome faster and more precisely than years of selective breeding can achieve, what's the harm? Others can see the potential, even with the example of glowing tropical fish, of something much more sinister.

In 2014 a German laboratory built the first transgenic honeybee queens. The work was replicated by scientists in Tokyo some years later. There now exist, therefore, a blueprint to build a GM bee, which some have called a *Frankenbee*. As the name suggests, the GM bee, like Frankenstein, has the potential to unleash sinister consequences.

Wouldn't it be a good thing if we could achieve Brother Adam's holy grail of the perfect bee; one that could resist specific diseases, pathogens, parasites, or predators such as the Asian hornet? What if it could also be made resistant to varroa and agricultural pesticides? Think of a lifetime's work that might be achieved in only a few years. It sounds like progress until you think it through, but then it appears to be a really stupid idea. If the big agrochemical companies know that bees can be made immune to pesticides, wouldn't that be a green light to putting even more pesticides on the market and for farmers to make even wider use of them? While the Frankenbee might not be affected, what about the numerous other species of pollinators that aren't immune? More worryingly, what if the big agricultural companies decide to build a Frankenbee and corner the market, the same way a company patented the glowing fish? They could then corner the market in bees, forbidding anyone else to breed them and selling them back to farmers in the same way agricultural companies now sell seed.

The commercialisation of the a transgenic bee would privatise the huge market of pollination, the only area of agriculture not yet dominated by the big agricultural companies. And it's worth billions a year. Imagine, they might say (recalling the fruit trees during Isle of Wight disease that were not pollinated), how many problems of food security this might solve for a hungry world. Think also of areas of China where there are no honeybees left and pollination of fruit trees has to be done by hand. Think of the scientists trying to build pollination drones, who could abandon artificial methods in favour of something more natural, efficient and cost-effective. Everyone buys what's natural too; there's always a better market for what's natural.

If we made a mess of honeybees during the First Industrial Revolution, imagine the potential for disaster looming from a Fourth Industrial Revolution in which there will be some winners who win it all and many losers who might lose everything. If you're on board with big agriculture and are willing to adjust to a new economic ideology perhaps, you'll likely be a winner; if you're not, the future looks quite sinister. That's social Darwinism for you. In the world of the new commercial Frankenbee, however, there might be no room for small hobbyists with their interest in locally adapted bees or unique subspecies and ecotypes whose gene pool will be diluted by the spread of the dominant and ubiquitous laboratory-designed GM bee. It would all be done in the name of a new humanism of the universal brotherhood of man or a Globalist utopian vision with the worthy ambition of feeding the world and saving the planet. Pollination, however, would be privatised. Good luck with that, but all totalitarian regimes in history gained traction by offering the common good. The truth is that if scientists engineer, patent and commercialise a GM honeybee, today's hobbyist beekeeper would probably be consigned to history along with the extinct cottager skeppist. Both will be remembered as no more than primitive forms of beekeeping.

This whole area of GMOs is difficult to navigate morally. There isn't an obvious or easy answer, but that isn't an excuse for moral relativism either, in which we simply say that it is acceptable in one circumstance but not another, or that ends justify means. Clearly, we have to weigh up the advantages against the risks involved in each case, but always rooted in a sound human ecology; the study of how we as individuals and groups interact with our environment and each other. The problem with this starting point is achieving an agreed definition of the human person. If we can't agree about who or what we are, we can't build a truly common human ecology. Issues such as GMOs and transgenic organisms then remain morally disorientating.

My own view, having given it much thought, is that the idea of inserting a gene from one organism into another outrages nature. We know it worked with the Hawaiian papaya, but it went very wrong with bt cotton. Fluorescent danios have also escaped into the wild. The fact that GMOs are a gamble and that we don't know the long term repercussions or how what goes right today might go horribly wrong tomorrow or in the wrong hands leads me to believe that in all cases and at all times GMOs are probably unethical. There must be limits to our ability, and they are defined by moral absolutes. If something has been proven to go wrong and to be unpredictable and dangerous, it should not be done.

*Orientated* – meaning to be adjusted or located in relation to our surroundings or circumstances. How located and adjusted are we beekeepers in relation to the circumstances in which we find ourselves and our beekeeping? A century ago beekeepers were grappling with the crises of their own times and attempting to adjust themselves to those circumstances, despite not fully understanding them. As I see it, from researching their arguments, they were trying to become located in relation to the diseases of foul brood and Isle of Wight disease by exploring those diseases' apparent connections with contemporary beekeeping methods. But there does appear to have been a blind spot in their efforts at orientation: that a key component of their circumstances was the increasing importation and cross-breeding of the different races of honeybee. Another blind spot was human greed – the desire to maximise production and profit from beekeeping.

I suggest that we won't begin to debate the contentious issue of queen importing and the continued use of different races in our apiaries until we redefine what progress in beekeeping means and until we abandon the dead-end pursuit of a perfect bee. Until then we remain, to some extent, as disorientated as our forebears. Potentially that risks not ending well, as history teaches us, because if we don't know where we are and how we got here, we can hardly do better than drift into the future repeating the mistakes of the past.

# MINDING THE BEES

## Chapter 19
### Ground-Level

' "DOOLITTLE' STRAIN", GOLDEN QUEENS.
Virgins 1s 6d. Fertiles 5s.
Customer writes: "Your queens head the best colonies I have."
D.TAYLOR. Ilminster.' (1)

*British Bee Journal 1907*

Early August. 6.00 am – a small flock of several jackdaws *jack-jack* as they fly over from the direction of St. Mary's church. Wood pigeons clap and rise into the sky... and fall, like missiles shot between the trees. The pheasants, recently released by local gamekeepers, fire volleys of staccato calls in the first light. I enjoy sitting in front of the Abbey church at this time of day, before Matins. It's deliciously cool – crisp as an apple from the fridge. In the distance an autumnal light gilds and flares through the spaces between foliage. I listen to the white noise of wind in the waiting trees.

Evening: these last sultry days of mid August are languorous and still long enough to sit outside and watch the sun coming up and going down. Everything seems to move slowly, the early sunlight thick and yellow as honey. I notice one afternoon that the sound of grasshoppers has ceased on the bleached meadow. I watch a fox cub curled up asleep one morning in the long grass of the library garden. The pink barley fields around the meadow have been harvested, and bales of rolled straw sit in vigil across the fields washed in amber light, waiting for the change in season that's on its way. In only a few months they'll be iced like Christmas puddings in a freezing world, but for now they are warmed by sinking evening sun. A roe deer and her fawn move through late shade under the firs, between the apiary and the field beyond. The doe freezes and stares in my direction. I turn and retreat calmly. I notice honeybees concentrating in large numbers on a bed of *Geranium rozanne* in our guest garden, as available forage begins to dwindle.

I look up. A number of trees are stippled with yellow against blue sky. There's a wind too, the first soft breath of autumn. I notice the light has changed, as I watch the sunshine on the hives; that low, slanting, autumnal brightness I associate with windfall, fermenting apples in wet grass, pears freckled as clay-coloured frogs, and the last brightly enamelled dragonfly adorning the glinting pond's lapel like a brooch.

By the end of the month it's dark before Matins. Evenings are drawing in too. *Ker-wit*. I hear the distant, haunting, hollow sound-scape of tawny owls fluting high in the black, ruffled trees. Red apples; polished cricket balls fallen silent at the season's end, adorn the trees across the Abbey grounds. Acorns turn from green to tan. I forage for fallen cob nuts along the edge of the meadow and salvage a handful not yet nibbled by mice. A whistling of frantic wings as a pair of Mallard hurry overhead. Sitting in the guest garden before Compline during our annual community retreat week, I notice darkness falling by 7.30 pm. Mottled pears ripen in the nearby pleached, yellow trees. Peels of church bells rise and fall sadly through gentle evening breezes scented with smoke of a barbeque in Abbey Gardens.

'A crossing breeze cuts a pause in its outrollings

Till they rise again as they were a new bell's boom.' (2)

It's hard not to think of endings as I sit on the far side of the uncut meadow in the evening with Banjo, my pet Call duck at my feet, watching the sun flare and fall behind the limes or back-lighting the tall, dried, singing grasses. Birdsong falls silent until only a yellowhammer calls from a high hedge. Summer is slipping away. Soon we'll hear the burr of the tractor cutting the blonde meadow. The wood-man will begin to lay the hedges across the lane from the Abbey, filling the bonfire-bright afternoons of early autumn with the *hack-hack* of his blade and the splitting of green wood at its base. It's a time that makes me pensive as I contemplate the end of another season with the bees. But all things pass, along with monks and bees. In the great scheme even the life of a monk is as short as the days of a forager bee. As the psalmist sings, 'For they wither quickly like grass.' (3)

I've been looking at old black and white photographs again: Biddy standing in the apiary in his habit, white-haired and bearded like Jack Hargreaves, holding a frame of honey; or bending down, rosary hanging from his hand, inspecting the entrance of a hive on the lawn between the 1923 building and the calefactory. He's in his habit, but wearing a veil. In the book *Beekeeping at Buckfast Abbey*, by Brother Adam, there are even older photographs - ghostly images: a hundred white hives in rows on the Moors in August 1920; two monks standing next to the hives in the spring of 1910; a vast hillside of hives on Dartmoor that was a mating station established in 1925. All gone. Gone too are William Woodley, Brother Adam and Fr. Robert Biddulph.

*Biddy at the hives*

It was only ninety years ago that Douai's first swarm arrived, and Fr. Anthony Baron, the apiarist at that time, was recorded in our annals a year or two later standing on a roof, catching a swarm. In that time the apiary has moved its site in the grounds of the Abbey, and beekeepers and monks have come and gone. The bees have come and gone too, with the passing of monks or the arrival of varroa, but like that swarm in the air brick over the bursar's office, they keep returning, as though they know better than we that monks and bees share a common, unbroken tradition.

Traditionally, however, August is really the start of the beekeeping year, rather than the end. It's now that preparations begin for the long winter sleep. After harvesting the honey surplus in a good year beekeepers would normally now be treating their bees with miticides such as thymol, to counter the build up of varroa mites at this time of year. In this way they hope to send the bees into winter with as low a mite load as possible, giving their stocks a better chance of surviving the winter.

No chemicals for our stocks though. I rely on other methods now. One management strategy is that by splitting the stocks into nucleus stocks, to raise new queens, the brood break in the split will lower its mite load. Between making the split and the production of new brood on which the varroa reproduce there is a brood break of at least five to six weeks, on average. In the parent stock the split lowers the mite load to some extent by the removal of brood to the nucleus stocks. Moving a nucleus, to bleed off foragers into another stock, can also lower the mite load in the nucleus.

We are now in year four since I stopped chemical treatments, and according to research this is about the time it takes for a colony to fully develop resistance. This means that more of our stocks should survive the winter into next spring. I must still expect some losses, but not as many year on year. In particular, these losses are likely to remain high in the swarms taken each year, especially if they are from managed stocks that have been treated. I must expect these stocks to go into winter with an increased mite load and that this could finish them off over the winter in year one, or at least in year two or three. Some swarms taken in my bait hives, however, are likely to be wild stocks and some of these might be resistant, if they have been in the wild for a few years or more. I won't know, except by monitoring them to see which survive each year. Not that resistance to varroa is my only goal; treatment-free bees, like wild bees, are also more likely to have higher levels of immunity to a range of other diseases.

Although once again I'm not treating the bees with chemicals this autumn, there are other things I must still do to give them a fighting chance of survival. The most important, especially in a lean year like this, is to feed the stocks. They are light in stores, having used most of the stored honey when they were unable to forage in poor weather this summer. This raises the thorny issue of feeding syrup. There are beekeepers (and I've been one of them in my time) who are sniffy about feeding bees granulated sugar. Sugar isn't honey, after all, and it lacks many of the nutritional elements in honey, but my attitude is that *needs must*. Sugar might be the equivalent of feeding me porridge every day instead of wholesome home cooking, but if I were starving I'd be jolly grateful for porridge.

Interestingly, if you give the bees a honey frame out in the open when there is a nectar flow on, they'll ignore it. Similarly, if you give them sugar syrup next to a honey frame they'll favour the syrup. The reason is that bees are hard-wired to collect nectar, which is sucrose, and sugar syrup mimics this by being sucrose. In other words, although it might not be nutritionally the ideal diet for them all the time, sugar syrup is closer to nectar than honey, so you can rely on them to take it down in a dearth. It's another reason why syrup encourages the queen to lay, because it mimics a nectar flow, and queens will increase laying in a nectar flow. It then has the water content reduced in the same way as nectar, and is stored as honey.

The strongest stocks have a super of honey at the moment, which I have decided to leave, letting them overwinter on this brood plus a super (aka brood and a half). It will also be an interesting experiment, to see how they fare and how they build up again next spring. As for the other stocks, they are very light. This is a problem not only for winter stores but also for the build up of winter bees that should be happening now and into September and October. Without feeding, the queen won't lay well (if at all) and the colonies will not raise enough brood to provide the winter bees needed to keep the cluster warm. Syrup will stimulate the queens to lay and will increase the production of winter bees. Although I have little control over the varroa or the kind of winter ahead, these two considerations – the number of winter bees and the amount of stores, are factors affecting the survival of the stocks over which I do have some control. The decision is simple then -  feed them syrup:1 part sugar to 1.5 parts water initially; then 1: 1 as the autumn arrives and the stocks begin to shut down. At the same time, I want to push the bees down into brood and a half for the winter, which the cluster will find easier to keep warm.

I spend some afternoons in the workshop, repainting the three nucleus boxes I inherited from Biddy's beekeeping days. I adapted them to the specification of the equipment we now use here, and with a coat of paint they are very serviceable too. I repaint them a terracotta colour and they look as good as new. I decide to make some extra nucleus hives with some tongue and groove planks I salvaged from a large packing case in which some kitchen equipment arrived some years ago. With a crowbar and hammer I manage to take most of it apart, with minimal damage to the wood, and I put together three new six-frame nuc boxes, three planks high, which leaves space beneath a national brood frame. Although the bees will build down from the bottom bars of the frames, the extra depth in the box will make them ideal as swarm boxes. From my experience this year, swarms seem to prefer a deep box rather than a wide one, and they like some free space beneath any frames that are put in.

I buy some ceramic floor tiles for roofs and bases that work out at just a few pound each, and paint them the same colour as the boxes, with masonry paint (*Harrow*). With the prices of beekeeping equipment rising, along with everything else, and poly nuc boxes having doubled in price since the pandemic, I'm feeling rather pleased with myself to have produced three new boxes for no more than a few pounds each. It's the age of *make-do-and-mend* again, and I'm all for it. I stand back and survey my boxes, old and new. They begin to symbolise a continuity with the past and with a long tradition of monasticism and beekeeping in which I have my own tremulous stay.

*Home-made nucs.*

This season, as my beekeeping has continued to change direction, it's been hard not to question the contribution of Brother Adam. Some might think that because I have rejected the Buckfast bee and some of the assertions and assumptions proposed by Brother Adam I must necessarily be critical of the whole enterprise of his career in apiculture. This is not the case, not least of all because his project to produce a superbee was interesting and in the finest tradition of other monastics and religious who have developed our understanding of genetics and animal husbandry or the production of garden cultivars in horticulture. I see Brother Adam in that admirable tradition of Mendel, the father of genetics, with his sweet peas; Martin Silverrudd, the Swedish monk in the 1970s who developed many of the varieties of autosexing chickens; and Brother Stefan Franczak, the Warsaw Jesuit clematis breeder. I can only admire and rather envy the dedication of these men whose achievements took a life time of patient work. If nothing else, there is a lesson in their example of sheer perseverance.

Despite this, I now regard the Buckfast bee as symbolic of a cul de sac down which apiculture began to turn in the mid nineteenth century. *Cul-de-sac* – French for *bottom of the sack*, first used in 1738 to describe how a dead-end street looked on a map. Since 1800 it has been used in English to describe a dead-end street. Allow me to elaborate, to develop an analogy: the cul de sac is a product of urbanisation and the invention of the motor car, which allowed home owners to live in isolated enclaves that provided the compromise of the best of urban life along with the advantages of rural life. Living on a cul de sac freed people from having to live near or on public transport routes. The road became a kind of moat, making its residents feel safe in their own little castle.

Urbanists are no longer so enchanted by the cul de sac, despite their continuing popularity with home owners. Firstly, they force us into an increased use of the motor car when policy-makers would have us use them less. Secondly, they minimise route choices, locking us into using the same route every day, which can also cause traffic congestion. Lastly, they might make us feel more secure if we live in one, but their isolation actually makes homes less safe. In short, the cul de sac actually restricts our freedom and even makes us less free.

When I suggest that beekeeping went down a dead-end street towards and during the twentieth century it is for many of the same reasons that cul de sacs themselves are now viewed as a dead-end in urban planning. People who keep alien races or strains of bee perhaps have the same disconnection with the wider world of beekeeping that cul de sac home owners might have with the rest of the population, keeping their bees in isolated enclaves from which it is almost impossible to free themselves. This kind of beekeeping looks like freedom, but it isn't.

Even cul de sacs might lead somewhere though, and if we exit them by the same way we go in, perhaps we emerge at least having learnt something. The fact is, I do see the Buckfast bee as a cul de sac in apiculture and a road that doesn't take us where we need to be going in the future. Even if something is deemed a failure, what's the worst that can come out of it? We learn something! Not least of all, Brother Adam remains a fine example of someone who championed honeybees because he was passionate about them.

Not only did he document the many subspecies he visited in their native conditions during his journeys in search of the best strains, often observing the traits of ecotypes and subspecies that were on the brink of extinction, but he was passionate in his day to day handling of the bees at Buckfast: 'True idealism and economic interests do not conflict but are complementary...' (4) He was also careful to point out that: 'The constant emphasis on the economic aspects of beekeeping may possibly have conveyed the impression that at Buckfast no value is placed on the aesthetic aspects of beekeeping. But anyone familiar with our endeavours will know that this is not so.' (5)

Do I accept his underlying assumption that the honeybee can be domesticated? No. Do I believe the native Black bee became extinct in Britain after World War I and Isle of Wight disease? Based on the evidence I've found, I doubt it very much. Do I think his assertion correct that there is no such thing as the perfect honeybee? I don't. The definition of a perfect bee will always be entirely subjective anyway. Do I believe that maximising the honey crop is the bee-all and end-all of apiculture? I regard this approach as no better than keeping chickens in batteries or pigs in tiny pens, because we compromise the health and well-being of animals when we treat them badly, and we diminish ourselves as stewards of Creation who are given dominion rather than domination of the created order. Do I accept that environmental factors make no difference to honeybees? No, and there is plenty of evidence now in the field of epigenetics to support that answer.

More evidence is emerging that our native Black bee has survived in certain enclaves, and is now thriving. In the fifty acre estate of Blenheim Palace, Oxfordshire, fifty colonies of bees have recently been discovered living in its ancient oak woodland. They appear to be native bees, unadulterated by contact with managed stocks, as there are no bees kept on the estate or nearby, while the estate is a closed environment surrounded by humid, damp valleys forming a physical barrier to entry by hybrid stocks. This seems to have kept the population pure. They are furry, with little banding, have distinct wing veins, and can forage in temperatures as low as 4C, swarming with multiple queens – all traits of our native Black bee. It is believed that one of their nest sites is at least two hundred years old. These bees are also resistant to varroa and are, no doubt, also immune to many other viruses and diseases. Could it be that they are among the last survivors of William Woodley's Olde English bee? It seems very likely.

Despite this, Brother Adam's work on the selection of desirable traits is still valuable for beekeepers who want to improve their stocks, and especially for those who want to help their stocks develop immunity to emerging or reemerging diseases and parasites. His study of the different subspecies of the western and eastern honeybee are also useful in helping us understand that these subspecies are also ecotypes with subtly differing characteristics that are minutely attuned to their country of origin, their locality and environment. It's precisely based on this knowledge that I have developed my own criticism of crossing the different subspecies and importing queens of different races.

As for the Douai apiary – when I know that it probably won't outlive me (at least in its present form) it can be tempting to question why I bother to go to these lengths, especially when I stare into those old photographs of Brother Adam's former Buckfast apiaries or Biddy and the bees on the lawns of Douai Abbey.'Vanity of vanities' (6) I hear in my head sometimes.

Quoeleth, the voice of *Ecclesiastes*, confronts us with his sobering question about the ultimate purpose or meaning of life in which everything of this material world is transient.

Why then don't I just treat our bees, buy in some Buckfast queens and enjoy the extra honey? The answer is for the same reason that I'm a monk – I'm part of something bigger, to which I'm connected in ways that might not always be obvious. Beekeepers and monks come and go, and apiaries come and go; even the bees have come and gone from this place, but something larger than them all continues, just as monasticism endures down the centuries. In a similar way bees and humans coexist and depend upon each other in the vast created order. This means that how we keep bees and how we look after the natural world matters. Minding the bees is much more than looking after them – it's also *looking out* for them, if we want to look out for ourselves and our food security.

Who knows but a century from now someone reading this might argue that I'm wrong, as perhaps William Woodley might have been partly wrong to resist Bee Disease legislation, or Brother Adam might have been wrong about the underlying principles of his Buckfast bee. They will interpret our own situation from their own social, historical and cultural perspectives, however those might have changed; or they might look at the situation in their own age and conclude, as I do, that the times remain the same and that there's nothing essentially new under the sun. Beekeeping might still be struggling with the same issues today that it faced a hundred years ago. All any of us can do is live in our own times, try to learn lessons from the past and attempt to avoid making the same mistakes in the future. Alongside the science and the craft, therefore, we also need the history.

As things stand this season, the bees have done well, despite a terrible winter, a late spring and a disappointing summer and the losses to be expected with treatment-free beekeeping. With a little help the survivor stocks have recovered and I'm in a good position heading into autumn. I remind myself that losses are not only to be expected, but are to be welcomed. Losses remove the weakest stocks, so that year on year the survivor stocks are left through natural selection. According to one treatment-free beekeeper on Youtube, he knew of several treatment-free beekeepers some years ago whose winter losses were 41% at their worst, compared with losses of 31% for a similar sample of beekeepers in the same winter who used miticides. Conclusion: there are losses even if you treat, but the greatest loss, in my view, is that your bees never reach the stage at which they can cope with varroa, while the mite becomes increasingly resistant. Neither do your bees get the chance to develop immunity to other diseases. That doesn't make any sense. Even if my losses are somewhat higher than the losses of beekeepers who treat, I'm gaining in the long run because I have bees that are increasingly able to survive varroa mites and are healthier generally.

At a Douai function a beekeeper from the Midlands approaches me and we begin talking about the bees. He tells me that a well-known scientist researching bee resistance to *varroa* was unable to work abroad during Covid and asked if he could examine local beekeepers' frames in the Midlands. Many of the beekeepers, he tells me, had stopped treating their bees with miticides. The scientist found that their bees had developed the same strategies for managing varroa that he had seen abroad. The bees were uncapping brood, removing affected larvae and biting the legs off the mites. Bees in this area of the Midlands were adapting and surviving.

I can't help drawing parallels between the poet and novelist Thomas Hardy and William Woodley. Both were Victorians who remembered and still referred to old *Wessex* and the country cottagers who laboured in a changing agricultural world at a time of enormous social and economic change. Just as Hardy revered the country folk and traditions of his own rural Wessex, William Woodley lived alongside and reverenced the agricultural workers on the edge of the Berkshire Downs whose bee gardens and skeps were like his great aunt's at Stanmore where he had grown up *mindin' the bees* and catching swarms under the direction of the blind old bee man.

Indeed, Brother Adam, it seems, also had the heart of a poet when it came to the old craft, as evidenced by his enjoyable article 'Mead', first published in *Bee World*, in which he recalled the simple rustic cottager of a world that had passed, and delighted in the '...similar contentment' (7) of a solitary traveller that was once shared by every cottager in England:

'For many a year there stood on Salisbury Plain a shepherd's hut on wheels, among bushes of gorse, and surrounded by hives. Few passers-by would notice the hives among the gorse, and possibly not even see the hut, unless the gale blew ribbons of smoke across the traveller's path and drew his attention to it. In that hut lived a solitary man who had travelled far, but had found anchorage at last among his bees on Salisbury Plain. He sold his honey, and made his candles, and brewed his mead.' (8)

As for Mr. Woodley, how much were his opinions and ideas formed by a similarly simple life lived in a small house in a tiny village in the country parish of Beedon, among many simple folk who were usually born and died in the same house? What conversations with his cottager neighbours did he have about bees and skeps and country traditions like *telling the bees* over a picket fence or an ancient hedge that made him respect their hard working lives and the importance of beekeeping to the cottager economy? How many skep stocks did he pick up when another old bee master died in Beedon, or a neighbour lost tenure of his cottage and land? Many of these rustic folk were not unlike Brother Adam's traveller, left stranded by history and progress on Salisbury Plain, the relics of another age.

He'd have known the travellers of his own time passing through, and the rural neighbours like the Blacksmith and the Cooper, the Shoemaker like Mr. Martin in his Cotswold apiary and workshop, and every cottager in the village, many of whom would have been fellow beekeepers. He'd have watched the Woodman (whose father and grandfather had probably also been Woodmen) working an ancient coppice, splitting the hazel rods with his *billhook* or *bill hook*. The Woodman might have called it a *hand bill, a hook bill, pruning hook* or a *hack*. In the West Country it is still called a *hacker*, closer to the original German *hackbeil*, meaning *chopping tool*. He might have discussed the Woodman's bees as he watched him splitting rods or making the traditional ash bar sheep hurdles, his L-shaped *froe* (halfway between an axe and a knife) and mallet cleaving the ash or hazel. Perhaps on stormy winter nights with the wind under the door Mr. Woodley had sipped mead with a neighbour: '...with a wax candle on the table and a fire of logs on the earthen hearth...the home was simple, but the fare was rich; the talk around the fire most companionable, and the liquor most stimulating.' (9)

The times remain the same; just as William Woodley and Thomas Hardy looked back to an era that was passing away as they were swept along by the tide of progress,  I find myself looking back at a world that has changed dramatically even since Brother Adam stopped keeping his bees at Buckfast Abbey and Biddy was running the apiary here at Douai. It was a binary world that now seems wonderfully innocent, simple, and free. It was a world in which every individual retains their importance and dignity in history. The recent speaker at the 2023 World Economic Forum who stated that we have to stop regarding people as mysterious souls and start thinking that every human being can now be hacked surely outrages human nature and the foundations of human civilisation.

Partly I miss that world of the old country, but I can also see that we're far better off these days than we were then: there were only three channels on terrestrial television before the so-called progress of the vast array of stations available today; sometimes less is more. There were test-cards before the day's programmes began, and transmission finished late at night with the national anthem. We watched the Apollo Missions in grainy black and white. I recall broken pictures interrupted by interference and the white noise of Mission Control, and countdowns backwards from ten to lift off that we mimicked in the playground with American accents. The television aerial was temperamental in our digital world. On Saturday evening we watched *Dad's Army*. It was bath night before central heating. After school we watched *Old Country* and a white-haired Jack Hargreaves in his shed asking, 'Do you know what these are?' as he explained to us the history of fishing rods, or showed us a piece of hornbeam, an old-pattern horseshoe, or related how a Dorset river found its name.

The telephone was fixed to the wall and all its conversations happened on the bottom step of the stairs. People on trains and buses had conversations with each other. There were red telephone boxes on every street and a thick phone book beside each payphone. You dialled with a forefinger, clockwise, a whirr of the slow dial you had to wait for between each number. You could talk to the operator. I did once. She phoned me; she was Susan, a girl in my class who left school at sixteen and worked at the exchange, connecting calls manually. Now you often can't even talk to someone when you ring a direct number.

The Douai Abbey telephone exchange has been replaced this year and seems to be a more complicated operation than we expected; we have so many rooms and numbers that monks have taken with them to other rooms as they've moved over the years, that our system has to be unravelled like a plate of spaghetti. We have the option to make it *modular*, they say. It's all completely over my head! As for options, how free are we? Is freedom having a smart phone that absorbs your attention at every moment and logs your actions and interests with endless algorithms? Are we free when we can hardly speak to real people any more when we telephone the utility companies or the NHS? Back then the doctor came out on call-outs. These days it's a job to get near a doctor even when you make the effort.

When I was a boy the milk float made the daily delivery round from Express Dairies based at Belmont circle, clattering and whirring along the slow, frosty Avenue as I lay in bed, a lad as feral as a Victorian chimney sweep collecting the empties and jumping onto the empty crates as the float whirred to the next house. When I was getting up to go to school that lad was a symbol of freedom.

My teachers had a job for life and there were many other careers for life. It was still the same employment world that Mr. and Mrs. Dyer, the beekeepers of Compton, had inhabited, who came to live at the crossing on the DN&S line at Compton in 1905, retiring in 1930. By then Sarah had given forty-two years service to the railway and her husband, who had helped build the line, had given forty-five years.

A girl from New York landed in my class in the brown fog of a winter dawn when I was ten Her exotic accent flowed with the freedom of a great river, like something from the colourful escapism of cinema. When we went to the *flicks* there was an intermission, and ice creams for sale with little wooden spoons (we had the freedom of three cinemas in Harrow), ushers with torches who told you off before children dared talk back to adults or barely knew of worse crimes.

We had Jack Frost on the inside of the windows in winter, and rags of breath in freezing air as we walked to school after a cooked breakfast. I aspired to eating porridge for breakfast so that I might have the glowing aura of the boy in the porridge adverts, but we always had eggs, bacon and toast. I was twelve the year of the great heatwave summer of 1976: the year of three popes, the year we had central heating installed and the year I learnt how to fish for newts with a worm tied to a piece of cotton in that old pond next to my primary school where I'd first been fascinated by the goldfish that had regressed to bronze.

There were hedgehogs snuffling through every Metroland garden, and flocks of sparrows before they became as novel as a nuthatch. I've been at Douai Abbey fifteen years and have never seen a hedgehog here, despite looking. Children were more free then too. I walked miles to Stanmore pond (Middlesex) on summer afternoons where I caught tiddlers and tadpoles in the days when it was still safe for children to go feral for the day. I read *Just William* and free wheeled through the woods of Bentley Priory or Kenton park with my friend, Adrian. A white-whiskered great Uncle in Portsmouth related stories of his war at sea and of chasing the *Turpitz* halfway round the free world, illustrated by black and white photos of *HMS Ramelles*. At the time it was like listening to the adventures of a Dover sole, but I now wish I could have the time again. I appreciate black and white photos now.

You gave your reel of camera film to *Boots* the chemist and picked up your developed photos a week later. It was something you looked forward to and it taught you patience. You chose your pictures carefully because film was expensive and you only had twelve, twenty-four or thirty-six exposures. Now taking photos is as throwaway as the rest of the culture; when I was on holiday with my sister recently I watched a woman photograph her pancakes with her phone. My father did his own developing in the kitchen. He once took a photograph of me sitting on the doorstep of Number10 Downing Street in short trousers, in black and white. You were free to walk along Downing Street in those days. Back then boys wore shorts until they were teenagers. Children were allowed to be children.

It was the Cold War, but everyone knew where they were. It was a world in which you knew where you were, and that has all but gone, just as Mr. Woodley's world passed away with Victorian England in the tide of progress after the Great War. The change is every bit as seismic as William Woodley and Thomas Hardy experienced. I miss aspects of that world, but on the whole I admit that we are better off for the technological progress that has lengthened life-expectancy, improved our standard of living, our health and our education.

I'm sure Mr. Woodley sat often at the kitchen table in his later years, lamenting not only the extremes of weather that happened even back then and the season's effects on his bees, but also *Isle of Wight disease*, the First World War and the indomitable march of progress. He probably looked back (as I do now) and wondered where the slow, rural life of the Berkshire Downs had gone since the innocent days when his only concern was mindin' the bees for his great aunt. Perhaps he noticed the old copses left uncut, the Woodman's sudden absence with the labourers who left for the towns and factories. Though his friends, Mr. and Mrs. Dyer, had arrived with the new railway at Compton, and the countryside had begun to roar with the occasional motor car like his own Benz Volo, he must have been aware of the fragile vestiges of the old country around him in the villages of Stanmore, Worlds End, Chieveley and the surrounding parishes.

Despite the mechanisation of agricultural life, there would still have been some horses left on the farms. I found a statistic that in England in 1894 there were still 1,176,248 working horses in agriculture. Doubtless there was someone locally who still made cider in the traditional way too, pressing the layers of crushed apple in sandwiched layers, folded into biscuits, in West Country hessian corn sacks. Out in the fields at harvest time William Woodley probably saw the last workers stop to tilt and swig a small wooden *costrel* of cider over their forearm. He doubtless knew what hornbeam was and that hornbeam cogs turned the creaking ancient mill wheels and fashioned the ninepins played for centuries in Wessex ale houses. Maybe there were just a few last cottagers alive who knew the art of eel-catching; a ring of green hazel sewn into the mouth of a barn sack, baited and filled with straw, with three strings from the opening tied to a taught line on the bank and weighted with a brick. Every cottager and farmer in those times did-it-themselves; there's nothing new under the sun about do-it-yourself. It's as old as the Berkshire Downs. It was as old as the skeppist, but by then even they had mostly gone. Like us, Mr. Woodley was caught in a current that moves us all inevitably onward, and he probably felt the same uneasiness about it that we do.

Progress and technology bring advantages nevertheless, and so does Capitalism. It is a dangerous thing to disown all technological progress and to espouse an apocalyptic environmentalism that despises humanity. Some activists who berate the world for not immediately achieving zero $CO_2$ emissions forget that the reason we have more trees and green spaces than a hundred years ago is because we started using coal. They forget that when people are pulled out of poverty they begin to care about the environment, so that if you push up energy prices and limit energy resources not only will it be at the expense of the poor but it will probably make them care even less about the environment and even less able to cope with extremes of weather and natural disasters. They forget that the developing world will also be the first to suffer for our virtue-signalling. At the moment, for example, the developing world talks about green colonialism and energy apartheid, as they can't get loans to develop their own energy plants because its too politically incorrect for the west whose policy makers are dismantling our own fossil fuel infrastructure. Ironically, the same people in the west who criticise oppressive western culture are the very ones imposing it on the developing world, and keeping them poor.

It's a warning to me, and a lesson drawn from William Woodley, to be very suspicious of the policy-makers and the agenda they can often bolt on to otherwise noble causes like the environment. For example, it's become fashionable in our times to blame everything on global warming, especially if politicians want to signal their virtue with voters. An example of this is that in 2019 the UN Climate Change website ran an article claiming that climate breakdown is causing an increase in domestic violence. Again and again, crouching behind the issues of environmentalism and global warming are policy makers who will be quick to use the narrative of crisis and apocalyptic doom to force an agenda of other issues. Interestingly, their other worthy goal of feeding the world (which should rank top) consistently comes lower in their priorities than global warming.

William Woodley resisted government interference in beekeeping for a similar reason; because the policy makers in the Beekeeping Associations had an agenda to consign the skep to history, and bee diseases was the issue they used to do it. The greatness of Mr. Woodley's character, whatever faulty arguments he might have had, was his inability to separate the skep from the poor of his day. I see increasingly that it was more than minding the bees; it was an issue of social justice.

Beekeepers today should beware the policy makers and government interference too, because an issue can be used as much as it always has been to drive another agenda; whether that be political, ideological or economic. It's why beekeepers need to work at ground level to solve the problems we can in beekeeping and not look too readily to those who offer easy answers such as GM bees.

Admittedly, without the freedom of the new technology and knowledge that came to apiculture and the world neither would Mr. Woodley have made a career as a bee farmer or an income from selling bees and honey, or his writing in the *British Bee Journal*. Despite his criticism of the motor car, Mr. Woodley must have also seen its advantages in his own life along with the steam train that sent his bees, his section-honey and his correspondence around the country. Indeed Charles Heap wrote of his visit to Beedon in the summer of 1910 and explained that William Woodley found a car necessary to reach his out apiary two miles away at Worlds End which he'd established in 1890. William was forty-five in those days and probably thought nothing of the two mile walk. By 1910 he was sixty-five and perhaps beginning to feel his advancing years. We know he used the car to reach the apiary from his *Notes By The Way* in 1911: 'On Saturday, January 28th, we had a beautiful day, more like May or June than January, every hive in the home apiary was in full force, some gathering natural pollen, and after lunch I ran the car up to my out apiary and found every stock alive and on the wing in goodly number.' (10) Despite embracing modern technological developments, however, perhaps at the end he was left, like the cottagers and the skeppists and Brother Adams' traveller, stranded by history and progress, the relic of another age.

William Woodley's demise has all the more pathos because it seems to have run parallel with the demise of the native Black bee and traditional cottage beekeeping. How many bee gardens and orchards had he known in Beedon parish that were left silent and empty of hives? Maybe he recalled wistfully the memory of his own grand apiaries when he and the native bee were both in their prime – an idyll evoked by Tickner Edwardes in his *Bee-Master of Warrilow*:

'We stopped in the centre of an old orchard. Overhead the swelling fruit-buds glistened against the blue sky. Merry thrush-music rang out far and near. Sun and shadow, the song of the bees, laughing voices, a snatch of an old Sussex chantie, the perfume of violet-beds and nodding gilly-flowers, all came over to us through the lichened tree-stems, in a flood of delicious colour and scent and sound.' (11)

Having sold off much of his equipment, downsized and closed the out apiary at Worlds End, Mr. Woodley was seventy-seven by 1922. I imagine him leaning, like Thomas Hardy, on his garden gate at the cottage in Stanmore, staring at the last of his frosted and lifeless hives under a desolate cloudy sky at the close of the year; Woodley himself frail, gaunt and small in his darkling days. His time was drawing to an end.

By 1923 he was already less well known when his obituary appeared in the Journal, 'Although Mr. Woodley is not so well-known to present bee-keepers...' (12) In September he had visited his cousin, Mr. A. D. Woodley, in Reading. He fell ill and refused a doctor, according to the obituary by his friend, Charles Heap. He returned home, but his condition weakened. William Woodley was eventually admitted to Newbury hospital and underwent a serious operation. He did not recover, but died on 8 October, aged seventy-eight. On 13 October his funeral was conducted at St. Nicholas' Church, Beedon, where he had regularly attended services. He was buried in the grave yard with his wife, Annie, their names punctuated on the gravestone by an engraving of a large queen bee.

A friend of mine contacts me from half a mile away in Midgham. I haven't seen him since before the first Covid lockdown. I taught him beekeeping several years ago. He tells me he's been in touch with a bee farmer who has offered to mentor him in queen rearing next year. He's spoken before about an interest in bee farming. I'm afraid I'm less than enthusiastic about the prospect of a bee farm in such close proximity to the Abbey, flooding the area with drones from Buckfast queens (most likely). It's the worst news I've heard since the pandemic because it threatens to undermine all my efforts to advocate locally adapted bees. My heart sinks at the thought of it.

I wonder if perhaps I am becoming intolerant of the diverse reasons people keep bees and the diverse ways in which they are kept? I hope not. It's just that I am increasingly convinced that we, like the beekeepers of a hundred years ago, have gone down a dead-end in apiculture and that we are now less free because of it. A hundred years ago, arguably, legislation would not have helped counter Isle of Wight disease, for which there was no known cure, because there was no understanding of the complex situation; a hundred years ago the cottager couldn't afford the transition to frame hives. A century ago there was also no option but to begin importing bees, to counter the losses of our native stocks and to ensure food security. We, on the other hand still have some freedom left to take responsibility and to make sensible choices.

Today the small hobbyist has superseded the rustic cottager, and because they can't afford to keep losing their bees to varroa many elect to use miticides and other chemical treatments. I can understand how modern hobbyists have become locked into this habit of treatment, just as the old skeppists were locked into their traditional ways of keeping bees. In contrast with the situation a century ago, however, further legislation today might go some way to getting us out of this cul de sac. Ideally, if we could make the choice ourselves, and not rely on legislation, to stop importing foreign queens we would reduce the possibility of introducing new pests and diseases that threaten to add to the issue of varroa mites; and we would give our native stocks a chance to reselect traits for survival and the *Amm* genes latent in the DNA of our locally adapted and near-native bees.

Some might accuse me of asserting my own subjective opinion of the perfect bee and progress in apiculture. My criteria of perfection, however, are no more than those traits that promote survival in the first instance. And that would certainly be progress for bees. After that, beekeepers will always select for other traits that suit their own purposes, such as docility, honey production and disease-resistance. If a beekeeper wants to select for a low propensity for swarming or reduced propolis production, I have no argument with these ambitions. To those, however, who argue that no native or near-native bee can compete with the Buckfast or Italians or Carniolans for the honey crop they produce, I would simply quote Tickner Edwardes:

'The English Black bee is a more generous honey-maker in indifferent seasons; she does not swarm so determinedly, under proper treatment, as the Ligurians or Carniolans; and, above all, though she is not so handsome as some of her Continental rivals, she comes of a hardy northern race, and stands the ups and downs of the British winter better than any of the fantastic yellow-girdled crew from overseas.' (13)

If some locally adapted bees over time will tend to have reselected near-native genes, they will eventually conform more or less to the description given above by Mr. Edwardes. All the beekeeper need do from then on in their own apiary is select and refine their bees as Brother Adam did until they have achieved their own definition of the locally adapted bee that works best for them.

The issue of legislation is, of course, a thorny one, even today. It is inextricably bound up with our freedoms. But is freedom an absolute value? During lockdown we relinquished various freedoms to take responsibility for the common good – to stop the spread of the virus and to ease the burden on the Health Service. Similarly, before D-Day in 1944 those who knew about the Normandy invasion were not free to disclose what they knew. On the other hand, freedom doesn't always serve the common good and can have disastrous consequences; think of the controversial gun law in the US and the repeated murderous rampages in US schools by teenage gunmen who are free to own guns but lack any responsibility for their actions. Think of Haber and his irresponsible desire for no limits but his ability and how that resulted in the birth of chemical warfare in 1915.

William Woodley has made me ponder this issue of freedom regarding Bee Disease legislation in the early twentieth century and to wonder when legislation becomes necessary. On balance I think history has proved Woodley right and that it was better then not to legislate about Isle of Wight disease. Today I'm not so sure if doing nothing in this area remains the right approach. I ask myself frequently if the time has come, for the common good, for new legislation to ban the importation of foreign bees. Would that outrage the freedom of beekeepers to keep the kind of bees they want, in the way they want to approach beekeeping? Or are our present fears about tightening legislation an equivalent of the gun law in the US? But if we continue doing nothing in this regard, do we simply have another accident waiting to happen in apiculture? Freedom is not an absolute value and we are not free to do absolutely anything we want as individuals or as a society, especially if we outrage nature and our own human nature. Would limited legislation now in this area of beekeeping protect us in the long run from an even greater interference if and when the next beekeeping disaster happens?

I'm all for freedom, but also responsibility. I resist the tyranny of ideologies that outrage my freedom of speech and beliefs. On balance, however, I'm also against the freedom to keep a fluorescent GM pet fish in an aquarium (if you like that sort of thing), since it has already led to these fish escaping into the wild. I am not, on the other hand, against the use of the bee genome to help us understand why bees are struggling in the modern world, if it helps us to help our bees. At the root of it all I'm concerned for the hobbyist beekeeper who enjoys their bees and wants to take a few pounds of honey from a craft they enjoy. That represents for me a freedom and a joy beyond measure, and we do well to be aware that many of our freedoms are at risk from the tyranny of a new kind of progress that dismisses the importance of the individual and empties human life of its dignity, vision, potential and responsibility for all that is meaningful, good and true.

I'm not against progress or technology, unless it diminishes or dismisses the importance, dignity and responsibility of the individual person. That's what we're losing in an increasingly post human world. It's why I'm against a certain ideology of environmentalism promoted by unelected globalists and irrational teenage tantrums whose underlying assumptions, beliefs and ideas express a worryingly antithetical sentiment in relation to humanity. Some of its core beliefs are, in their logical conclusions, almost genocidal. What I miss most about the Old Country is its humanity. It reflected a world that still had some belief in humanity.

*In* the next Industrial Revolution, which has begun, there will be a fusion of our new technologies, the digital revolution and biology. With our smart phones we are already in the foothills of trans humanism, while some are already talking about cyber physical systems and gene editing. A Chinese rogue doctor has already illegally edited twins' DNA for HIV immunity and discovered, alarmingly, that it also made them more intelligent. It is not too much of a stretch of the imagination that someone (a Technocrat, Big Business or government) will eventually aspire to create and patent a GM bee and that this would eventually supersede any of the subspecies or strains kept by hobbyists or commercial beekeepers today. If we make another mess in beekeeping it'll be almost guaranteed, because it will be posed as the only solution. And when it all goes wrong, the way Africanised bees, bt cotton and fluorescent fish have gone wrong, and when we need the genetic diversity of wild strains of bee, what will we do? We will search for the best strains of honeybee, as Brother Adam did, only to find them gone for ever. And that really will be the bottom of the sack.

Locally adapted bees and ground-level beekeeping might be the only traction remaining to lobby against the looming possibility of a GM bee in a future that has never looked more promising and yet more potentially dangerous. On the other hand, some bee keepers, already locked into chemical treatments and who want the desirable traits of foreign races, would probably welcome a Frankenbee. Imagine a bee (they might say) that is immune to varroa and viruses, that yields a heavy crop of honey and that perhaps doesn't even sting. It would seems to be the answer to all our beekeeping problems, but it would be grossly irresponsible. At what point, for example, does the harmless freedom to keep a fluorescent GM fish in an aquarium become part of a larger dystopian vision that might include a globalist Frankenbee? Beekeeping would then be a monoculture, a single strain owned by a few multinational agrochemical companies, to which even the hobbyist would have to submit, as the skeppist was once forced to submit to modern methods. That would be when beekeeping ceases to embody our greatest freedoms and when a scourge will be unleashed far worse than Isle of Wight disease or the varroa mite. If it happens I doubt that we would even enjoy the freedom of debate that William Woodley and others were able to have in their old bee journal for so many years. Indeed the hobbyist beekeeper might well by then be as despised as the straw skep, as blamed and disparaged as the old cottager skeppists were for holding back progress and the enlightened ways of a new kind of beekeeping. People like me might even end up as scourged as William Woodley. But I should console myself that I would be in very good company, if that were the case.

# MINDING THE BEES

## Epilogue
### Read, Mark, Learn and Digest

'DUTCH BEES. 4-frame Nuclei, May and June delivery,
also six Dutch-Italian Nuclei; £3 3s;
cash with orders. Carriage paid – SEALE,
Hardumont, Oatlands Drive, Weybridge, Surrey.' (1)

*British Bee Journal 1919*

October. A bleak northwesterly is blowing. It's colder than usual for this time of year. In choir we've begun wearing our pleated cowls again over our habits and are singing the winter hymns at Vespers in the dark. Leaning on the gate to the peaceful graveyard of St. Nicholas' Church, Beedon, lead-grey clouds enclosing the afternoon, I feel the faltering of the season's pulse. The creaking trees are haunted with the cold lament of death. Hooded hordes of jackdaws are restless among limp leaves. Weak sunshine winks wearily through tangled stems and coloured leaves and rests upon trembling, blackened flowers, their hard, dry germs dangling along mossy, weedy walls. Autumn will soon be laid to rest, and the beekeeping season will fall silent, a creeping spectre through the crypt of winter's long, dark sleep.

I look up at the timeless sky and the old, bent trees. A chesty pheasant coughs nearby. I look at my watch, as much to check the date as the time. We have passed the autumn equinox, and the tide of time is carrying us into the cold months when the bee man retires to his workshop and his bench, to meditate, to read and to gather together the accumulating wisdom of his craft.

'But the flash and quiver of wings, and the drowsy song of summer days, were gone in the iron-bound January weather; and the bee-master was lounging idly to and fro in the great main-way of the waxen city, shot-gun under arm, and with apparently nothing more to do than to meditate over past achievements, or to plan out operations for the season to come.' (2)

We had the late Queen Elizabeth II's magnificent funeral some weeks ago. The whole nation tuned in to watch, as though to to remind ourselves who we are and where we are in this moment of history. The event reminded us how we came to that day and where we're going. I read that her beekeeper at Clarence House and Buckingham Palace had tied black ribbons to her beehives in the old traditional way and had told the bees the old queen had died. William Woodley would have approved of that. So do I. It's more than just a quaint superstition, as a letter pointed out in the *British Bee Journal* in 1913:

'Some years ago I asked an old Worcestershire bee-keeper what was his belief on the subject of "Telling the Bees," and he replied that there is a great amount of common sense in the old superstition; because the people who took enough interest in the bees to tell them of their master's death generally looked after them in other ways, and naturally the bees benefited.' (3)

Now we have had a new Prime Minister, the resignation of a Prime Minister, yet another Prime Minister, a new Monarch and a new reign, and at the Abbey we have a new Abbot too. It is a new era in many ways. Two seasons of beekeeping have passed since I last stood here at the grave of William Woodley. At that time we were in the thick of the Covid pandemic and another lockdown and we were all wondering how we had got into that mess. Now we have more crises: war in Europe and the threat of nuclear weapons, an energy and cost of living crisis, an NHS in crisis, inflation and strike action. A winter of discontent looks likely. People are still asking, as deep as an existential crisis, how did we get to this?

The times really do remain the same, it seems. But I resist the narrative of apocalyptic gloom. A century ago there was World War and Spanish Flu and then a crisis in food production after the war. Beekeeping was emerging from Isle of Wight disease and changing direction with the increasing importations of foreign queens from Holland, France and Italy and the development of the cross-breeding of different honeybee races. Beekeeping had changed almost beyond recognition in less than a century. So had everything else. Beekeeping had survived though, and Isle of Wight disease was coming to an end, like the Great War. Brighter days lay ahead, for beekeepers and for the world.

When the poet and novelist, Thomas Hardy, leaned on that coppice gate at the threshold of 1900 he knew he was at a turning point much like our own and that an era was passing with the old century's fading pulse. A thrush sang from the tree overhead,

'In a full-hearted evensong

Of joy illimited;' (4)

and in that darkling thrush the old poet found meaning and hope to face the uncertainties of the future.

In a country graveyard there is a real sense of time standing still, held in tension with the poignant reminder at our feet of the relentless cycle of birth and death, and of the constant turning of the seasons. I hold this moment as I might hold an old pocket watch whose hands tick-tick-tick around its yellowing face.

Another reign of queens bred this year from survivor stocks have taken their thrones in the stocks of Douai's apiary. Last year was a difficult season, and I lost all but one of the captured swarms over the following winter, along with a few nucs and one operational hive; but five of the six production hives and a few other nucs survived and built up again in a favourable spring and the fine summer of 2022, giving us a splendid honey crop this year. The one surviving swarm did well too, the hive giving us a surplus of about forty pounds of honey.

The colony above the bursar's office died out last winter, though they were flying as late at mid-February, but a new swarm arrived in May. I put five bait hives up on the patio of the unused novitiate area, some twenty feet off the ground. I baited them with the pheromone of a couple of dead queens I'd found in the hives the previous winter. I preserved them in some alcohol stored in a vial until I needed to bait the lures. Swarms arrived in four of the bait hives. (I need to try more experiments with preserved queens in alcohol – perhaps baiting bee-bobs in the trees, the way William Woodley did. I'm told Biddy did it too, with little blue rags.

My first swarm of 2022 arrived on the first official day of summer:

*Novitiate area and swarms.*

313

### 21ˢᵗ June.(The First Day of Summer)

24 C. Sunny. No wind. The apiary has that distinctive sound on a torrid afternoon of hives in top gear now that the main nectar flow is hitting its stride. The freedom of an unrestrained apiary in a nectar flow is a tangible joy.

On the novitiate terrace the favoured nuc box is getting more attention again, but not a serious number of scouts. The other boxes are ignored.

It's a white nuc on an old, leaky green water-tank near the poultry that looks the most promising. It has the largest number of scouts visiting that I've seen at any bait hive this season. It crosses my mind that a swarm might already have arrived, but the frenzy around the entrance suggests scout activity – and that the site is being taken seriously.

I decide not to touch the box, in case a swarm's arrival is imminent. Patience! Could this be my first swarm this season?

By 5.00 pm the traffic to and fro around the white box has increased. Up on the terrace there are a lot of bees too, investigating the boxes sited up there.

7.00 pm. Supper time. As I arrive at the refectory for supper, there are bees all over the terrace, criss-crossing in large numbers. A swarm has landed on one of the boxes previously ignored, and the two boxes on the wall are both busy with scouts too.

After night prayer the terrace has settled down and I have a single prime swarm. But it's in a box that was ignored until today. The swarm has surprised me. Near the poultry the white nuc has gone quiet too, a couple of scouts still hanging around the entrance. On the meadow the bucket hive in the small beech has the attention of a small group of scouts.

I walk to the far side of the meadow. A few startled female pheasants explode from the long grass. The limes are flowering – creamy white bunches of small, starry flowers. The main nectar flow is on, the weather is holding and my first swarm has arrived. It doesn't get much better than this in beekeeping.

### Sunday 3ʳᵈ July. 20 C. Sunny.

Before Matins I listen to the midsummer sound-scape of the surrounding countryside. Sheep are calling, and yellowhammers, wood pigeons, then a distant male tawny owl, and the calls of cock pheasants – like the stuttering of a reluctant starter motor on a winter morning.

At the top of a beech near the Abbey gates two red kites, a few metres apart, survey their dawn kingdom. One preens, the blade of its kite tail pink-orange in early sunshine. In a lower branch a jay scrambles from leafy bough to bough.

At St Bernadette's Church I park the car, arriving for mass, and am greeted by three collared doves on the lawn.

*Before Midday prayer the Prior informs me that a swarm of bees went past his open window at 11.15am, making a lot of noise. I ask where they were heading. Towards the far end of South Block, following the line of the buildings, he tells me.*

*They're heading in the direction of the terrace and my bait hives...*

*I rush up there at lunchtime and see two swarms entering two of the boxes. After lunch I put on my bee suit and check the apiary – none of our hives have swarmed, so today's swarms are from elsewhere. I take a quick look at them under the hive lids. Both look large enough to be prime swarms. Elation!*

*On the meadow - the ticking and whirring of grasshoppers through long grass. In the apiary there's the sound of static as the stocks reach their peak strength.*

*It would appear that the novitiate terrace and Swarm Alley are hot-spots for swarm catching. Even my two experimental bait hives, made crudely from two polystyrene packing cases taped together, have proved suitable. I made two and both have attracted a swarm.*

*Swarm Alley*

*I now have three captured swarms to add to the one that survived last winter – all stocks I can monitor for health and survival, that could prove to be resistant to varroa, especially if they are wild bees. I note that most of them look very dark, which is a good sign.*

*9.30pm – walking on the meadow and listening to the birds falling silent. This time of the evening in high summer is pure freedom. In the gloaming I stand at the gate beyond the apiary, looking at a field of barley. Dark ears appear in the middle of the field, then disappear. Hares? A head. It's a roe deer stag. Other smaller doe ears appear, then vanish in the field. The light is vanishing too. A single bat circles over me against silvery sky between the silhouetted boughs of the great mothy oaks.*

*Nightfall. In the apiary bees are bearding the hive entrances. The apiary calms to a contented purr under midsummer stars.*

### 8ᵗʰ July. Sunny. 27C.

*6.00am. I go up to the terrace. There's activity at two more of the bait hives up there, which suggests that scouts have staked them out overnight and more swarms might be on the way. After breakfast it's quieter by mid-morning. I'm guessing the scouts have gone back to trigger the swarm.*

*I'm right! After lunch the boxes are both very busy.*

*I inspect the apiary in very hot weather at 2.00 pm, then check the boxes on the terrace at 3.00 pm while I'm suited. There's a prime swarm in one of the two boxes and what appears to be a smaller cast in another box.*

*It's good news, but the cast is a sign that from now on swarms will probably be smaller.*

*After Compline, I walk on the meadow until 9.30 pm as darkness falls. Up in my room I notice a tawny owl has landed on the lawn in the garden and is just sitting there. It stays there for a few minutes. I look away for a minute or so and when I next glance out it's gone – into the freedom of the night.*

### 9ᵗʰ July. 27 C.

*As evening cools I walk out onto the brown meadow. A roe deer stag with short antlers stops in long grass a hundred feet away, looking over his shoulder at me. It's a scene from the African bush. I look away, trying not to spook him, but when I look again he's pronking away across the grassland.*

*I reach the bait bucket trap hanging in the small tree on the meadow, scattering yellow rattle seed from papery seed-heads as I walk. The bucket is busy. Are they scouts? If so, there are a lot of them, and a swarm might be coming tomorrow? I incline my ear towards the bucket that's about six feet off the ground. It doesn't sound as though the bucket is full of bees yet.*

*10<sup>th</sup> July. 27 C.*

*After lunch I walk through long meadow grass ticking and whirring with grasshoppers, to inspect the bucket trap in the young beech tree.*

*Meadows take you back in time, I'm thinking. They belong to an analogue age.*

*I notice honeybees foraging Birdsfoot trefoil flowers stitched into the tapestry of the meadow.*

*Grasshopper eggs remain underground for ten months, emerging in spring as nymphs. Several weeks later they moult into adult grasshoppers. Grasshoppers spend most of their lives as buried treasure.*

*There are still plenty of bees at the hanging bucket. I watch their behaviour – there's traffic going to and fro, but some bees are returning with yellow pollen bags. That's not scout activity – it suggests foraging, which means a swarm has arrived. But how big? From the level of activity and the sound at the entrance I speculate that it's probably a cast, especially as we're now into July.*

*Boniface told me recently that Biddy never touched swarms in July.*

In Late September an email arrived from BeeBase, the online database of the National Bee Unit of the Animal and Plant Health Agency (APHA), designed for beekeepers. It told me I'd been selected for a survey. The questions were as I expected : would I recognise European and American  foul brood? Do I treat for varroa? Do I import queens or raise my own? Have I done any recent training?

In particular I'm interested in the question asking why I keep bees. It's a question I have often asked in the last few years, and my answer is complex. I suspect what's behind the question is that there are now so many reasons why people keep bees, which affects how they keep them and their attitudes to the other questions in the survey. Diversity seems to be one of the modern world's most important values, and there's certainly more diversity in beekeeping today than there would have been in William Woodley's day in the age of the cottager economy.

Perhaps the humble cottager has been replaced now by the small hobbyist. Though their reasons for keeping bees might differ and be more varied, I'd be the first to sympathise with hobbyists with only a couple of stocks who have to treat their colonies with miticides because they can't sustain the kind of losses I can with the freedom of a larger apiary for the four or five years of losses required to select for resistance to varroa and other diseases.

I suspect that many small hobbyists today are as locked in to treating with miticides as the old cottagers were with their skeps, and for the same reason; that the alternative seems beyond their means. Other aspects of beekeeping, however, aren't entirely beyond the ambitions of the average hobbyist, such as raising our own queens from local stock and perhaps keeping one or two captured swarms a year that are treatment-free alongside treated hives. Perhaps beekeepers who treat might only do so when they need to, rather than as a matter of routine. A change of direction towards treatment-free, locally adapted bees is a long term goal, but seems to me achievable if enough beekeepers are prepared to work towards it. We would end up with a bee which would have reselected at least some of the latent *Amm* genes in many of our wild and managed stocks, producing a bee that in some places would be, to all intents and purposes, as close as we can get to the native old British Black bee under pressure of natural selection for survival. Whatever the genetics, however, our bees would be increasingly selected for survival in our local conditions.

I'm heartened by a letter in the latest edition of the *BBKA News,* by someone congratulating the magazine for running a number of articles in July this year (2022) on treatment-free beekeeping. He relates in his letter calls he received in 2015-16 about bees in abandoned hives, log piles and trees, which made him ask if colonies in the wild were evolving to live with varroa. After finding his own hives with very low mite loads, he stopped treating in 2017. He still uncaps drone brood in spring, to check mite levels, which remain low or even zero. The writer of that letter is convinced, as I am, that bees are evolving to cope with varroa, and that we can keep our bees without medicating them.

There are many reasons why I keep bees: I know where I am with the bees' ordered world that seems somehow much simpler than the chaos of an ever stranger world. I like producing honey for a monastic community and people who come to the monastery wanting to buy something produced by a monk's simple manual work; I enjoy working outside where I notice the cycle of the changing seasons and the wildlife around me; I have a wonderful sense of standing in some small way in the venerable tradition and heritage of monastic beekeepers of the past who were better craftsmen than I will ever be, both at Douai Abbey and in the wider monastic world; I am also endlessly fascinated by honeybees and excited by the swarm season. But added to all these reasons, I have a stronger sense, as I get older, of our broader cultural heritage and why we preserve buildings and art, traditions and old orchards. If we lose them, we lose something of ourselves. We know from history that the stripping of sacred altars, the destruction of beautiful churches, art and language, the murder of monarchies and the ruin of ancient orchards and meadows are all acts of vandalism with consequences far beyond themselves. Ultimately we lose the old country of who we are.

Telling the story of the demise of William Woodley's Olde English bee, and the demise of William Woodley and the traditional skep beekeeper, is important because it's the story of our beekeeping heritage. It's also part of the story of our basic freedoms and of the ordinary men and women who come and go unnoticed, and of the ordinary folk before them, the skeppists, who handed us their ancient and noble craft. And that points to who we are and to our future. It's as necessary to us as a state funeral for a Queen and an era that has passed, so that in the same breath we are free to sing with optimism for the future: *God save the king!*

My beekeeping began to change for ever one midsummer evening, standing near the beehives of King Charles III at Highgrove House. It was more than something poetic about the name *Anglesey Bees*. Something resonated with me - like the poignant passing of a queen.

I walk towards William Woodley's grave and crouch down, to examine at close quarters again that cold, carved monarch, her thorax marked with coloured lichen, commissioned no doubt by William himself when Annie died. She is also now a touching memorial to the Black bee's former reign. I gaze up briefly into the sunlit trees, recalling the frail, gaunt poet who stood at the coppice gate, thinking sadly that there is little cause for my own caroling either; when it seemed to me suddenly that this forgotten little carving was in fact a cause for great celebration, written not on terrestrial things afar, but simply graven there on an old headstone,

'That I could think there trembled through

(Her) happy good-night air

Some blessed Hope, whereof (she) knew

And I was unaware.' (5)

William Woodley has long since gone, but his treasured Black bees are still here after all; hidden as treasure somewhere in the freedom of a high, ruffled woodland and the tolling church steeples of Wessex and the old, beloved country; or clustered in a crooked orchard, or hunkered in a hidden hole in some derelict Berkshire barn; or perhaps concealed as yet within the mystery of a chanting monastery's darkling bees....waiting....

It's a long, hard wait for them and for me before they reemerge next spring, and I'm always a little sad as another year latches its gate and the drowsy bees cluster against their white-capped honey stores. As the old cottagers sat through winter evenings by crackling hearths, weaving or repairing their skeps, I will retire to the old cobwebby workshop of our 1918 building, making new swarm boxes and rendering the year's fragrant wax. I will dream through frost and fog and days of black, wet boughs scribbled upon the dead, dreary months ahead just as the ancient bee men dreamed in centuries past; of that first day of spring when we hear again the hidden *crump* of each stock igniting like a boiler's pilot-light deep in their box. That moment in beekeeping is as near perfect as anything in this life can be. It's a liberating moment, like hearing the first cuckoo.

The bee-master of Warrilow was asked, 'Can you...after all these years of experience, lay down for beginners in beemanship one royal maxim of success above any other?' (6) To which the bee-master replied: 'Well, they might take warning...and beware the foreign feminine element. Let British bee-keepers cease to import queen bees from Italy and elsewhere, and stick to the good old English Black.' (7)

His Black bees, he assured the writer, were mostly from, '...one pure original Sussex stock.' (8) No doubt that was one of the few survivor stocks after Isle of Wight disease that still reigned in some murmuring bee garden or high in the knotty boughs of an ancient woodland. And they haven't completely gone after all. Brother Adam would surely have been as intrigued as he would be surprised. William Woodley, I have little doubt, would smile. What might he say to me now, along with Biddy and Brother Adam and all the bee men of yore recorded in the old, yellow Journals and serif-font minutes of their BKAs who have already passed this way?

With the bee-master of Warrilow, perhaps they might agree:

' 'Twas before you were born, likely as not; and bee science has seen many changes since then. In those days there were nothing but the old straw skeps, and most bee-keepers knew as little about the inner life of their bees as we do of the bottom of the South Pacific. Now things are very different; but the improvement is mostly in the bee-keepers themselves. The bees are exactly as they always have been, and work on the same principles as they did in the time of Solomon. They go their appointed way inexorably, and all the bee-master can do is to run on ahead and smooth the path a little for them.' (9)

Ironically, I find a truth in Brother Adam's prophetic words, though his definition of progress in apiculture differed greatly from my own: 'It is in the bee itself that we foresee the most profound and far-reaching progress – progress of a kind that will prove almost as revolutionary as the great technical and mechanical developments that have taken place in bee culture in the last hundred years, or possibly even more so.' (10)

It's the end of another beekeeping season. I hope perhaps that there has been some slow improvement in me these last few seasons, both as a beekeeper and a monk. I know that the most profound progress is indeed happening with the bees in the Douai Abbey apiary in any case, where I hope that bee culture will continue long after I have gone.

The light will soon be fading and the birds' full-hearted evensongs will cease. I touch the stone bee a final time, decoding her Braille. Turning away, I hear, faintly, autumn's first light winds chanting like children in distant chapels. I blink in late sunlight that blinds me momentarily like smoke in the eyes of the mind. I smile to myself. Perhaps once more, before winter calls down the final, fading leaves of another season, I'll kneel to tell the bees that I'm still alive, and that they are still very much alive too; that together we are all survivors. And I shall dream my dreams of spring through the cold, dark months ahead of that bright day to come when the first cuckoo calls again and the bees are set free by a flood of warm sunshine across the bright new meadow. On that day all the hopes and dreams of the beekeeper are liberated anew.

But now I must return to the hushed Abbey and the gentle stillness of its silent waxen city among the ferns, to my own evening psalms along the wondrous ways and recollected cloisters of this beautiful world and the quiet joy of my slow and ancient craft. The words of William Woodley in one of his last *Notes By The Way* from 1 February 1917 remain as pertinent and as wise now as they were then, because they are so full of hope:

'Surely there are brighter and happier days for beekeepers coming in the future. Read, mark, learn and digest... W. Woodley, Beedon, Newbury.' (11)

# Bibliography and Useful Links

**British Bee Journal & Bee-Keepers Adviser**
Free Internet Archive
https://archive.org/details/britishbeejournal

*The Bee-Master of Warrilow* by Tickner Edwardes
Published by Kessinger's Legacy Reprints (originally Published by Pall Mall Press – 1907)

*Breeding The Honeybee* by Brother Adam - 1982
Published by Northern Bee Books

*In Search of the Best Strains of Bees* by Brother Adam – 1983
Published by Northern Bee Books

*Beekeeping at Buckfast Abbey* by Brother Adam – first Published 1975
Fourth Ed Published 1987 by Northern Bee Books

*Swarming Its Control and Prevention* by L.E. Snelgrove 1934
Seventeenth Edition Published by Northern Bee Books 2014

*The Lives of Bees The Untold Story of the Honey Bee in the Wild* by Thomas D. Seeley - 2019
Published by Princeton University Press

*The Compleat Fisherman (The Contemplative Man's Recreation)* by Izaak Walton
First Published by Richard Marriot, London. 1653

Wordsworth Classics. Published by Wordsworth Editions Ltd. 1996

*Isle of Wight Disease: It's Historical and Practical Aspects* by Brother Adam
Published 1 March 1968 in *Bee World*

*Isle of Wight Disease: The Origin and Significance of The Myth* by Leslie Bailey 1964
Published online: 31 Jul 2015 - Originally published *Bee World* Vol 45, No 1
(tandfonline.com)

*H is for Hawk* by Helen McDonald, Vintage Books (Penguin) 2014
Published by Vintage Books, London. 2014

*The Man Who Planted Trees* by Jean Gione 1953 Abebooks

*Green Tyranny Exposing the Totalitarian Roots of the Climate Industrial Complex* by Rupert Darwall, Encounter Books, USA 2019

*Georgics* by Virgil translation by Project Gutenberg Ebooks
https://www.gutenburg.org/files/232/232-h/232-h.htm

**Wikipedia** *For A Swarm of Bees*

*Telling The Bees*
 https://wwwpoetryfoundation.org.poems45491/telling-the-bees

*Beekeeping at Buckfast – Past and Present (BIBBA)*
https://www.youtube.com/watch?=9Tg6zXh0hqw

*BeeBase* – Beekeeping Information resource for Beekeepers (nationalbeeunit.com)
https://www.nationalbeeunit.com

*Eva Crane Trust*
https://www.evacranetrust.org

*In Your Garden* Pathe Films 1944
https://www.youtube.com/watch?v=71b22Td50o4

*Jack Asks Have the Conservationists got it right & Where are the Cow Horns?*
https://www.youtube.com/watch?v=0OPr1_-SJio

*Old Country*
https://www.youtube.com/watch?v=YAgom

*Nat BIP website*
https://bibba.com/bip/

*Bee Disease Bill*
https://api.parliament.uk/historic-hansard/commons/...

*William Woodley: An Introduction*
 https://youtube.com/watch?v=8YDTnmMn0

*William Woodley: Early Years At Stanmore* (documentary) (Beehive yourself)
https://www.youtube.com/watch?v=-LUInUBaE-w

*The Apiary at Railway Crossing Cottage, Compton* (Beehive Yourself)
https://www.youtube.com/watch?v=b8M7RUHVPCc

*The Monk and the Honeybee* DVD 2005

*Detectorists*

https://www.imdb.com/title/tt4082744/

***Heathland Beekeeping***
https://www.bing.com/video/
search?&q=youtube+skep+beekeeping+in+lower+axony&docid=

# References

## Prologue

1.  *British Bee Journal* internet archive advert April 1896
    https://archive.org/details/britishbeejournal1896

2.  *British Bee Journal* internet archive  advert July 1921
    https://archive.org/details/britishbeejournal1921

3.  *British Bee Journal* internet archive  March 1915
    https://archive.org/details/britishbeejournal1915

4.  *British Bee Journal* internet archive  April 1916
    https://archive.org/details/britishbeejournal1916

5.  *British Bee Journal* internet archive  April 1915
    https://archive.org/details/britishbeejournal1915

## Chapter 1

1.  *British Bee Journal* internet archive advert March 1885
    https://archive.org/details/britishbeejournal1885

2.  *British Bee Journal* internet archive 'Superstitions about bees' by George D. Leslie
    October 1896
    https://archive.org/details/britishbeejournal

3.  ibid

4.  Edwardes Tickner *The Bee-Master of Warrilow* 1907 Kessinger's Legacy reprints p5

5.  *British Bee Journal* internet archive Woodley William 'Notes By The Way' June 1908
    https://archive.org/details/britishbeejournal1908

6.  ibid

7.   *British Bee Journal* internet archive 'Bristol Somerset & Gloucester Association Minutes' March 1893
     https://archive.org/details/britishbeejournal1893

8.   ibid

9.   *British Bee Journal* internet archive Woodley William 'Notes By The Way' January 1905
     https://archive.org/details/britishbeejournal1905

10.  *British Bee Journal* internet archive Woodley William 'Notes By The Way' June 1905
     https://archive.org/details/britishbeejournal1905

11.  *British Bee Journal* internet archive Woodley William 'Notes By The Way' May 1913
     https://archive.org/details/britishbeejournal1913

12.  ibid

13.  Cobbett William *The Cottage Economy and the Poor Man's Friend* 1821
     https://www.gutenberg.org/files/32863/32863-h/32863-h.htm

14.  ibid

15.  *British Bee Journal* internet archive April 1896
     https://archive.org/details/britishbeejournal1886

16.  ibid

17.  *British Bee Journal* internet archive June 1898
     https://archive.org/details/britishbeejournal1886

18.  *British Bee Journal* internet archive Woodley William 'Notes By The Way' June 1908
     https://archive.org/details/britishbeejournal1908

19.  *British Bee Journal* internet archive Woodley William 'Notes By The Way' August 1908
     https://archive.org/details/britishbeejournal1908

20.  ibid

21.  *British Bee Journal* internet archive Woodley William 'Notes By The Way' July 1896
     https://archive.org/details/britishbeejournal1896

22.  *British Bee Journal* internet archive Woodley William 'Notes By The Way' June 1905
     https://archive.org/details/britishbeejournal1905

23.  ibid

24.  *British Bee Journal* internet archive Woodley William 'Notes By The Way' May 1913

https://archive.org/details/britishbeejournal1913

25. ibid

26. *British Bee Journal* internet archive Woodley William 'Notes By The Way' February 1913
https://archive.org/details/britishbeejournal1913

27. *British Bee Journal* internet archive Woodley William 'Notes By The Way' June 1910


## Chapter 2

1. *British Bee Journal* internet archive advert May 1890
https://archive.org/details/britishbeejournal

2. *Adam Brother Breeding The Honeybee* 1982 Northern Bee Books p115

3. Cobbett William Introduction to *The Cottage Economy and the Poor Man's Friend* 1821
https://www.gutenberg.org/files/32863/32863-h/32863-h.htm

4. Charles David from article on 'BBKA and the Making of Experts'
https://www.bbka.org.uk/bbka-history-making-of-experts

5. ibid

6. *British Bee Journal* internet archive May 1873
https://archive.org/details/britishbeejournal1873

7. *British Bee Journal* internet archive Woodley William 'Notes By The Way' May 1913
https://archive.org/details/britishbeejournal1913

8. *British Bee Journal* internet archive November 1911
https://archive.org/details/britishbeejournal1911

9. *British Bee Journal* internet archive July 1912
https://archive.org/details/britishbeejournal1912


## Chapter 3

1. *British Bee Journal* internet archive advert June 1910

https://archive.org/details/britishbeejournal1910

2. *Douai Magazine* 1934

3. Thistleton Dyer T.F. *The Folk Law of Plants* 1889 p 119
   https://archive.org/details/folkloreplants00thisgoog/page/n7/mode/2up

4. Adam Brother *Breeding The Honeybee* 1982 Northern Bee Books p116

5. Adam Brother *Breeding The Honeybee* 1982 Northern Bee Books p 96

6. Adam Brother *In Search of the Best Strains of Bees* 1983 Northern Bee Books p17

7. *British Bee Journal* internet archive April 1900
   https://archive.org/details/britishbeejournal1900

8. *British Bee Journal* internet archive May 1900
   https://archive.org/details/britishbeejournal1900

9. *British Bee Journal* internet archive June 1915
   https://archive.org/details/britishbeejournal1915

10. Adam Brother *Breeding The Honeybee* 1982 Northern Bee Books p 97

11. ibid

12. *British Bee Journal* internet archive January 1910
    https://archive.org/details/britishbeejournal1910

13. *British Bee Journal* internet archive June 1910
    https://archive.org/details/britishbeejournal1910

14. *British Bee Journal internet* archive 'Breeding The British Golden Bee In Ripple Court Apiary' December 1909
    https://archive.org/details/britishbeejournal1909

15. Adam Brother *In Search of the Best Strains of Bees* 1983 Northern Bee Books p199

16. ibid p200

17. ibid p200

18. Adam Brother *Breeding The Honeybee* 1982 Northern Bee Books p 57

19. *British Bee Journal* internet archive 1889
    https://archive.org/details/britishbeejournal1889

20. Adam Brother *In Search of the Best Strains of Bees* 1983 Northern Bee Books p 200

21. ibid p200

22. Adam Brother *Breeding The Honeybee* 1982 Northern Bee Books p 62

23. ibid p 64

24. *British Bee Journal* internet archive December 1909
    https://archive.org/details/britishbeejournal1909

25. *British Bee Journal* internet archive June 1915
    https://archive.org/details/britishbeejournal1915

26. ibid

27. ibid

28. ibid

29. House of Commons 17 July 1913
    https://api.parliament.uk/...

30. ibid

31. Adam Brother *Breeding The Honeybee* 1982 Northern Bee Books p12

32. ibid

## Chapter 4

1. *British Bee Journal* internet archive advert July 1896
   https://archive.org/details/britishbeejournal1896

2. Adam Brother *In Search of the Best Strains of Bees* 1983 Northern Bee Books p12

3. Adam Brother *Breeding The Honeybee* 1982 Northern Bee Books p116

4. Adam Brother *In Search of the Best Strains of Bees* 1983 Northern Bee Books p11

5. ibid

6. Adam Brother *Breeding The Honeybee* 1982 Northern Bee Books p 6

7. ibid p45

8. Adam Brother *In Search of the Best Strains of Bees* 1983 Northern Bee Books p14

9. ibid

10.  Adam Brother *Breeding The Honeybee* 1982 Northern Bee Books p 54

11.  ibid p22

12.  ibid

13.  ibid

14.  *British Bee Journal* internet archive C. Heap 'Random Jottings' February 1915
     https://archive.org/details/britishbeejournal

**Chapter 5**

1.  *British Bee Journal* internet archive advert July 1901
    https://archive.org/details/britishbeejournal1901

2.  Edwardes Tickner *The Bee-Master of Warrilow* 1907 Kessinger's Legacy reprints 1907 p5

3.  ibid

4.  ibid

5.  Adam Brother *Breeding The Honeybee* 1982 Northern Bee Books p13

6.  Adam Brother *In Search of the Best Strains of Bees* 1983 Northern Bee Books p167

7.  ibid p168

8.  ibid p172

9.  ibid p173

10. ibid p25

11. Adam Brother *Breeding The Honeybee* 1982 Northern Bee Books p115

12. ibid p6

13. Adam Brother *In Search of the Best Strains of Bees* 1983 Northern Bee Books p195

14. Adam Brother *Breeding The Honeybee* 1982 Northern Bee Books p98

15. ibid p100

16. ibid p83

17. ibid

18.  Edwardes Tickner *The Bee-Master of Warrilow* 1907 Kessinger's Legacy reprints p 6

19.  ibid

20.  ibid

21.  ibid

22.  Adam Brother *Beekeeping at Buckfast Abbey* 1975 Northern Bee Books p 29

23.  ibid p11

24.  ibid p11-12

25.  ibid p12

26.  ibid p16

27.  ibid

28.  Edwardes Tickner *The Bee-Master of Warrilow* 1907 Kessinger's Legacy reprints p 30

29.  ibid p11

## Chapter 6

1.  *British Bee Journal* internet archive advert April 1910
https://archive.org/details/britishbeejournal1910

2.  *British Bee Journal* internet archive Woodley William 'Notes By The Way' February 1917
https://archive.org/details/britishbeejournal1917

3.  *British Bee Journal* internet archive April 1916
https://archive.org/details/britishbeejournal1916

4.  *British Bee Journal* internet archive February 1917
https://archive.org/details/britishbeejournal1917

5.  *British Bee Journal* internet archive October 1888
https://archive.org/details/britishbeejournal1888

6.  *British Bee Journal* internet archive 'Obituary Notice' of William Woodley by Charles C. Heap October 1923
https://archive.org/details/britishbeejournal1923

7.  *British Bee Journal* internet archive October 1888
    https://archive.org/details/britishbeejournal1888

8.  Woodley William *Letter to landlord* March 23 1917 Lockinge Estate

9.  *British Bee Journal* internet archive C. Heap 'Random Jottings' 1912

10. *British Bee Journal* internet archive William Woodley 'Notes By The Way' February 1917
    https://archive.org/details/britishbeejournal1917

11. Edwardes Tickner *The Bee-Master of Warrilow* 1907 Kessinger's Legacy reprints p 33

12. ibid

13. Snelgrove L.E. *Swarming Its Control and Prevention* reprint 2014 Northern Bee Books p10

14. Adam Brother *Beekeeping at Buckfast Abbey* 1975 Northern Bee Books p 46

15. Adam Brother *In Search of the Best Strains of Bees* 1983 Northern Bee Books p 173

16. Adam Brother *Beekeeping at Buckfast Abbey* 1975 Northern Bee Books p14

17. ibid p15

18. Edwardes Tickner *The Bee-Master of Warrilow* 1907 Kessinger's Legacy reprints  p 36

19. ibid

20. ibid p 31

21. ibid

**Chapter 7**

1.  *British Bee Journal* internet archive advert March 1915
    https://archive.org/details/britishbeejournal1915

2.  Edwardes Tickner *The Bee-Master of Warrilow* 1907 Kessinger's Legacy reprints p31

3.  ibid p32

4.  ibid p34

5.  ibid

6.  ibid p34-35

7.  ibid p32

8.  Adam Brother Breeding The Honeybee 1982 Northern Bee Books p 7

9.  ibid p19


**Chapter 8**

1.  *British Bee Journal* internet archive advert May 1907
    https://archive.org/details/britishbeejournal1907

2.  White Gilbert *A Natural History of Selborne* 'June 28th' 1789 reprinted Cambridge
    University Press 2014

3.  Edwardes Tickner *The Bee-Master of Warrilow* 1907 Kessinger's Legacy reprints p 44

4.  Gray Thomas Elegy *Written in a Country Churchyard* 1751

5.  Woodley William letter to landlord March 23 1917 Lockinge Estate Archive

6.  *British Bee Journal* internet archive William Woodley 'Notes By The Way' February
    1917

    https://archive.org/details/britishbeejournal1917

7.  *British Bee Journal* internet archive 'Our Prominant Beekeepers' November 1909
    https://archive.org/details/britishbeejournal1909

8.  Adam Brother *Beekeeping at Buckfast Abbey* 1975 Northern Bee Books p 32

9.  Edwardes Tickner *The Bee-Master of Warrilow* 1907 Kessinger's Legacy reprints p15

10. ibid


**Chapter 9**

1.  *British Bee Journal* internet archive advert May 1904
    https://archive.org/details/britishbeejournal1904

2.  *British Bee Journal* internet archive January 1911

https://archive.org/details/britishbeejournal1911

3.  Edwardes Tickner *The Bee-Master of Warrilow* 1907 Kessinger's Legacy reprints p4

4.  *British Bee Journal* internet archive April 1905
    https://archive.org/details/britishbeejournal1905

5.  *British Bee Journal* internet archive William Woodley 'Notes By The Way' December 1911
    https://archive.org/details/britishbeejournal1911

6.  *British Bee Journal* internet archive March 1911
    https://archive.org/details/britishbeejournal1911

7.  *British Bee Journal* internet archive William Woodley 'Notes By The Way' May 1911
    https://archive.org/details/britishbeejournal1911

8.  *British Bee Journal* internet archive William Woodley 'Notes By The Way' May 1905
    https://archive.org/details/britishbeejournal1905

9.  *British Bee Journal* internet archive William Woodley 'Notes By The Way' February 1911
    https://archive.org/details/britishbeejournal1911

10. *British Bee Journal* internet archive William Woodley 'Notes By The Way' March 1919
    https://archive.org/details/britishbeejournal1919

11. ibid

12. *British Bee Journal* internet archive William Woodley 'Notes By The Way' March 1911
    https://archive.org/details/britishbeejournal1911

13. *British Bee Journal* internet archive William Woodley 'Notes By The Way' January 1911
    https://archive.org/details/britishbeejournal1911

14. ibid

15. *British Bee Journal* internet archive October 1903
    https://archive.org/details/britishbeejournal 1903

16. *British Bee Journal* internet archive William Woodley 'Notes By The Way' February 1911
    https://archive.org/details/britishbeejournal 1911

17. *British Bee Journal* internet archive William Woodley 'Notes By The Way' June 1911
    https://archive.org/details/britishbeejournal 1911

18. ibid

19. *British Bee Journal* internet archive William Woodley 'Notes By The Way' July 1911
https://archive.org/details/britishbeejournal 1911

20. ibid

21. *British Bee Journal* internet archive William Woodley 'Notes By The Way' October 1911
https://archive.org/details/britishbeejournal 1911


**Chapter 10**

1. *British Bee Journal* internet archive advert March 1915
https://archive.org/details/britishbeejournal1915

2. Edwardes Tickner *The Bee-Master of Warrilow* 1907 Kessinger's Legacy reprints p 24

3. *British Bee Journal* internet archive 'Prominant Beekeepers' November 1909
https://archive.org/details/britishbeejournal1909

4. *British Bee Journal* internet archive William Woodley 'Notes By The Way' May 1913
https://archive.org/details/britishbeejournal1913

5. *British Bee Journal* internet archive William Woodley 'Notes By The Way' June 1898
https://archive.org/details/britishbeejournal1898

6. *British Bee Journal* internet archive William Woodley 'Notes By The Way' May 1913
https://archive.org/details/britishbeejournal1913

7. *British Bee Journal* internet archive William Woodley 'Notes By The Way' June 1898
https://archive.org/details/britishbeejournal1898

8. ibid

9. *British Bee Journal* internet archive William Woodley 'Notes By The Way' May 1913
https://archive.org/details/britishbeejournal1913

10. ibid

11. *British Bee Journal* internet archive William Woodley 'Notes By The Way' February 1917
https://archive.org/details/britishbeejournal1917

12. *British Bee Journal* in https://archive.org/details/britishbeejournalternet archive

William Woodley 'Notes By The Way' February 1919

13. *British Bee Journal* internet archive 'Editorial' November 1919
    https://archive.org/details/britishbeejournal1919


## Chapter 11

1.  *British Bee Journal* internet archive advert June 1910
    https://archive.org/details/britishbeejournal1910

2.  Edwardes Tickner *The Bee-Master of Warrilow* 1907 Kessinger's Legacy reprints p15

3.  ibid p6

4.  ibid

5.  ibid p5

6.  ibid p6

7.  ibid p6

8.  ibid p5

9.  ibid

10. ibid

11. ibid

12. ibid p5-6

13. *British Bee Journal* internet archive January 1914
    https://archive.org/details/britishbeejournal1914

14. ibid

15. Edwardes Tickner *The Bee-Master of Warrilow* 1907 Kessinger's Legacy reprints p 6

16. *British Bee Journal* internet archive January 1913
    https://archive.org/details/britishbeejournal1913

17. ibid

18. *British Bee Journal* internet archive September 1913
    https://archive.org/details/britishbeejournal1913

19. NATBIP

*Nat BIP website*

https://bibba.com/bip/

20. ibid

21. ibid

22. CoLoss Report in Journal of Apicultural Research 2014 Vol 53, Issue 2

*Nat BIP website*

https://bibba.com/bip/

23. Edwardes Tickner *The Bee-Master of Warrilow* 1907 Kessinger's Legacy reprints p6

24. ibid

25. *British Bee Journal* internet archive William Woodley 'Notes By The Way' August 1912
https://archive.org/details/britishbeejournal 1912

26. *British Bee Journal* internet archive January 1914
https://archive.org/details/britishbeejournal 1914


## Chapter 12

1. *British Bee Journal* internet archive advert April 1919
https://archive.org/details/britishbeejournal1919

2. *British Bee Journal* internet archive 'On Swarming' August 1913
https://archive.org/details/britishbeejournal1913

3. *British Bee Journal* internet archive 'Charm for Catching a Swarm' February 1905
https://archive.org/details/britishbeejournal1905

4. Wikipedia *For a Swarm of Bees* (Anglo Saxon Text discovered by John Mitchell Kemble in nineteenth century.)

5. Wikipedia *Lorsch Bee Blessing* from Apocalyse of Paul (Vatican Library)

6. *British Bee Journal* internet archive from 'Essex County Chronicle' report by Inspector Wapling June 1902
https://archive.org/details/britishbeejournal1902

7.  Whittier John Greenleaf *Telling The Bees* 1858  (Poetry Foundation)
    https://www.poetryfoundation.org.poems45491/telling-the-bees

8.  Edwardes Tickner *The Bee-Master of Warrilow* 1907 Kessinger's Legacy reprints p33

9.  ibid

10. ibid

11. ibid

12. Edwardes Tickner *The Bee-Master of Warrilow* 1907 Kessinger's Legacy reprints p34

13. Edwardes Tickner *The Bee-Master of Warrilow* 1907 Kessinger's Legacy reprints p37

14. *British Bee Journal* internet archive William Woodley 'Notes By The Way' July 1902
    https://archive.org/details/britishbeejournal1902

15. *British Bee Journal* internet archive William Woodley 'Notes By The Way' April 1904
    https://archive.org/details/britishbeejournal1904

16. *British Bee Journal* internet archive William Woodley 'Notes By The Way' July 1902
    https://archive.org/details/britishbeejournal1902

17. Edwardes Tickner *The Bee-Master of Warrilow* 1907 Kessinger's Legacy reprints p 34

18. ibid p35

19. *British Bee Journal* internet archive William Woodley 'Notes By The Way' May 1891
    https://archive.org/details/britishbeejournal1891

20. ibid

21. *British Bee Journal* internet archive July 1891
    https://archive.org/details/britishbeejournal1891

22. *British Bee Journal* internet archive William Woodley 'Notes By The Way' May 1891
    https://archive.org/details/britishbeejournal1891

23. *British Bee Journal* internet archive May 1891
    https://archive.org/details/britishbeejournal 1891

## Chapter 13

1. *British Bee Journal* internet archive advert 1912
   https://archive.org/details/britishbeejournal1912

2. Adam Brother *Beekeeping at Buckfast Abbey* 1975 Northern Bee Books p 24

3. *British Bee Journal* internet archive December 1915
   https://archive.org/details/britishbeejournal1915

4. *British Bee Journal* internet archive September 1915
   https://archive.org/details/britishbeejournal1915

5. *British Bee Journal* internet archive October 1915
   https://archive.org/details/britishbeejournal1915

6. *British Bee Journal* internet archive November 1915
   https://archive.org/details/britishbeejournal1915

7. *British Bee Journal* internet archive Mr. Ryder quoting Mr. Woodley in the 'Record' November 1915
   https://archive.org/details/britishbeejournal1915

8. *British Bee Journal* internet archive 'Report of Annual Meeting of Northumberland BKA' of 13 April 1916
   https://archive.org/details/britishbeejournal1916

9. *British Bee Journal* internet archive William Woodley 'Notes By The Way' February 1917
   https://archive.org/details/britishbeejournal1917

10. *British Bee Journal* internet archive June 1879
    https://archive.org/details/britishbeejournal1879

11. *British Bee Journal* internet archive April 1916
    https://archive.org/details/britishbeejournal1916

12. ibid

13. *British Bee Journal* internet archive William Woodley 'Notes By The Way' January1902
    https://archive.org/details/britishbeejournal1902

14. ibid

15. ibid

## Chapter 14

1.  *British Bee Journal* internet archive advert July 1904
    https://archive.org/details/britishbeejournal1904

2.  *British Bee Journal* internet archive William Woodley 'Notes By The Way' July1909
    https://archive.org/details/britishbeejournal1909

3.  *British Bee Journal* internet archive 'Homes of the Honeybee' December 1905
    https://archive.org/details/britishbeejournal1905

4.  *British Bee Journal* internet archive William Woodley 'Notes By The Way' May1913
    https://archive.org/details/britishbeejournal1913

5.  *British Bee Journal* internet archive C. Heap 'Obituary Notice' for William Woodley
    October 1923
    https://archive.org/details/britishbeejournal1923

6.  *British Bee Journal* internet archive October 1907
    https://archive.org/details/britishbeejournal1907

7.  *British Bee Journal* internet archive C. Heap 'Obituary Notice' for William Woodley
    October 1923

8.  ibid

9.  Woodley William letter to landlord 1917 Lockinge Estate

10. *British Bee Journal* internet archive William Woodley 'Notes By The Way' February1917
    https://archive.org/details/britishbeejournal1917

11. ibid

12. *British Bee Journal* internet archive C. Heap 'Obituary Notice' for William Woodley
    October 1923
    https://archive.org/details/britishbeejournal1923

13. ibid

14. *British Bee Journal* internet archive William Woodley 'Notes By The Way' April 1902
    https://archive.org/details/britishbeejournal1902

15. ibid

16. ibid

## Chapter 15

1. *British Bee Journal* internet archive advert May 1912
   https://archive.org/details/britishbeejournal1912

2. *British Bee Journal* internet archive William Woodley 'Notes By The Way' June 1902
   https://archive.org/details/britishbeejournal1902

3. ibid

4. ibid

5. *British Bee Journal* internet archive January 1920
   https://archive.org/details/britishbeejournal1920

6. *British Bee Journal* internet archive 'Editors' October 1921
   https://archive.org/details/britishbeejournal1921

7. *British Bee Journal* internet archive 'Editors' November 1921
   https://archive.org/details/britishbeejournal1921

8. *British Bee Journal* internet archive March 1921
   https://archive.org/details/britishbeejournal1921

9. ibid

10. ibid

11. *British Bee Journal* internet archive 'Editors' October 1921
    https://archive.org/details/britishbeejournal1921

12. *British Bee Journal* internet archive January 1920
    https://archive.org/details/britishbeejournal1920

13. ibid

14. ibid

15. ibid
    https://archive.org/details/britishbeejournal

16. *British Bee Journal* internet archive 'Lover of Bees' January 1920
    https://archive.org/details/britishbeejournal

17. *British Bee Journal* internet archive January 1920

18. ibid

19. ibid

20. *British Bee Journal* internet archive 'Editors' January 1920
    https://archive.org/details/britishbeejournal1920


**Chapter 16**

1. *British Bee Journal* internet archive advert April 1907
   https://archive.org/details/britishbeejournal1907

2. *British Bee Journal* internet archive 'Bee Paralysis: Is The Cause Known?' February 1906
   https://archive.org/details/britishbeejournal1906

3. ibid

4. *British Bee Journal* internet archive 'Editorial' August 1906
   https://archive.org/details/britishbeejournal1906

5. ibid

6. ibid

7. *British Bee Journal* internet archive 'Editorial' July 1906
   https://archive.org/details/britishbeejournal1906

8. ibid

9. ibid

10. ibid

11. ibid

12. *British Bee Journal* internet archive Mr Silver's Report July 1906
    https://archive.org/details/britishbeejournal1906

13. *British Bee Journal* internet archive 'Bee Epidemic: Isle of Wight Scourge' June 1907
    https://archive.org/details/britishbeejournal1907

14. ibid

15. *British Bee Journal* internet archive 'Editorial' June 1907
    https://archive.org/details/britishbeejournal1907

16. ibid

17. ibid

18. *British Bee Journal* internet archive reference to Dr. E. Assmuss 1865 Editorial  June 1907
    https://archive.org/details/britishbeejournal1907

19. Dr. Malden lecture 1909

20. Adam Brother *Isle of Wight or Acarine Disease: Its Historical and Practical Aspects* 1968 p 9

21. ibid p10

22. Dr. Rennie 1919

23. Adam Brother *Isle of Wight or Acarine Disease: Its Historical and Practical Aspects* 1968 p17

24. ibid p10

25. ibid p9

26. ibid p9

27. Bailey L. *Isle of Wight Disease: The Origin and Significance of the Myth* 1964 p 4

28. ibid p5

29. ibid p2

30. ibid p4-5

31. ibid p5

32. ibid p8

33. *British Bee Journal* internet archive Rev. Donald Moore December 1902
    https://archive.org/details/britishbeejournal1902

34. ibid

35. ibid

36. ibid

37. ibid

38. *British Bee Journal* internet archive William Woodley 'Notes By The Way' June 1904
    https://archive.org/details/britishbeejournal1904

39. ibid

## Chapter 17

1. *British Bee Journal* internet archive advert  May 1886
   https://archive.org/details/britishbeejournal1886

2. Adam Brother *Beekeeping at Buckfast Abbey* 1975 Northern Bee Books p16

3. Paraphrasing from Walton Izaak *The Compleat Angler (The Contemplative Man's Recreation)* first published 1653 by Richard Marriot, London. Published by Wordsworth Editions Ltd 1996

## Chapter 18

1. *British Bee Journal* internet archive advert  November 1921
   https://archive.org/details/britishbeejournal1921

2. *Douai Magazine* VIII 1, 1934

3. *British Bee Journal* internet archive 'Homes of the Honeybee' May 1898
   https://archive.org/details/britishbeejournal1898

4. ibid

5. *Douai Magazine* XVI, 3, 1951

6. Tennyson Alfred Lord *Eleanor* (first printed 1833)
   https://americanliterature.com/author/alfred-lord-Tennyson/poem/eleanor

## Chapter 19

1. *British Bee Journal* internet archive advert May 1907
   https://archive.org/details/britishbeejournal1907

2. Hardy Thomas *Afterwards* (first published in 'Moments of Vision' 1917)
   https://poets.org/poem/afterwards

3. Psalm 36:2 *The Grail Psalms* (Collins)

4. Adam Brother *Beekeeping at Buckfast Abbey* 1975 Northern Bee Books p110

5. ibid p 109

6. *Ecclesiastes* 1:2

7.  Adam Brother *Beekeeping at Buckfast Abbey* 1975 Northern Bee Books p112 (first published in 'Bee World' 1953)

8.  ibid p111

9.  ibid

10. *British Bee Journal* internet archive William Woodley 'Notes By The Way' February 1911

11. Edwardes Tickner *The Bee-Master of Warrilow* 1907 Kessinger's Legacy reprints p27-28

12. *British Bee Journal* internet archive October 1923
    https://archive.org/details/britishbeejournal

13. Edwardes Tickner *The Bee-Master of Warrilow* 1907 Kessinger's Legacy reprints p15

## Epilogue

1.  *British Bee Journal* internet archive advert  April 1919
    https://archive.org/details/britishbeejournal 1919

2.  Edwardes Tickner *The Bee-Master of Warrilow* 1907 Kessinger's Legacy reprints p11

3.  *British Bee Journal* internet archive March  1913
    https://archive.org/details/britishbeejournal1913

4.  Hardy Thomas *The Darkling Thrush* First published 29 December 1900
    https://www.poetryfoundation.org/poems/44325/the-darkling-thrush

5.  ibid (paraphrasing)

6.  Edwardes Tickner *The Bee-Master of Warrilow* 1907 Kessinger's Legacy reprints p15

7.  ibid

8.  ibid

9.  ibid

10. Adam Brother *In Search of the Best Strains of Bees* 1983 Northern Bee Books p17

11. *British Bee Journal* internet archive William Woodley 'Notes By The Way' February 1917
    https://archive.org/details/britishbeejournal1917

www.ingramcontent.com/pod-product-compliance
Lightning Source LLC
Chambersburg PA
CBHW050104220326
41598CB00043B/7380